Topics in Applied Physics
Volume 84

Available Online

Topics in Applied Physics is part of the Springer LINK service. For all customers with standing orders for Topics in Applied Physics we offer the full text in electronic form via LINK free of charge. Please contact your librarian who can receive a password for free access to the full articles by registration at:

http://link.springer.de/orders/index.htm

If you do not have a standing order you can nevertheless browse through the table of contents of the volumes and the abstracts of each article at:

http://link.springer.de/series/tap/

There you will also find more information about the series.

Springer
Berlin
Heidelberg
New York
Barcelona
Hong Kong
London
Milan
Paris
Tokyo

Physics and Astronomy ONLINE LIBRARY

http://www.springer.de/phys/

Topics in Applied Physics

Topics in Applied Physics is a well-established series of review books, each of which presents a comprehensive survey of a selected topic within the broad area of applied physics. Edited and written by leading research scientists in the field concerned, each volume contains review contributions covering the various aspects of the topic. Together these provide an overview of the state of the art in the respective field, extending from an introduction to the subject right up to the frontiers of contemporary research.

Topics in Applied Physics is addressed to all scientists at universities and in industry who wish to obtain an overview and to keep abreast of advances in applied physics. The series also provides easy but comprehensive access to the fields for newcomers starting research.

Contributions are specially commissioned. The Managing Editors are open to any suggestions for topics coming from the community of applied physicists no matter what the field and encourage prospective editors to approach them with ideas.

See also: http://www.springer.de/phys/books/tap/

Managing Editors

Dr. Claus E. Ascheron

Springer-Verlag Heidelberg
Topics in Applied Physics
Tiergartenstr. 17
69121 Heidelberg
Germany
ascheron@springer.de

Dr. Hans J. Kölsch

Springer-Verlag Heidelberg
Topics in Applied Physics
Tiergartenstr. 17
69121 Heidelberg
Germany
koelsch@springer.de

Assistant Editor

Dr. Werner Skolaut

Springer-Verlag Heidelberg
Topics in Applied Physics
Tiergartenstr. 17
69121 Heidelberg
Germany
skolaut@springer.de

Mathias Fink   William A. Kuperman
Jean-Paul Montagner   Arnaud Tourin (Eds.)

Imaging of Complex Media with Acoustic and Seismic Waves

With 162 Figures

Springer

Prof. Mathias Fink
Lab. Ondes et Acoustiques ESPCI
Université Paris VII-Denis Diderot
10, rue Vauquelin
75231 Paris Cedex 05
France
mathias.fink@espci.fr

Prof. William A. Kuperman
Marine Physical Laboratory
University of California, San Diego
9500 Gilman Drive
92093 La Jolla, CA
USA

Prof. Jean-Paul Montagner
Departement de Sismologie
Institut de Physique du Globe
4, Place Jussieu
75252 Paris Cedex 05
France
jpm@ipgp.jussieu.fr

Prof. Arnaud Tourin
Lab. Ondes et Acoustiques ESPCI
Université Paris VII-Denis Diderot
10, rue Vauquelin
75231 Paris Cedex 05
France
arnaud.tourin@espci.fr

Library of Congress Cataloging-in-Publication Data

Imaging of complex media with acoustic and seismic waves / Mathias Fink ... [et al.].
   p. cm. -- (Topics in applied physics ; v. 82)
   Includes bibliographical references and index.
   ISBN 3540416676 (alk. paper)
   1. Acoustic imaging. I. Fink, Mathias, 1945- II. Series.

TA1770 .I43 2002
620.2'8--dc21

2001032277

Physics and Astronomy Classification Scheme (PACS):
43.30.+m, 81.70, 87.57.G, 91.30.-f, 02.30.Zz, 43.58.+z, 43.60.+d

ISSN print edition: 0303-4216
ISSN electronic edition: 1437-0859
ISBN 3-540-41667-6 Springer-Verlag Berlin Heidelberg New York

This work is subject to copyright. All rights are reserved, whether the whole or part of the material is concerned, specifically the rights of translation, reprinting, reuse of illustrations, recitation, broadcasting, reproduction on microfilm or in any other way, and storage in data banks. Duplication of this publication or parts thereof is permitted only under the provisions of the German Copyright Law of September 9, 1965, in its current version, and permission for use must always be obtained from Springer-Verlag. Violations are liable for prosecution under the German Copyright Law.

Springer-Verlag Berlin Heidelberg New York
a member of BertelsmannSpringer Science+Business Media GmbH

© Springer-Verlag Berlin Heidelberg 2002
Printed in Germany

http://www.springer.de

The use of general descriptive names, registered names, trademarks, etc. in this publication does not imply, even in the absence of a specific statement, that such names are exempt from the relevant protective laws and regulations and therefore free for general use.

Typesetting: DA-TEX Gerd Blumenstein, Leipzig
Cover design: *design & production* GmbH, Heidelberg

Printed on acid-free paper   SPIN: 10749907   56/3141/mf   5 4 3 2 1 0

Preface

The objective of the workshop held in Cargèse from the 26 April to the 8 May 1999 was to bring together scientists of different communities that were interested in "acoustic and seismic wave imaging of complex media," a subject which covers many areas of applied research. Indeed, acoustic and elastic wave propagation is being investigated in media such as the ocean, the earth, biological tissues and solid materials. In these different areas, many specific techniques have been developed which differ by the wavelength (sound, ultrasound, seismic waves), polarization and the instrumentation used. The various communities have traditionally worked in an independent fashion, communicating only at specific and focused workshops, so that the interactions between underwater acousticians, geophysicists, medical scientists and researchers in nondestructive evaluation have been very limited up to now.

Today, improvements in multi-element sensor technology and computer capacity make possible the transfer of migration and tomography techniques used in seismics to medical imaging or nondestructive evaluation. The adaptive focalization techniques, first developed in optics, have also appeared in acoustics; such very various methods are open to comparison. Especially, the ultrasonic time-reversal mirror approach has some similarities but also significant differences with phase conjugation methods used in underwater acoustics. New, very promising fields such as anisotropic media studies (seismology and nondestructive testing) or imaging based on speckle correlation techniques (seismology and medical imaging) are developing quickly. Finally, medical imaging scientists have been interested for a few years by studies concerning low-frequency elastic wave propagation through the human body (elastography), and analogies with seismic problems become important.

Thus, it appeared that it had become necessary for physicists, geophysicists and engineers to gather at a meeting devoted to an interdisciplinary program. The different contributions of the lecturers are now gathered in this book.

Although four main fields of research are represented in this book, we did not organise it in four distinct parts, since some contributions actually cover several subjects. The book begins with five contributions dealing with the connections between time reversal, imaging and the inverse problem from both theoretical and practical points of view. Then, in the two next parts, classical imaging and detection techniques are presented in the context of medical imaging and nondestructive testing. The last contributions concern more specifically the resolution of the inverse problem based on the study of elastic wave propagation. Despite the diversity of the propagation media (the human body, the earth or polycrystals) and of the wavelengths involved, the goal remains the same: determine the elastic parameters, velocities structures or even nature of the source in order to elaborate an elastic model of the tested medium as reliably as possible. More specifically, the role of polarization effects and anisotropy are discussed in both seismology and NDT. As for the medical imaging field, elastography techniques are elaborated upon which present great similarities with the ones used in seismics. This work was made possible by financial support from the Centre National de la Recherche Scientifique (CNRS), the Groupement de Recherches "Propagation et imagerie en milieu aléatoire" (GDR PRIMA), the Délégation Générale de l'Armement (DGA, Ministère de la Défense), the Collectivité Territoriale de Corse and finally the Organization of Naval Research (ONR), to whom we owe special thank. The personal of the "Institut d'Etudes Scientifiques de Gargése" has made our stay very enjoyable. We thank all of them.

Finally, we thank Dr. Elisabeth Dubois-Violette, Director of the Cargese Institute, where the ideas leading to this work originated.

Paris, November 2001

Mathias Fink
William A. Kuperman
Jean-Paul Montagner
Arnaud Tourin

Contents

Time-Reversal Invariance and the Relation between Wave Chaos and Classical Chaos
Roel Snieder 1

1. Time-Reversal Invariance of the Laws of Nature 1
2. Wave Chaos and Particle Chaos 4
3. Instability of Particle Trajectories 6
4. Instability of Wave Propagation 7
5. Numerical Examples 10
6. Discussion 14

Acoustic Time-Reversal Mirrors
Mathias Fink 17

1. Introduction 17
2. Time-Reversal Cavities and Mirrors 17
   2.1. The Time-Reversal Cavity 18
   2.2. The Time-Reversal Mirror 20
3. Time-Reversal Experiments 21
   3.1. Time Reversal through Random Media 21
   3.2. Time Reversal in Waveguides 27
   3.3. Time Reversal in Chaotic Cavities 32
4. Applications of Time-Reversal Mirrors 37
5. Conclusion 40

Ocean Acoustics, Matched-Field Processing and Phase Conjugation
William A. Kuperman and Darrel R. Jackson 43

1. Review of Ocean Acoustics 43
   1.1. Qualitative Description of Ocean Sound Propagation 43
   1.2. Sound Propagation Models 49
   1.3. Quantitative Description of Propagation 53
2. Matched-Field Processing 56

3. Phase Conjugation in the Ocean 60
    3.1. Basic Properties of Phase Conjugation 60
    3.2. Background Theory and Simulation
         for Phase Conjugation/Time-Reversal Mirror in the Ocean 65
    3.3. Implementation of a Time-Reversal Mirror in the Ocean 71
    3.4. Summary of Time-Reversal Mirror Experiments 75
4. The Range-Dependent Ocean Waveguide 75
5. The Effect of Ocean Fluctuations on Phase Conjugation 83
    5.1. Time-Independent Volume Scattering 83
    5.2. Time-Dependent Scattering by Surface Waves 85
    5.3. Time-Dependent Scattering by Internal Waves 87
6. Conclusions 90
7. Appendix A: Parabolic Equation (PE) Model 91
    7.1. Standard Parabolic Equation Split-Step Algorithm 91
    7.2. Generalized or Higher-Order Parabolic Equation Methods 92
8. Appendix B: Units 93

Time Reversal, Focusing and Exact Inverse Scattering
James H. Rose 97

1. Introduction 97
2. Direct and Inverse Scattering Problems 98
    2.1. The Forward Problem 99
    2.2. Inverse Scattering Problem 100
3. Physics of the Newton–Marchenko Equation 100
4. Discussion and Summary 104

Detection and Imaging in Complex Media
with the D.O.R.T. Method
Claire Prada 107

1. Introduction 107
2. Basic Principle of the D.O.R.T. Method 109
    2.1. The Transfer Matrix 109
    2.2. Invariants of the Time-Reversal Process and Decomposition
         of the Transfer Matrix 110
    2.3. Transfer Matrix for Point-Like Scatterers 111
    2.4. Decomposition of K for Well-Resolved Scatterers 112
    2.5. The D.O.R.T. Method in Practice 113
3. Selective Focusing Through an Inhomogeneous Medium
   with the D.O.R.T. Method 114
4. Highly Resolved Detection and Selective Focusing in a Waveguide ... 116
    4.1. Selective Highly Resolved Focusing in a Waveguide 118
    4.2. Detection Near the Interface 120
    4.3. Detection in a Nonstationary Waveguide 121

5. Inverse-Scattering Analysis and Target Resonance 122
   5.1. Experiment 123
   5.2. Invariants of the Time-Reversal Process 125
   5.3. Resonance Frequencies of the Shell 127
6. The D.O.R.T. Method in the Time Domain 128
   6.1. Construction of the Temporal Green's Functions 129
   6.2. Selective Focusing in the Pulse Mode 131
7. Conclusion 132

Ultrasound Imaging and Its Modeling
Jørgen A. Jensen 135

1. Fundamental Ultrasound Imaging 135
2. Imaging with Arrays 138
3. Focusing 142
4. Ultrasound Fields 144
   4.1. Derivation of the Fourier Relation 144
   4.2. Beam Patterns 146
5. Spatial Impulse Responses 149
   5.1. Fields in Linear Acoustic Systems 149
   5.2. Basic Theory 150
   5.3. Geometric Considerations 153
   5.4. Calculation of Spatial Impulse Responses 154
   5.5. Examples of Spatial Impulse Responses 156
   5.6. Pulse–Echo Fields 157
6. Fields from Array Transducers 159
7. Examples of Ultrasound Fields 161
8. Summary 164

Nondestructive Acoustic Imaging Techniques
Volker Schmitz 167

1. Introduction 167
2. The Nondestructive Testing Task 168
3. The Inverse Problem 170
4. Special Features of SAFT 172
   4.1. Lateral Resolution 173
   4.2. Signal-to-Noise Ratio Improvement by SAFT 175
   4.3. Localization Accuracy 175
   4.4. Pulse–Echo/Pitch-and-Catch Reconstruction 177
   4.5. Acoustic Imaging in a 3-dimensional CAD Environment 180
5. Imaging in Transversally Isotropic Material – Ray Tracing 183
   5.1. Sound Propagation Through a V Weld with Defects 184
   5.2. A 10-Layer Approximated Austenitic Weld 185
6. Summary 188

Seismic Anisotropy Tomography
Jean-Paul Montagner 191

1. Introduction 191
2. The Anatomy of Seismograms 192
   2.1. Progress in Instrumentation 192
   2.2. Body Waves, Surface Waves and Normal Modes 195
   2.3. Normal-Mode Theory 198
3. An Anisotropic Earth 202
   3.1. Seismic Anisotropy at All Scales 202
   3.2. First-Order Perturbation Theory in the Planar Case 205
4. Tomography of Anisotropy 212
   4.1. Forward Problem 212
   4.2. Inverse Problem 216
   4.3. Practical Implementation 218
   4.4. Geophysical Applications 218
5. Conclusions 224

Elastic-Wave Propagation in Random Polycrystals: Fundamentals and Application to Nondestructive Evaluation
Bruce R. Thompson 233

1. Introduction 233
2. Simple Polycrystals 235
   2.1. Background 235
   2.2. Theory 236
   2.3. Randomly Oriented, Equi-axed Polycrystals 240
   2.4. Equi-axed Polycrystals with Preferred Orientation 241
   2.5. Randomly Oriented Polycrystals with Grain Elongation 243
   2.6. Polycrystals with Both Preferred Orientation
        and Grain Elongation 244
3. Complex Microstructures 244
   3.1. Background 244
4. Effects on Imaging 251
5. Conclusions 253

Imaging the Viscoelastic Properties of Tissue
Mostafa Fatemi and James F. Greenleaf 257

1. Introduction 257
2. Theory of the Radiation Force 260
3. Radiation-Force Methods 261
   3.1. Transient Method 262
   3.2. Shear-Wave Methods 262
   3.3. Vibro-Acoustography 263
4. Capabilities and Limitations 272
5. Summary 274

Estimation of Complex-Valued Stiffness
Using Acoustic Waves Measured with Magnetic Resonance
Travis E. Oliphant, Richard L. Ehman and James F. Greenleaf 277

1. Introduction 277
2. Measurement Model 278
   2.1. Acoustic Model 278
   2.2. Displacement Measurement with Magnetic Resonance 281
3. Estimating Material Properties 284
   3.1. Algebraic Inversion of the Differential Equation (AIDE) 285
   3.2. Other Inversion Methods 287
4. Examples 289
   4.1. Experimental Phantom 289
5. Conclusion 292

A New Approach for Traveltime Tomography
and Migration Without Ray Tracing
Philippe O. Ecoublet and Satish C. Singh 295

1. Introduction 295
2. The Traveltime Function 296
   2.1. Traveltime as a Series Expansion 298
   2.2. The Eikonal Equation 299
   2.3. The Equations of Constraint 300
3. Tomography 301
   3.1. The Misfit Function 301
   3.2. The Initial Model 302
   3.3. Optimization 302
   3.4. Slowness Image Reconstruction 302
4. Error and Resolution Analyses 303
5. Prestack Depth Migration 303
   5.1. Computation of the Incidence Angle of the Ray 304
6. Conclusions 305

Simple Models in the Mechanics of Earthquake Rupture
Shamita Das 311

1. Introduction 311
2. Brief Derivation of the Underlying Equations 312
3. The Finite Circular Shear Fault 321
4. Spontaneous Faults 322
   4.1. Fracture Criterion 324

Index 333

Time-Reversal Invariance and the Relation between Wave Chaos and Classical Chaos

Roel Snieder

Department of Geophysics and Center for Wave Phenomena,
Colorado School of Mines, Golden/Colo./CO/ 80401-1887, USA
rsnieder@mines.edu

Abstract. Many imaging techniques depend on the fact that the waves used for imaging are invariant for time reversal. The physical reason for this is that in imaging one propagates the recorded waves backward in time to the place and time when the waves interacted with the medium. In this chapter, the invariance for time reversal is shown for Newton's law, Maxwell's equations, the Schrödinger equation and the equations of fluid mechanics. The invariance for time reversal can be used as a diagnostic tool to study the stability of the temporal evolution of systems. This is used to study the relation between classical chaos and wave chaos, which also has implications for quantum chaos. The main conclusion is that in classical chaos perturbations in the system grow exponentially in time [$\exp(\mu t)$], whereas for the corresponding wave system perturbations grow at a much smaller rate algebraically with time ($\sqrt{t}$).

1 Time-Reversal Invariance of the Laws of Nature

Most laws of nature are invariant for time reversal. The only exceptions are the weak force that governs radioactive decay and equations that describe statistical properties such as the heat equation. This means that when we let the clock run backwards rather than forwards, the deterministic laws that govern the macroscopic world do not change. Mathematically, time reversal implies that the time t is replaced by $-t$. By making the substitution $t \to -t$ and by verifying whether the equation under consideration changes, one can verify whether the physical law is unchanged under time reversal.

As a first example let us consider Newton's third law which governs the motion of bodies in classical mechanics:

$$m\frac{d^2 r}{dt^2} = F(r) \,. \tag{1}$$

In this expression F denotes the force that acts on a particle with mass m at location r. Under the substitution $t \to -t$, Newton's law does not change because the second time derivative is insensitive to the multiplication with the factor $(-1)^2$ that follows from this substitution. Mathematically this can be expressed by stating that Newton's law transforms as $md^2 r/dt^2 = F(r) \to md^2 r/d(-t)^2 = F(r)$, which is identical to the original law (1). This means that when $r(t)$ is a solution, then so is $r(-t)$. Physically this means that when a particle follows a trajectory then when one reverses the velocity of the particle at one point it will retrace its original trajectory.

Pressure waves in an acoustic medium satisfy the acoustic wave equation:

$$\rho \nabla \cdot \left(\frac{1}{\rho}\nabla p\right) - \frac{1}{c^2}\frac{\partial^2 p}{\partial t^2} = 0 \ . \qquad (2)$$

Because of the invariance of the second time derivative under time reversal, pressure waves in an acoustic medium are invariant for time reversal as well. This means that when $p(\mathbf{r}, t)$ is a solution then the time-reversed wavefield $p(\mathbf{r}, -t)$ is also a solution.

At this point you may think that the invariance for time reversal is due to the occurrence of the second time derivative in the equations. This is not necessarily the case. In classical electromagnetism the electric field $\mathbf{E}$ and the magnetic field $\mathbf{B}$ obey in vacuum Maxwell's equations, which contain only the first time derivative:

$$\nabla \cdot \mathbf{E} = 4\pi\rho/\varepsilon_0 \quad , \quad \mu_0 \nabla \times \mathbf{B} - \frac{\varepsilon_0}{c}\frac{\partial \mathbf{E}}{\partial t} = \frac{4\pi}{c}\mathbf{J} \ ,$$

$$\nabla \cdot \mathbf{B} = 0 \quad , \quad \nabla \times \mathbf{E} + \frac{1}{c}\frac{\partial \mathbf{B}}{\partial t} = 0 \ . \qquad (3)$$

In this expression ρ is the charge density, $\mathbf{J}$ is the electrical current density, ε_0 is the electrical permittivity and μ_0 is the magnetic susceptibility. Under time reversal $(t \to -t)$ Maxwell's equations transform to

$$\nabla \cdot \mathbf{E} = 4\pi\rho/\varepsilon_0 \quad , \quad \mu_0 \nabla \times (-\mathbf{B}) - \frac{\varepsilon_0}{c}\frac{\partial \mathbf{E}}{\partial (-t)} = \frac{4\pi}{c}(-\mathbf{J}) \ ,$$

$$\nabla \cdot (-\mathbf{B}) = 0 \quad , \quad \nabla \times \mathbf{E} + \frac{1}{c}\frac{\partial (-\mathbf{B})}{\partial (-t)} = 0 \ . \qquad (4)$$

Note that these expressions are identical to the original Maxwell's equations (3) with the exception that the magnetic field $\mathbf{B}$ and the current $\mathbf{J}$ have changed sign. This is due to the fact that when one changes the direction of time the velocity of the charges changes sign; hence the associated current changes sign as well: $\mathbf{J} \to -\mathbf{J}$. Since the electric current is the source of the magnetic field, the magnetic field therefore also changes sign under time reversal $(\mathbf{B} \to -\mathbf{B})$. However, the Lorentz force $\mathbf{F} = q(\mathbf{E} + \mathbf{v} \times \mathbf{B})$ does *not* change sign because under time reversal both the magnetic field $\mathbf{B}$ and the velocity $\mathbf{v}$ change sign. This means that under time reversal the magnetic field $\mathbf{B}$ is not invariant because it changes sign. However, the functional form of the transformed equations (4) and the original equations (3) is identical and the imprint of the associated fields on charges is unaffected because the Lorentz force does not change under time reversal. For this reason one can state that the laws of classical electromagnetism are invariant under time reversal.

In quantum mechanics the wave-character of a particle is described by Schrödinger's equation:

$$i\hbar \frac{\partial \psi}{\partial t} = -\frac{\hbar^2}{2m}\nabla^2 \psi + V\psi , \tag{5}$$

where $V(\boldsymbol{r})$ denotes a real potential. When t is replaced by $-t$ and when one takes the complex conjugate, this equation transforms to

$$(-i)\hbar \frac{\partial \psi^*}{\partial(-t)} = -\frac{\hbar^2}{2m}\nabla^2 \psi^* + V\psi^* . \tag{6}$$

This equation is identical to (5) because the minus signs in the first term cancel each other. This implies that when $\psi(\boldsymbol{r},t)$ is a solution of Schrödinger's equation, then $\psi^*(\boldsymbol{r},-t)$ is a solution as well. Since in quantum mechanics only the absolute value $|\psi|^2$ leads to observable effects, one can state that the Schrödinger equation is invariant for time reversal.[1]

The time-reversal invariance of Maxwell's equations is related to the fact that the current and the magnetic field reverse sign under time reversal. From this you may have concluded that time-reversal invariance only holds for linear equations. To show that this is not true, we consider as a last example a fluid that is exposed to a body force $\boldsymbol{F}$. The equation of motion is given by

$$\rho \frac{\partial \boldsymbol{v}}{\partial t} + \rho \boldsymbol{v} \cdot \nabla \boldsymbol{v} = \boldsymbol{F} . \tag{7}$$

Under time reversal, $t \to -t$, the velocity changes sign because $\boldsymbol{v} = \mathrm{d}\boldsymbol{r}/\mathrm{d}t \to \mathrm{d}\boldsymbol{r}/\mathrm{d}(-t) = -\boldsymbol{v}$, so that this expression changes under time reversal to

$$\rho \frac{\partial(-\boldsymbol{v})}{\partial(-t)} + \rho(-\boldsymbol{v}) \cdot \nabla(-\boldsymbol{v}) = \boldsymbol{F} . \tag{8}$$

This expression is identical to the original expression (7) because the minus signs cancel. This implies that when $\boldsymbol{v}(\boldsymbol{r},t)$ is a solution of the equation of motion of the fluid, then $-\boldsymbol{v}(\boldsymbol{r},-t)$ is a solution as well. This has been demonstrated in a beautiful experiment by *Chaiken* et al. [1], who mixed white paint through black paint by rotating a cylinder in the paint. When the cylinder was rotated in the reverse direction the paint "un-mixed" again and the white paint contracted to a localized blob in the black paint.

It should be noted that when one adds dissipation to the equation, the invariance for time reversal is lost. For example, when viscosity is added to expression (7), this expression changes to $\rho \partial \boldsymbol{v}/\partial t + \rho \boldsymbol{v} \cdot \nabla \boldsymbol{v} = \mu \nabla^2 \boldsymbol{v} + \boldsymbol{F}$. Under time reversal this expression transforms to $\rho \partial(-\boldsymbol{v})/\partial(-t) + \rho(-\boldsymbol{v}) \cdot \nabla(-\boldsymbol{v}) =$

---

[1] This is in fact a slight over-simplification. In quantum mechanics the expectation value of Hermitian operators are the only observable quantities. One can show that for such an operator the expectation value does not change when one takes the complex conjugate of the wavefunction ψ.

$\mu\nabla^2(-\boldsymbol{v}) + \boldsymbol{F}$, which is *not* identical to the original expression because the viscous term $\mu\nabla^2\boldsymbol{v}$ changes sign. In general, dissipation implies a direction of time because energy is lost from the system (with time). Therefore, invariance for time reversal can only be expected in the absence of dissipation.

The results of this section imply that in the absence of dissipation the laws of classical mechanics, acoustic wave propagation, classical electromagnetism, quantum mechanics, and fluid mechanics are invariant under time reversal. Yet in our daily life we clearly experience a "direction" of time. When a vase falls on the floor we see it break and all the parts fly around, yet we never see pieces of pottery suddenly assemble themselves to a vase which then flies upward in the air. With the years our body ages, the phrase "growing older" encapsulates a notion that time moves in one direction. The fact that the basic natural laws are invariant for time reversal, but that we clearly experience a direction of time, is called the paradox of the "arrow of time" [2]. Detailed accounts of this issue are given by *Coveny* and *Highfield* [3] and by *Price* [4].

The invariance of the basic equations in physics for time reversal forms the basis of many applications that are described in this book where waves are re-emitted from receivers so that they propagate back to the original source. A clear example is given by *Derode* et al. [5], who propagated acoustic waves though a dense assemblage of scatterers to an array of receivers. The recorded signals are digitized and then time-reversed in a computer after which they are re-emitted from the receivers. These waves focus after a certain time at the original source. This process of time-reversed propagation is very stable to errors in re-emitted signal [6]. The stability of the back-propagation of strongly scattered waves to the addition of random noise is explained by *Scales* and *Snieder* [7].

2 Wave Chaos and Particle Chaos

At the end of the 19th century, scientists saw the universe as a clockwork that obeyed the laws of classical physics. If one would know the initial position and initial velocity of all the particles, one could predict the future evolution of the universe with great accuracy. Quantum mechanics shattered this mechanistic dream (or nightmare?) because chance or probability forms and integral part of this theory. In the 20th century it became apparent that very simple dynamical systems showed such a strong sensitive dependence on perturbations in the initial conditions that the prediction of the temporal evolution of such a system is practically impossible. A famous example is the Lorenz system which accounts for the air flow in a simplified model for the atmosphere [8]. In chaotic dynamical systems the perturbations in the solutions grow exponentially with time $[\exp(\mu t)]$ so that errors in the initial condition quickly lead to a very different solution. The factor μ is called the *Lyapunov exponent*. A detailed and clear account of classical chaos is given by *Tabor* [9].

Suppose one has a system where the classical equations of motion lead to chaotic behavior. How does the corresponding wave system then behave? This question is highly relevant because quantum mechanics is the wave extension of classical mechanics, and one may wonder how the solution of the Schrödinger equation behaves under a perturbation of the system that for the classical system leads to chaos. This has led to the formulation of "quantum chaos" [9]. It is not obvious that the wave system shows the same dependence to perturbations of the system as the classical system, because waves carry out a natural smoothing [10,11]. The finite extent of the wave field ensures that a wave samples space in a more extended way than a particle. This smoothing effect could render wave propagation much more stable for perturbations of the system than particle propagation [12].

In this chapter the stability of wave propagation is not addressed with solutions of the Schrödinger equation, but with a numerical simulation of waves that are analogous to the acoustic waves used in the experiments of *Derode* et al. [5]. The geometry of the system is shown in Fig. 1. A numerical simulation of the experiment of *Derode* et al. [5] is well suited to study the stability of the propagation of waves or particles to perturbations. The idea is that waves (or particles) propagate from a source through the system and are recorded at receivers. The system is then perturbed, and the waves (or particles) are re-emitted backward in time from the receivers. Because of the invariance of the system for time reversal, the waves (or particles) should converge back onto the source at time $t = 0$. When the system is perturbed, this focussing of waves (or particles) onto the original source position is degraded. This degradation of the focusing on the source position can be used as a measure of the sensitivity of system to perturbations.

In the numerical experiment waves or particles are emitted from a source and then propagate through an assemblage of isotropic point scatterers. In the wave experiments the waves are recorded on 96 equidistant receivers on the

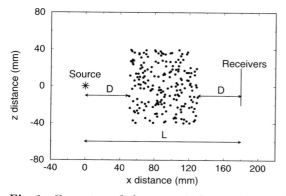

Fig. 1. Geometry of the numerical experiment with time-reversed propagation. *Dots:* scatterers

line marked "receivers." For the particles a particle is detected when it crosses the receiver line. For the time-reversed particle propagation the velocity of the particle at the receiver location is reversed, and the particle is re-emitted at time $-t$ towards the scatterers. For the time-reversed propagation of the waves the recorded wave field is numerically time-reversed and re-emitted from the receivers. It should be noted that the particles do not interact with each other, they are only scattered by the scatterers of the system.

In order to make a fair comparison, the isotropic point scatterers have the same cross-section for the waves and the particle simulation. The numerical values of the parameters are shown in Table 1. For these parameters the system can be considered to be a strongly scattering system because the size of the scatterer array (approximately 80 mm) is much larger than the mean free path (15.56 mm). As discussed by *Scales* and *Snieder* [13] such strongly scattered waves show aspects of diffusive behavior.

It should be noted that the equations of particle propagation in classical mechanics are formally equivalent to the equation of kinematic ray-tracing that governs the trajectories of rays. Therefore the comparison between wave propagation and particle propagation not only has a bearing on the relation between classical chaos and quantum chaos, it is also relevant for the relation between ray-geometric solutions and full-wave solutions.

Table 1. Numerical values of parameters in numerical experiment

Symbol	Property	Value
σ	Scattering cross-section	1.592 mm
l	Mean free path	15.56 mm
λ	Dominant wavelength	2.5 mm

3 Instability of Particle Trajectories

In this section the stability properties of particle trajectories are treated. A full derivation of the results presented in this section and the next can be found in *Snieder* and *Scales* [14] and in *Snieder* [15]. It is shown in these references that when a trajectory has the initial perturbation Δ_{in} the perturbation after n scattering events is on average given by

$$\Delta_{\text{out}} = \left(\frac{2\pi l}{\sigma}\right)^n \Delta_{\text{in}}, \tag{9}$$

where l is the mean free path and σ is the scattering cross-section. On average, a particle encounters a scatterer after each time interval l/v, where v is the velocity of the particle. This means that after a time t the number of encountered scatterers is on average given by

$$n = \frac{vt}{l}. \tag{10}$$

Using this in (9) one finds that the perturbation $\Delta(t)$ in trajectories grows exponentially with time:

$$\Delta(t) = e^{\mu t}\Delta(0), \qquad (11)$$

where the Lyapunov coefficient μ is given by

$$\mu = \frac{v}{l}\ln\left(\frac{2\pi l}{\sigma}\right). \qquad (12)$$

This implies that perturbations in the trajectories grow exponentially with time, which is one of the characteristics of chaotic behavior.

When the growth in the perturbation of a trajectory is comparable to the cross-section σ of the scatterers, a particle may suddenly miss a scatterer that it encountered on its unperturbed trajectory. Under this condition the particle will follow a fundamentally different trajectory. The associated critical length scale for the perturbation of the scatterer positions is derived by *Snieder* and *Scales* [14] and is given by

$$\delta_c^{\text{part}} = \left(\frac{\sigma}{2\pi l}\right)^n \frac{\sigma}{2}. \qquad (13)$$

Using (10) and (12) one finds that this critical length scale decreases *exponentially* with time:

$$\delta_c^{\text{part}} = \frac{\sigma}{2}e^{-\mu t}. \qquad (14)$$

Equation (13) states that the critical length scale is proportional to the cross-section σ multiplied by the dimensionless number $(\sigma/2\pi l)^n$. Since for the employed parameter setting the mean-free path l is much larger than the scattering cross-section σ (see Table 1), the critical length scale decreases dramatically as a function of the number of scatterers encountered (see Table 2). Note that even for a limited number of scatterer encounters n the critical length scale becomes much smaller than any of the characteristic dimensions of the system (which are all of the order of millimeters).

4 Instability of Wave Propagation

The wave field that propagates through a system of isotropic scatterers can be computed in a relatively simple way using the method of *Groenenboom* and *Snieder* [10], which is described in great detail by *Snieder* [15]. The wave field recorded at a receiver in the middle of the receiver array is shown in Fig. 2. The wave field consists of a long extended wave train of multiply scattered waves. This is a result of the fact that the mean free path is much less than the propagation distance of the waves (see Table 1).

Table 2. Critical error δ_c for different numbers of scattering encounters. Also indicated is the employed machine precision

n	δ_c (mm)
1	0.0129
2	2.11×10^{-4}
3	3.43×10^{-6}
4	5.60×10^{-8}
5	9.11×10^{-10}
6	1.48×10^{-11}
7	2.41×10^{-13}
8	3.93×10^{-15}
Machine precision	0.22×10^{-15}
9	6.41×10^{-17}

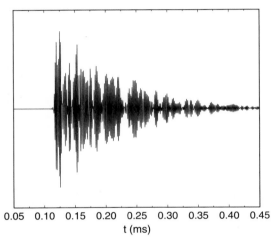

Fig. 2. Wave field at a receiver located in the middle of the receiver array

The recorded waves can be separated into the *ballistic wave* and the *coda*[2]. The ballistic wave is the wave that travels more or less along the line of sight from the source to the receiver. This wave is only affected by multiple forward scattering and consists of the early part of the wave train in Fig. 2. The detour of the multiple-scattered waves compared to the un-scattered direct wave is by definition less than a fraction of the wavelength. This means that the waves that comprise the ballistic wave are scattered within the first Fresnel zone. (The concept of the Fresnel zone and its implications are described in great detail by *Kravtsov* [16].) The coda consists of the multiply scattered waves later in the signal. These waves have been scattered in all directions.

---

[2] The term "coda" comes from music, where it denotes the closing part of a piece of music.

A critical length scale can be defined for the average perturbation of the locations of the scatterers that leads to a perturbed wave field that is uncorrelated with the unperturbed wave field. The physics of wave propagation for the coda waves and the ballistic wave is fundamentally different because the ballistic wave is mostly sensitive to the average structure of the medium within the first Fresnel zone [10,15]. For this reason, the critical length scale is different for the ballistic wave and for the coda waves. As shown by *Snieder* [15] the critical length scale for the coda waves is given by

$$\delta_c^{coda} = \frac{\lambda}{4\sqrt{2n}}, \qquad (15)$$

while the critical length scale for the ballistic wave is given by

$$\delta_c^{ball} = \frac{\sqrt{\lambda L}}{\sqrt{12(n+1)}}. \qquad (16)$$

In these expressions λ is the wavelength and L is the source–receiver separation as shown in Fig. 1.

For the coda waves the critical length scale is given by the wavelength divided by the square-root of the number of scatterers encountered. Note that in contrast to the critical length scale (13) for the particles the critical length scale for the coda waves does not depend on the scattering cross-section. The reason for this is that for the coda waves the wavelength is the relevant length scale that determines the interference between the multiply scattered waves. For the ballistic wave the critical length scale is proportional to $\sqrt{\lambda L}$. This quantity gives the width of the Fresnel zone in a homogeneous medium [16]. Since the ballistic wave is only sensitive to the properties of the medium averaged over the Fresnel zone, the ballistic wave is only affected when scatterers are moved out of the Fresnel zone when they are displaced. For this reason the width $\sqrt{\lambda L}$ of the Fresnel zone is the relevant scale length for the ballistic wave.

Using (10) the critical length scale for the coda waves can be rewritten as

$$\delta_c^{coda} = \frac{\lambda}{4\sqrt{2vt/l}}. \qquad (17)$$

Note that the time dependence of this quantity is given by $1/\sqrt{t}$, which indicates an algebraic decay of the critical length scale with time. In contrast to this, the time dependence of the critical length scale (14) for the particle scattering decays as $\exp(-\mu t)$, which denotes an exponential decay with time. Since the algebraic decay of (17) implies a much slower decay with time than the exponential decay of the critical length scale of the particles, the wave propagation is much more robust for perturbations of the system than the particle propagation. *Snieder* and *Scales* [14] conjecture that this is due to the fact that the scattered waves travel along all possible trajectories between scatterers and continue to do so when the system is perturbed, whereas the particles may travel along a fundamentally different path when the scatterer locations are perturbed.

The difference in the time dependence for the wave system and the particle system (algebraic versus exponential) has also been noted for the periodically kicked rotator. In this system a particle (or wave) moves along a ring and is exposed after each time interval T to a kick in a fixed direction. Classically this system displays chaotic behavior and the initial perturbation of the particle along the ring grows exponentially with time. As shown by *Ballantine* and *Zibin* [17] the corresponding quantum system is sensitive to a critical perturbation in the initial angle that varies with time as $1/\sqrt{t}$, which is the same time dependence as in (17). It is striking that two different systems give rise to the same time dependence of the critical length scale for both the particles and the waves.

5 Numerical Examples

In this section numerical examples are used to illustrate the analytical results of Sects. 3 and 4. Let us first consider the time-reversed propagation of particles. When a particle is re-emitted at time $-t$ from the receiver line after its velocity is reversed ($\boldsymbol{v} \to -\boldsymbol{v}$) it should return to the original source position at time $t = 0$. This is illustrated in Fig. 3, where the location of the time-reversed particles is shown at time $t = 0$. In the top panel the location of the particles that are scattered less than or equal to 6 times is shown. In this figure several thousand particles are located at the original source position at $x = z = 0$. The middle panel shows the locations of the particles after time-reversed propagation for the particles that are scattered between 7 and 9 times. Although the particles cluster near the original source position at $x = z = 0$, it can be seen that the particles do not completely converge to this point. The bottom panel shows the location of the time-reversed particles that are scattered 10 or more times. In this case the particles are not concentrated near the original source location at all.

In the numerical simulation of Fig. 3 no explicit errors have been imposed. This implies that the only error is the round-off error of the numerical calculations. The behavior of the time-reversed particles in Fig. 3 can be understood by considering Table 2, where the critical length scale is shown as a function of the number of encountered scatterers. Also shown is the numerical precision of the machine employed to carry out the calculations. When round-off errors are the only source of error, particles that are scattered 6 times or less will refocus on the source after time reversal because the numerical errors are much smaller than the critical length scale (see Table 2). Conversely, the particles that are scattered 10 times or more have according to Table 2 a critical length scale that is much smaller than the round-off error. These particles do not return to the original source position after time reversal. The numerical simulations of Fig. 3 therefore agree well with the results of Table 2.

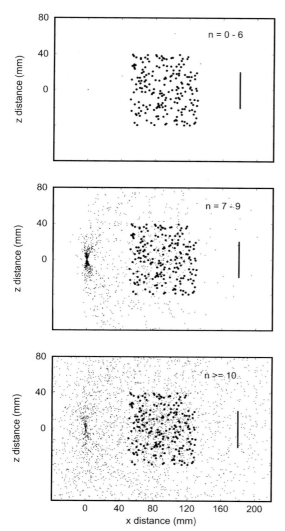

Fig. 3. Locations of the particles (*small dots*) at $t = 0$ after time-reversed propagation for particles that had 6 or less scatterer encounters (*top*), between 7 and 9 scatterer encounters (*middle*) and 10 or more scatterer encounters (*bottom*). In the *top panel* several thousand particles are located at the source position at $x = z = 0$. The original source position is indicated by a *star* in Fig. 1

In the next examples the particles are time-reversed after the scatterer locations have been randomly perturbed. The distance of the time-reversed particles to the source location is a measure of the error in the time-reversed propagation through the perturbed system. This error in converted to a number that measures the quality of the time-reversed propagation by the function $\exp(-\text{error}/D)$, where "error" is the mean distance of the time-reversed particles from the original source location and D is the typical length scale of the experiment. When all the particles refocus on the source, the error is zero and the exponential is equal to unity, while the function $\exp(-\text{error}/D)$ is much smaller than unity when the particles do not propagate back to the source.

The quality of the time-reversed propagation that is defined in this way is shown in Fig. 4 as a function of the root-mean-square value of the perturbation of the scatterer positions before the reversed propagation. The curves are shown for different numbers, n, of encountered scatterers. Note that the logarithmic scale along the horizontal axis spans 13 orders of magnitude! The critical length scale of (13) is indicated by the vertical arrows. Each of the curves shows a characteristic decay when the error in the scatterer location exceeds a certain critical value. The point at which the quality of the time-reversed propagation degrades agrees well with the critical length scale indicated by the vertical arrows.

The waves that are time-reversed refocus at time $t = 0$ at the original source position through a process of constructive interference. When the system is perturbed before time reversal the height of the interference peak decreases. Thus the ratio of the interference peak at the original source location $x = z = 0$ of the perturbed system to the same corresponding quantity of the unperturbed system is a measure of the accuracy of the time-reversed propagation. When this quantity is equal to unity, the time-reversed prop-

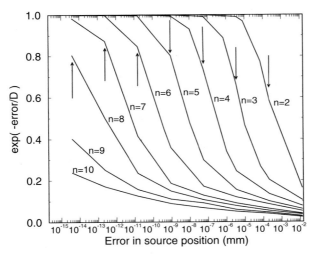

Fig. 4. Imaging quality defined as $\exp(-\text{error}/D)$ as a function of the perturbation of the initial position of the time-reversed particle. The analytical estimates of the the critical perturbation are indicated by *vertical arrows*

agation is optimal, whereas a small value of this quantity denotes degraded time-reversed propagation.

The quality of the time-reversed propagation of the waves that is defined in this way is shown in Fig. 5 both for the ballistic wave (dashed line) and for coda waves that are scattered approximately 10, 20 or 30 times (the solid lines of the left). Details of this numerical experiment are given by *Snieder* and *Scales* [14]. The analytical estimates of the critical length scales as given in (15) and (16) are shown by the vertical errors. Note that the critical length scales agree well with the numerical simulations.

The solid lines in the middle denote the quality of the time-reversed propagation when the source for the time-reversed propagation is perturbed for the ballistic wave and coda waves in three time intervals. In this case only the wave path from the source to the first scatterer changes; hence the critical length scale is given by $\lambda/4$, and this length scale does not depend on the time interval that is used for the time-reversed propagation [14].

The ballistic wave is much more robust for perturbations of the system than the coda waves. This is due to the fact that the critical length scale for the ballistic wave is given according to (16) given by the width of Fresnel zone, whereas the critical length scale of the coda waves is given according to (15) proportional to the wavelength. Since the width of the Fresnel zone is much larger than the wavelength (with a factor $\sqrt{L/\lambda}$), the critical length

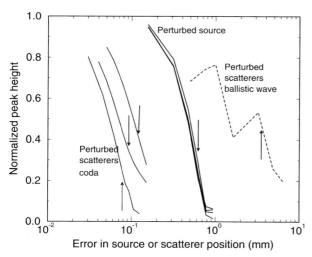

Fig. 5. Quality of time-reversed propagation of waves measured as the ratio of the peak height of the imaged section for the experiment with perturbed conditions compared to the peak height for the unperturbed imaged section. The *dashed line* represents the ballistic wave with perturbed scatterers. The *dotted lines* are for the three coda intervals for perturbed scatterers, with the latest coda interval on the left. The critical length scales from the theory are shown by *vertical arrows*

scale of the ballistic wave is much larger than the critical length scale for the coda waves. In addition, the number of scatterers encountered is much larger for the coda waves than for the ballistic waves. With the factor $1/\sqrt{n}$ in (15) this gives a further reduction of the critical length scale for the coda waves compared to the same quantity for the ballistic wave.

6 Discussion

The most striking feature in a comparison of Figs. 4 and 5 is the scale along the horizontal axis. For the particles the critical scale length ranges between 10^{-15} mm and 10^{-3} mm for particles that are scattered 8 or 2 times respectively. For the waves the critical length scale ranges between 10^{-1} mm and 4 mm for the coda waves that are scattered 30 times and the ballistic wave respectively. This indicates that wave propagation is vastly more robust than the propagation of particles. Mathematically this is related to the fact that the critical length scale decays exponentially with time $[\exp(-\mu t)]$ for the particles, while for the waves the critical length scales decay algebraically with time $(1/\sqrt{t})$. Physically there are two reasons that explain this difference. The particles are point-like objects with no scale, whereas the waves are associated with a wavelength. The wave field "feels" its environment on a scale that only depends on the wavelength; hence the wavelength determines a natural scale for the sensitivity of scattered waves for perturbations of the scatterer locations. In addition, the waves travel along all possible scattering paths, whereas the particles each travel along a unique path. When the scatterer locations are perturbed, the waves still travel along all possible scattering paths, whereas a particle may suddenly follow a fundamentally different trajectory. Both effects should form an essential element in accounting for the differences between classical chaos and wave chaos.

References

1. J. Chaiken, R. Chevray, M. Tabor, Q. M. Tan, Experimental study of Lagrangian turbulence in a Stokes flow, Proc. R. Soc. Lond. A **408**, 165–174 (1986)
2. K. Popper, The arrow of time, Nature **177**, 538, (1956)
3. P. Coveney, R. Highfield, *The Arrow of Time* (Harper Collins, London 1991)
4. H. Price, *Time's Arrow and Archimedes' Point, New Directions for the Physics of Time* (Oxford Univ. Press, New York 1996)
5. A. Derode, P. Roux, M. Fink, Robust acoustic time reversal with high-order multiple scattering, Phys. Rev. Lett. **75**, 4206–4209 (1995)
6. A. Derode, A. Tourin, M. Fink, Ultrasonic pulse compression with one-bit time reversal through multiple scattering, J. Appl. Phys. **85**, 6343–6352 (1999)
7. J. Scales, R. Snieder, Humility and nonlinearity, Geophysics **62**, 1355–1358 (1997)
8. E. N. Lorenz, Deterministic nonperiodic flow, J. Atmos. Sci. **20**, 130–141 (1963)

9. M. Tabor, *Chaos and Integrability in Nonlinear Dynamics* (Wiley-Interscience, New York 1989)
10. J. Groenenboom, R. Snieder, Attenuation, dispersion and anisotropy by multiple scattering of transmitted waves through distributions of scatterers, J. Acoust. Soc. Am. **98**, 3482–3492 (1995)
11. R. Snieder, A. Lomax, Wavefield smoothing and the effect of rough velocity perturbations on arrival times and amplitudes, Geophys. J. Int. **125**, 796–812 (1996)
12. M. C. Gutzwiller, *Chaos in Classical and Quantum Mechanics* (Springer, Berlin, Heidelberg 1990)
13. J. Scales, R. Snieder, What is a wave?, Nature **401**, 739–740 (1999)
14. R. Snieder, J. A. Scales, Time reversed imaging as a diagnostic of wave and particle chaos, Phys. Rev. E **58**, 5668–5675 (1998)
15. R. Snieder, Imaging and averaging in complex media, in J. P. Fouque (Ed.) *Diffuse Waves in Complex Media* (Kluwer, Dordrecht 1999) pp. 405–454
16. Ya. A. Kravtsov, Rays and caustics as physical objects, Prog. Opt., **XXVI**, 227–348 (1988)
17. L. E. Ballantine, J. P. Zibin, Classical state sensitivity from quantum mechanics, Phys. Rev. A **54**, 3813–3819 (1996)

Acoustic Time-Reversal Mirrors

Mathias Fink

Laboratoire Ondes et Acoustique, Ecole Supérieure de Physique et de Chimie
Industrielle de la Ville de Paris, Université Denis Diderot,
UMR CNRS 7587, 10 Rue Vauquelin, 75005 Paris, France
mathias.fink@espci.fr

Abstract. The objective of this paper is to show that time-reversal invariance can be exploited in acoustics to accurately control wave propagation through complex media.

1 Introduction

In time-reversal acoustics [1,2,3,4] a signal is recorded by an array of transducers, time-reversed and then re-transmitted into the medium. The re-transmitted signal propagates back through the same medium and refocuses on the source. In a time-reversal cavity (TRC) the array completely surrounds the source, and thus the time-reversed signals propagate backwards through the medium and go through all the multiple scattering, reflections and refraction that they underwent in the forward direction. If the time-reversal operation is only performed on a limited angular area (a time-reversal mirror TRM), a small part of the field radiated by the source is captured and time-reversed, thus limiting reversal and focusing quality.

The basic principles and limitations of time-reversal acoustics are described in Sect. 2. Various time-reversal experiments conducted with TRMs are then discussed in Sect. 3. It will be shown that focusing quality is improved if the wave traverses random media or if the wave propagates in media with reflecting boundaries as waveguides or reverberating cavities. The focusing resolution may be much better than the resolution obtained in an homogeneous medium. Multiple scattering or multiple reflections allow one part of the initial wave to be redirected towards the TRM that normally misses the transducer array. TRM appears to have an aperture that is much larger than its physical size. It will be shown that, for a reflecting cavity with chaotic boundaries, a one-channel TRM is sufficient to ensure optimal focusing. Then differences between time reversal and phase conjugation are discussed. Finally, Sect. 4 includes a short description of the potential of TRM in various applications (medical therapy and non-destructive testing).

2 Time-Reversal Cavities and Mirrors

The basic theory employs a scalar wave formulation, $\phi(\boldsymbol{r},t)$, and, hence, is strictly applicable to acoustic or ultrasound propagations in fluid. However, the basic ingredients and conclusions apply equally well to elastic waves in a solid and to electromagnetic fields.

In any propagation experiment, the acoustic sources and the boundary conditions determine a unique solution, $\phi(\mathbf{r}, t)$, in the fluid. The goal, in time-reversal experiments, is to modify the initial conditions in order to generate the dual solution $\phi(\mathbf{r}, T - t)$, where T is a delay due to causality requirements. *Jackson* and *Cassereau* [4,5] have studied theoretically the conditions necessary to insure the generation of $\phi(\mathbf{r}, T - t)$ in the entire volume of interest.

2.1 The Time-Reversal Cavity

From an experimental point of view a TRC consists of a two-dimensional piezoelectric transducer array that samples the wavefield over a closed surface. An array pitch of the order of $\lambda/2$, where λ is the smallest wavelength of the pressure field, is needed to insure the recording of all the information on the wavefield. Each transducer is connected to its own electronic circuitry, which consists of a receiving amplifier, an analog-to-digital converter, a storage memory and a programmable transmitter able to synthesize a time-reversed version of the stored signal. Although reversible acoustic retinas usually consist of discrete elements, it is convenient to examine the behavior of idealized continuous retinas, defined by two-dimensional surfaces. In the case of a TRC, we assume that the retina completely surrounds the source. The basic time-reversal experiment can be described in the following way:

As a first step, a point-like source located at $\mathbf{r}_0$ inside a volume V surrounded by the retina surface S emits a pulse at $t = t_0 \geq 0$. The wave equation in a medium of density $\rho(\mathbf{r})$ and compressibility $\kappa(\mathbf{r})$ is given by

$$(L_\mathrm{r} + L_\mathrm{t})\phi(\mathbf{r}, t) = -A\delta(\mathbf{r} - \mathbf{r}_0)\delta(t - t_0),$$
$$L_\mathrm{r} = \nabla\left(\frac{1}{\rho(\mathbf{r})}\nabla\right), \quad L_\mathrm{t} = -\kappa(\mathbf{r})\partial_{tt}, \tag{1}$$

where A is a dimensional constant that insures the compatibility of physical units between the two sides of the equation; for simplicity, this constant will be omitted in the following. The solution to (1) reduces to the Green's function $G(\mathbf{r}, t|\mathbf{r}_0, t_0)$. Classically, $G(\mathbf{r}, t|\mathbf{r}_0, t_0)$ is written as a diverging spherical wave (homogeneous and free space case) and additional terms that describe the interaction of the field itself with the inhomogeneities (multiple scattering) and the boundaries.

We assume that we are able to measure the pressure field and its normal derivative at any point on the surface S during the interval $[0, T]$. As time-reversal experiments are based on a two-step process, the measurement step must be limited in time by a parameter T. In all the following, we assume that the contribution of multiple scattering decreases with time and that T

is chosen such that the information loss can be considered to be negligible inside the volume V.

During the second step of the time-reversal process, the initial source at r_0 is removed, and we create on the surface of the cavity monopole and dipole sources that correspond to the time reversal of those same components measured during the first step. The time-reversal operation is described by the transform $t \to T - t$ and the secondary sources are

$$\begin{cases} \phi_s(r,t) = G(r, T-t|r_0, t_0), \\ \partial_n \phi_s(r,t) = \partial_n G(r, T-t|r_0, t_0). \end{cases} \quad (2)$$

In this equation, ∂_n is the normal derivative operator with respect to the normal direction n to S, oriented outward. Due to these secondary sources on S, a time-reversed pressure field $\phi^{tr}(r_1, t_1)$ propagates inside the cavity. It can be calculated using a modified version of the Helmoltz–Kirchhoff integral:

$$\phi^{tr}(r_1, t_1) = \int_{-\infty}^{+\infty} dt \iint_S [G(r_1, t_1|r, t) \partial_n \phi_s(r, t) \\ - \phi_s(r, t) \partial_n G(r_1, t_1|r, t)] \frac{d^2 r}{\rho(r)}. \quad (3)$$

Spatial reciprocity and time-reversal invariance of the wave equation (1) yield the following expression:

$$\phi^{tr}(r_1, t_1) = G(r_1, T-t_1|r_0, t_0) - G(r_1, t_1|r_0, T-t_0). \quad (4)$$

This equation can be interpreted as the superposition of incoming and outgoing spherical waves centered on the initial source position. The incoming wave collapses at the origin and is always followed by a diverging wave. Thus the time-reversed field, observed as a function of time, from any location in the cavity, shows two wavefronts, the second one being an exact replica of the first, but multiplied by -1.

If we assume that the retina does not perturb the propagation of the field (free-space assumption) and that the acoustic field propagates in an homogeneous fluid, the free-space Green's function G reduces to a diverging spherical impulse wave that propagates with a sound speed c. Introducing its expression in (4) yields the following formulation of the time-reversed field:

$$\phi^{tr}(r_1, t_1) = K(r_1 - r_0, t_1 - T + t_0), \quad (5)$$

where the kernel distribution $K(r, t)$ is given by

$$K(r, t) = \frac{1}{4\pi |r|} \delta\left(t + \frac{|r|}{c}\right) - \frac{1}{4\pi |r|} \delta\left(t - \frac{|r|}{c}\right). \quad (6)$$

The kernel distribution $K(r, t)$ corresponds to the difference between two impulse spherical waves one converging to and one diverging from the origin of the spatial coordinate system, i.e., the location of the initial source.

Resulting from this superposition, the pressure field remains finite for all time throughout the cavity, although the converging and diverging spherical waves show a singularity at the origin. Note that this singularity occurs at time $t_1 = T - t_0$.

The time-reversed pressure field, observed as a function of time, shows two wavefronts, the second one being an exact replica of the first, but multiplied by -1. If we consider a wide-band excitation function instead of a Dirac distribution, $\delta(t)$, the two wavefronts overlap near the focal point, therefore resulting in a temporal distortion of the acoustic signal. It can be shown that this distortion yields a temporal derivation of the initial excitation function at the focal point.

If we now calculate the Fourier transform of (6) over the time variable t, we obtain

$$\tilde{K}(\boldsymbol{r},\omega) = \frac{1}{2\mathrm{j}\pi} \frac{\sin(\omega|\boldsymbol{r}|/c)}{|\boldsymbol{r}|} = \frac{1}{\mathrm{j}\lambda} \frac{\sin(k|\boldsymbol{r}|)}{k|\boldsymbol{r}|}, \qquad (7)$$

where λ and k are the wavelength and wavenumber, respectively. As a consequence, the time-reversal process results in a pressure field that is effectively focused on the initial source position, but with a focal spot size limited to one half-wavelength. The size of the focal spot is a direct consequence of the superposition of the two wavefronts and can be interpreted in terms of the diffraction limitations (loss of the evanescent components of the acoustic fields).

A similar interpretation can be given in the case of an inhomogeneous fluid, but the Green's function G now takes into account the interaction of the pressure field with the inhomogeneities of the medium. If we were able to create a film of the propagation of the acoustic field during the first step of the process, the final result could be interpreted as a projection of this film in the reverse order, immediately followed by a re-projection in the initial order.

The apparent failure of the time-reversed operation that leads to diffraction limitation can be interpreted in the following way: The second step described above is not strictly the time reversal of the first step. During the second step of an ideal time-reversed experiment, the initial active source (which injects some energy into the system) must be replaced by a *sink* (the time reversal of a source). An acoustic sink is a device that absorbs all arriving energy without reflecting it. De Rosny and Fink, using the source as a diverging wavefront canceller, have recently built such a sink in our laboratory and have observed a focal spot size quite below diffraction limits [6].

2.2 The Time-Reversal Mirror

This theoretical model of the closed time-reversal cavity is interesting, since it affords an understanding of the basic limitations of the time-reversed self-focusing process; but it has some limitations, particularly compared to an experimental setup:

- It can be proven that it is not necessary to measure and time-reverse both the scalar field (acoustic pressure) and its normal derivative on the cavity surface; measuring the pressure field and re-emitting the time-reversed field in the backward direction yields the same results, on the condition that the evanescent parts of the acoustic fields have vanished (propagation along several wavelengths) [7]. This comes from the fact that each transducer element of the cavity records the incoming field from the forward direction and retransmits it (after the time-reversal operation) in the *backward* direction (and not in the forward direction). The change between the forward and backward directions replaces the measurement and the time reversal of the field-normal derivative.
- From an experimental point of view, it is not possible to measure and re-emit the pressure field at any point on a 2-dimensional surface; experiments are carried out with transducer arrays that spatially sample the receiving and emitting surface. The spatial sampling of the TRC by a set of transducers may introduce grating lobes. These lobes can be avoided by using an array pitch smaller than $\lambda_{min}/2$, where λ_{min} is the smallest wavelength of the transient pressure field. In this case, each transducer senses all the wavevectors of the incident field.
- The temporal sampling of the data recorded and transmitted by the TRC has to be at least of the order of $T_{min}/8$ (T_{min} is the minimum period) to avoid secondary lobes [8].
- It is generally difficult to use acoustic arrays that completely surround the area of interest, and the closed cavity is usually replaced by a TRM of finite angular aperture. This yields an increase in the point spread function dimension that is usually related to the mirror angular aperture observed from the source.

3 Time-Reversal Experiments

3.1 Time Reversal through Random Media

Derode et al. [9] carried out the first experimental demonstration of the reversibility of an acoustic wave propagating through a random collection of scatterers with strong multiple-scattering contributions. In an experiment such as the one depicted in Fig. 1, a multiple-scattering sample is placed between the source and an array made of 128 elements. The whole setup is in a water tank. The scattering medium consists of a set of 2000 parallel steel rods (diameter 0.8 mm) randomly distributed. The sample thickness is $L = 40$ mm, and the average distance between rods is 2.3 mm. The source is 30 cm away from the TRM and transmits a short (1μs) ultrasonic pulse (3 cycles of a 3.5 MHz, Fig. 2a). Figure 2b shows the waveform received on the TRM by one of the elements. It spread over 250μs, i.e., ~ 250 times the initial pulse duration. A long incoherent wave is observed, which results

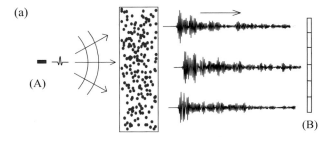

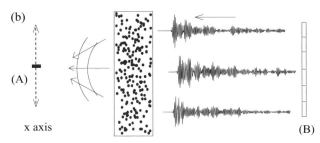

Fig. 1. Sketch of experiment (a) first step, the source sends a pulse through the sample, the transmitted wave is recorded by the TRM. (b) second step, the multiplay scattered signals are time reversed, they are retransmitted by the TRM and A records the reconstructed pressure field

from the multiply scattered contribution. As a second step to the experiment, the 128 signals are time-reversed and transmitted and an hydrophone measures the time-reversed wave around the source location. Two different aspects of this problem have been studied: the property of the signal recreated at the source location (time compression) and the spatial property of the time-reversed wave around the source location (spatial focusing).

The time-reversed wave traverses the rods back to the source, and the signal received at the source is represented in Fig. 2c; an impressive compression is observed, since the received signal lasts about 1 µs, in comparison to 250µs. The pressure field is also measured around the source, in order to obtain the directivity pattern of the beam emerging from the rods after time reversal, and the results are plotted in Fig. 3. Surprisingly, multiple scattering has *not* degraded the resolution of the system; indeed, the resolution is found to be six times finer (thick line) than the classical diffraction limit (thin line). However, this effect does not contradict the laws of diffraction. The intersection of the incoming wavefront with the sample has a typical size D. After time reversal, the waves travel on the same scattering paths and focus back on the source as if they were passing through a converging lens with size D. The angular aperture of this pseudo-lens is much wider than

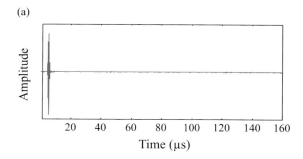

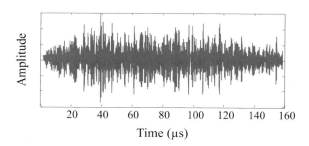

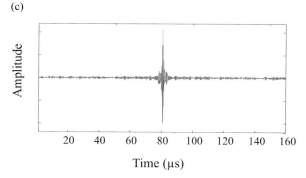

Fig. 2. (a) Signal transmitted in water and recieved on transducer 64. (b) Signal transmitted through the multiple scattering sample and recieved on transducer 64. (c) Signal recieved on the source

that of the array alone, and hence there is an improvement in resolution. In other words, because of the scattering sample, the array is able to detect higher spatial frequencies than it would in a purely homogeneous medium. High spatial frequencies that would have been otherwise lost are redirected towards the array, due to the presence of the scatterers in a large area.

This experiment also shows that the acoustic time-reversal experiments are surprisingly stable. The recorded signals are sampled with 8-bit analog-to-

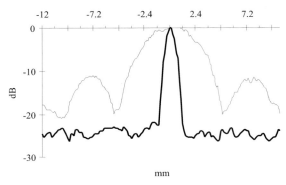

Fig. 3. Directivity patterns in water (*thin line*), through the multiple scattering medium (*thick line*)

digital converters that introduce quantization errors, but the focusing process still works. This has to be compared to time-reversal experiments involving particles moving like balls on an elastic billiard table of the same geometry. Computation of the direct and reversed particle trajectories moving in a plane among a fixed array of some thousand obstacles shows that the complete trajectory is irreversible. Indeed, such a system is a well-known example of a chaotic system that is highly sensitive to initial conditions. The finite precision that occurs in the computer leads to an error in the trajectory of the time-reversed particle that grows exponentially with the number of scattering encounters.

Recently, *Snieder* and *Scales* [10] performed numerical simulations to point out the fundamental difference between waves and particles in the presence of multiple scattering by random scatterers. In fact, they used time reversal as a diagnostic of wave and particle chaos; in a time-reversal experiment, complete focusing on the source will only take place if the velocity and positions are known exactly. The degree δ to which errors in these quantities destroy the quality of focusing is diagnostic of the stability of the wave or particle propagation. Intuitively, the consequences of a slight deviation δ in the trajectory of a billiard ball will become more and more obvious as time goes on and as the ball undergoes more and more collisions. Waves are much less sensitive than particles to initial conditions. In a multiple-scattering situation, the critical length scale δ that causes a significant deviation at a time t in the future decreases exponentially with time in the case of particles, whereas it only decreases as the square root of time for waves in the same situation.

Waves and particles react in fundamentally different ways to perturbations of the initial conditions. The physical reason for this is that each particle follows a well-defined trajectory, whereas waves travel along all possible trajectories, visiting all the scatterers in all possible combinations. While a small error in the initial velocity or position makes the particle miss one obstacle and completely change its future trajectory, the wave amplitude is much more stable because it results from the interference of all the possible trajectories; small errors in the transducer operations will sum up in a linear way for wave propagation resulting in only a small perturbation.

3.1.1 Time Reversal as a Time Correlator

As for any linear and time-invariant process, wave propagation through a multiple-scattering medium may be described as a linear system with different impulse responses. If a source located at r_0 sends a Dirac pulse $\delta(t)$, the jth transducer of the TRM will record the corresponding impulse response $h_j(t)$ for a point transducer located at r_j to the Green function $G(r_j, t|r_0, 0)$. Moreover, due to reciprocity, $h_j(t)$ is also the impulse response describing the propagation of a pulse from the jth transducer to the source. Thus, neglecting the causal time delay T, the time-reversed signal at the source is equal to the convolution product $h_j(t) * h_j(-t)$.

This convolution product, in terms of signal analysis, is typical of a *matched filter*. Given a signal as input, a matched filter is a linear filter whose output is optimal in some sense. Whatever the impulse response $h_j(t)$, the convolution $h_j(t) * h_j(-t)$ is maximum at time $t = 0$. This maximum is always positive and equals $\int h_j^2(t)dt$, i.e., the energy of the signal $h_j(t)$. This has an important consequence. Indeed, with an array of N elements, the time-reversed signal recreated on the source writes as a sum:

$$\phi^{\text{tr}}(r, t) = \sum_{j=1}^{j=N} h_j(t) * h_j(-t). \tag{8}$$

Even if $h_j(t)$ are completely random and apparently uncorrelated signals, each term in this sum reaches its maximum at time $t = 0$. Therefore, all contributions add constructively around $t = 0$, whereas at earlier or later times uncorrelated contributions tend to destroy one another. Thus the re-creation of a sharp peak after time reversal in an array of N elements can be viewed as an interference process between the N outputs of N matched filters.

The robustness of the TRM can also be accounted for through the matched filter approach. If, for some reason, the TRM does not exactly retransmit $h_j(-t)$ but rather $h_j(-t) + n_j(t)$, where $n_j(t)$ is an additional noise on channel j, then the re-created signal is written as follows:

$$\sum_{j=1}^{j=N} h_j(t) * h_j(-t) + \sum_{j=1}^{j=N} h_j(t) * n(t).$$

The time-reversed signals $h_j(-t)$ are tailored to exactly match the medium impulse response, which results in a sharp peak. However, an additional small noise is not matched to the medium and, given the extremely long duration involved, it generates a low-level, long-lasting background noise instead of a sharp peak.

3.1.2 Time Reversal as a Spatial Correlator

Another way to consider the focusing properties of the time-reversed wave is to follow the impulse response approach and treat the time-reversal process as a spatial correlator. If we consider $h'_j(t)$ to be the propagation impulse response from the jth element of the array to an observation point $\boldsymbol{r}_1$, not the source location $\boldsymbol{r}_0$, the signal recreated at $\boldsymbol{r}_1$ at time $t_1 = 0$ can be written:

$$\phi_j^{\mathrm{tr}}(\boldsymbol{r}_1, 0) = \int h_j(t) h'_j(t) \mathrm{d}t \tag{9}$$

Notice that this expression can be used as a way to define the directivity pattern of the time-reversed waves around the source. Now, due to reciprocity, the source S and the receiver can be exchanged, i.e., $h'_j(t)$ is also the signal that would be received at $\boldsymbol{r}_1$ if the source was the jth element of the array. Therefore, we can imagine that this array element is the source and that the transmitted field is observed at two points $\boldsymbol{r}_1$ and $\boldsymbol{r}_0$. The spatial correlation function of this wavefield would be $\left\langle h_j(t) h'_j(t) \right\rangle$, where the impulse-response product is averaged over different realizations of the disorder. Therefore, (9) can be viewed as an estimator of this spatial correlation function. Note that in one time-reversal experiment we have only access to a single realization of the disorder. However, the ensemble average can be replaced by a time average, a frequency average or a spatial average over a set of transducers. In that sense, the spatial resolution of the TRM (i.e., the $-6\,\mathrm{dB}$ width of the directivity pattern) is simply an estimate of the correlation length of the scattered wavefield [11].

This has an important consequence. Indeed, if the resolution of the system essentially depends on correlation properties of the scattered wavefield, it should become independent from the array's aperture. This is confirmed by the experimental results. Figure 4 presents the directivity patterns obtained

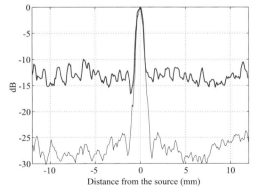

Fig. 4. Directivity patterns with $N = 122$ transducers (*thin line*) and $N = 1$ transducer (*thick line*)

through a 40-mm-thick multiple-scattering sample, using either one array element or the whole array (122 elements) as a TRM. In both cases, the spatial resolution at $-6\,\text{dB}$ is the same: $\sim 0.85\,\text{mm}$. In contrast to what happens in a homogeneous medium, enlarging the aperture of the array does not change the $-6\,\text{dB}$ spatial resolution. However, even though the number N of active array elements does not influence the typical width of the focal spot, it has a strong impact on the background level of the directivity pattern ($\sim -12\,\text{dB}$ for $N = 1$, $\sim -28\,\text{dB}$ for $N = 122$), as can be seen in Fig. 4.

Finally, the fundamental properties of time reversal in a random medium rely on the fact that it is both a space and time correlator, and the time-reversed waves can be viewed as an estimate of the space and time autocorrelation functions of the waves scattered by a random medium. The estimate becomes better as the number of transducers in the mirror is increased.

Moreover, the system is not sensitive to a small perturbation, since adding a small noise to the scattered signals (e.g., by digitizing them on a reduced number of bits) may alter the noise level but does not drastically change the correlation time or the correlation length of the scattered waves. Even in the extreme case where the scattered signals are digitized *on a single bit*, Derode has shown recently that the time and space resolution of the TRM were practically unchanged [12], which is striking evidence for the robustness of wave time reversal in a random medium.

3.2 Time Reversal in Waveguides

In the time-reversal cavity approach, the transducer array samples a closed surface surrounding the acoustic source. In the last section, we saw how the multiple-scattering processes in a large sample widen the effective TRM aperture. The same kind of improvement may be obtained for waves propagating in a waveguide or in a cavity. Multiple reflections along the medium boundaries significantly increase the apparent aperture of the TRM. The basic idea is to replace one part of the TRC transducers by reflecting boundaries that redirect one part of the incident wave towards the TRM aperture. Thus spatial information is converted into the time domain and the reversal quality depends crucially on the duration of the time-reversal window, i.e., the length of the recording to be reversed.

3.2.1 Ultrasonic Waveguide

Experiments conducted by Roux in rectangular ultrasonic waveguides have shown the effectiveness of time-reversal processing in compensating for multipath effects [13]. The experiment is conducted in a waveguide whose interfaces are plane and parallel. The length of the guide is $L \approx 800\,\text{mm}$ along the y axis, with a vertical water depth $H \approx 40\,\text{mm}$ along the x axis.

A point-like ultrasonic source is located on one side of the waveguide. On the other side, a TRM, identical to the one used in the multiple-scattering medium, is used. Ninety-six of the array elements are used, which corresponds to an array aperture equal to the waveguide aperture.

Figure 5a shows the field radiated by the source and recorded by the transducer array after propagation through the channel. After the arrival of the first wavefront corresponding to the direct path, we observe a set of signals, due to multiple reflections of the incident wave between the interfaces, that spread over 100μs. Figure 5b represents the signal received on one transducer of the TRM.

After the time-reversal operation of the 100μs signals, we observe the spatio-temporal distribution of the time-reversed field on the source plane (Fig. 6a) and we note a remarkable temporal compression at the source location (Fig. 6b). This means that multipath effects are fully compensated for. The signal observed at the source is nearly identical to the one received in a time-reversed experiment conducted in free space.

In this experiment, the transfer function of the waveguide has been completely compensated for by the time-reversal process. As with a multiple-scattering medium, the time-reversal process enables the realization of an

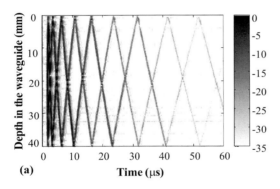

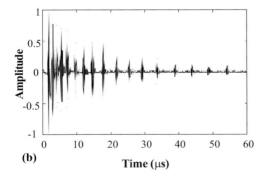

Fig. 5. (a) Field radiated by the source and recorded by the transducer array after propagation through the channel. (b) Signal recieved on one transducer of the TRM

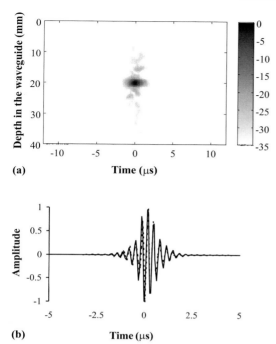

Fig. 6. (a) Spatio-temporal distribution of the time-reversed field on the source plane. (b) Temporal compression at the source location

optimal matched filter of the waveguide transfer function. Analysis of Fig. 6 shows that the ratio between the peak signal and the side lobe level is on the order of 45 dB.

Figure 7 shows the directivity pattern of the time-reversed field observed in the source plane. The time-reversed field is focused on a spot which is much smaller than the one obtained with the same TRM in free space. In our experiment, the −6 dB lateral resolution is improved by a factor of 9. This can be easily interpreted by the images theorem in a medium bounded by two mirrors. For an observer located at the source point, the 40 mm TRM appears to be accompanied by a set of virtual images related to multipath reverberation. The effective TRM is then a set of TRMs as shown in Fig. 8. When taking into account the first 10 arrivals, the theoretical effective aperture of the mirror array is 10 times larger than the real aperture. However, in practice, as the replicas arrive later, their amplitudes decrease. The angular directivity of the transducers leads to an apodization of the effective aperture of the TRM.

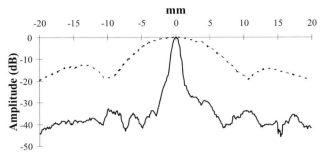

Fig. 7. Directivity pattern of the time-reversed field observed in the source plane in free space (*dotted line*), in the waveguide (*solid line*)

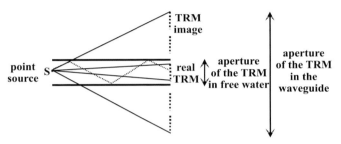

Fig. 8. Due to reverberation in the waveguide, the effective aperture of the TRM is increased

3.2.2 Underwater Acoustics

Acoustic waveguides are currently found in underwater acoustics, especially in shallow water, where multipath propagation limits the capacity of underwater communication systems. The problem arises because acoustic transmission in shallow water bounces off the ocean surface and floor, so that a transmitted pulse gives rise to multiple arrivals at the receiver.

To compensate for acoustic reverberation in the ocean, one-channel time reversal was first introduced in the early 1960s by *Parvulescu* and *Clay* [14,15]. They performed experiments in shallow water at sea with one transducer working in a time-reversed mode. They observed temporal compression but their experiments did not include the spatial focusing property of TRMs. Parvulescu's approach consists of considering the ocean as a correlator. *Jackson* and *Dowling* [4] developed a theoretical formalism to describe phase conjugation in the ocean. This formalism is based on the modal decomposition of the pressure field in an acoustic waveguide. Following this approach, in 1992 *Feuillade* and *Clay* [16] carried out numerical time-reversal experiments in shallow water. Since 1996, *Kuperman* et al. [17,18] have performed several underwater acoustics experiments in a 120-m-deep ocean waveguide. At frequencies of 500 Hz and 3.5 kHz, they used a 24-element TRM to accomplish time-reversal focusing and multipath compensation from 7 km up to 30 km.

Theoretically speaking, for one spectral component at frequency ω, the time-reversal operation consists of a phase conjugation of the incident field. For an incident field coming from a point source located at depth x_s in the plane $y = L$, the phase conjugation, performed in shallow water, from a vertical array of N discrete sources located in the plane $y = 0$ leads to the following time-reversed pressure field at observation point (x, y):

$$\phi^{\text{tr}}(x, y, t) = \sum_{j=1}^{N} G_\omega(x, y | x_j, 0) G_\omega^*(x_j, 0 | x_s, L) \exp(-i\omega t) , \quad (10)$$

where $G_\omega^*(x_j, 0 | x_s, L)$ is the conjugated monochromatic "Green's function" of the waveguide at frequency ω between a source at depth x_s and range L and a receiver at depth x_j and range 0. In other words, the phase-conjugated field in the plane of the source is the sum over the array elements of a product of two Green's functions: one describes the propagation from the source to the array, and the other describes the propagation from the array to the observation plane. Time reversal appears in the conjugation of the Green's function between the source and the array in the right term of (10). In a range-independent waveguide, the Green's function is expressed as follows:

$$G_\omega(x, 0 | x_s, L) = \frac{i}{\rho(x_s)(8\pi L)^{1/2}} \exp\left(-i\frac{\pi}{4}\right) \sum_n \frac{u_n(x_s) u_n(x)}{k_n^{1/2}} \exp(ik_n L) , \quad (11)$$

where n is the number of the propagating mode, $u_n(x)$ corresponds to the modal shape as a function of depth and k_n is the wavenumber. To demonstrate that $\phi^{\text{tr}}(x, y, t)$ focuses at the position of the initial source, we simply substitute (11) into (10), which specifies that we sum over all modes and array sources:

$$\phi^{\text{tr}}(x, y, t) \approx \sum_j \sum_n \sum_m \frac{u_m(x) u_m(x_j) u_n(x_j) u_n(x_s)}{\rho(x_j) \rho(x_s) \sqrt{k_m k_n y L}} \exp[i(k_m y - k_n L) - \omega t] . \quad (12)$$

For an array which substantially spans the water column, we approximate the sum of sources as an integral and invoke the orthonormality of the modes:

$$\int_0^\infty \frac{u_m(x) u_n(x)}{\rho(x)} dx = \delta_{nm} . \quad (13)$$

The sum over j selects mode $m = n$, and (12) becomes

$$\phi^{\text{tr}}(x, y, t) \approx \sum_m \frac{u_m(x) u_m(x_s)}{\rho(x_s) k_m \sqrt{yL}} \exp[ik_m(y - L) - \omega t] . \quad (14)$$

In the plane of the source at $y = L$, the closure relations which define the modes as a complete set $\left(\sum_m \frac{u_m(x) u_m(x_s)}{\rho(x_s)} = \delta(x - x_s)\right)$ can be applied under the assumption that k_n are nearly constant over the interval of the contributing modes. This leads to $\phi^{\mathrm{tr}}(x, L, t) \approx \delta(x - x_\mathrm{s}) \exp(-\mathrm{i}\omega t)$, which proves that the phase-conjugated field focuses back at the source.

Kuperman et al. have experimentally demonstrated in the ocean the robustness of time-reversal focusing, provided the array adequately samples the field in the water column. They have shown that temporal changes in the ocean due to surface waves and internal waves degrade the focus but that this degradation is tolerable if the average Green's function is not severely perturbed by these time variations [17]. Moreover they experimentally achieved a shift of the focal range on the order of 10% by shifting the central frequency of the TRM prior to retransmission [18].

3.3 Time Reversal in Chaotic Cavities

In this section, we are interested in another aspect of multiply reflected waves: waves confined in closed reflecting cavities such as elastic waves propagating in a silicon wafer. With such boundary conditions, no information can escape from the system and a reverberant acoustic field is created. If, moreover, the cavity shows ergodic properties and negligible absorption, one may hope to collect all information at only one point. *Draeger* and *Fink* [19,20,21] have shown experimentally and theoretically that in this particular case a time reversal can be obtained *using only one time-reversal channel* operating in a closed cavity. The field is measured at one point over a long period of time and the time-reversed signal is re-emitted at the same position.

The experiment is 2-dimensional and is carried out by using elastic surface waves propagating along a monocrystalline silicon wafer whose shape is a chaotic stadium. The shape of the cavity is of crucial importance. The chaotic stadium geometry insures that each acoustic ray radiated by the source will pass, after several reflections, sufficiently close to any point of the cavity. This ergodic property may be obtained for different geometries, and the geometry called the "D-shape stadium" was chosen for its simplicity.

Silicon was selected for its weak absorption. The elastic waves which propagate in such a plate are Lamb waves. An aluminum cone coupled to a longitudinal transducer generates these waves at one point in the cavity. A second transducer is used as a receiver. The central frequency of the transducers is 1 MHz, and its bandwidth is 100%. At this frequency, only three Lamb modes are possible (one flexural, two extensional). The source is isotropic and considered point-like because the cone tip is much smaller than the central wavelength. A heterodyne laser interferometer measures the displacement field as a function of time at different points on the cavity. Assuming that there is nearly no mode conversion between the flexural mode and other modes at the boundaries, we have only to deal with one field, the flexural-scalar field.

The experiment is a two-step process as described above: In the first step, one of the transducers, located at point A, transmits a short omnidirectional signal of duration 0.5μs into the wafer. Another transducer, located at B, observes a very long chaotic signal, which results from multiple reflections of the incident pulse along the edges of the cavity and which continues for more than 50 ms, corresponding to some one hundred reflections along the boundaries. Then, a portion of 2 ms of the signal is selected, time-reversed and re-emitted by point B. As the time-reversed wave is a flexural wave that induces vertical displacement of the silicon surface, it can be observed using the optical interferometer that scans the surface around point A (Fig. 9).

One observes both an impressive time recompression at point A and a refocusing of the time-reversed wave around the origin (Fig. 10), with a focal spot whose radial dimension is equal to half the wavelength of the flexural wave. Using reflections at the boundaries, the time-reversed wave field converges towards the origin from all directions and gives a circular spot, like the one that could be obtained with a closed time-reversal cavity covered with transducers. The 2-ms time-reversed waveform is the time sequence needed to focus exactly on point A.

The success of this time-reversal experiment is particularly interesting with respect to two aspects. Firstly, it proves again the feasibility of time

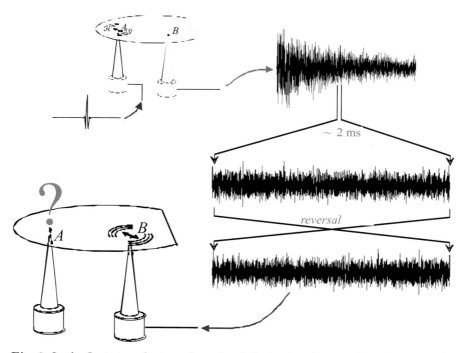

Fig. 9. In the first step, the transducer located at point A transmits a short omnidirectional signal $(0, 5\mu s)$ into the wafer. The transducer located at point B observes a very long chaotic signal which lasts more than 50 ms. In the second step, a portion of 2 ms of the signal is selected, time-reversed and re-emitted by point B

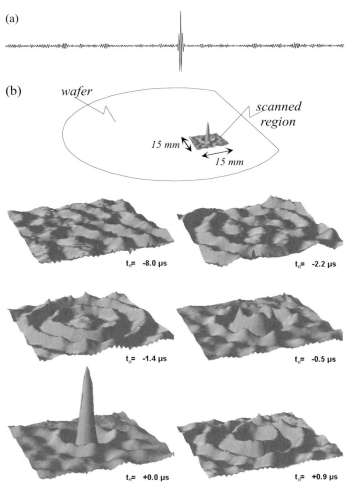

Fig. 10. (a) Time-reversed signal observed at point A. The observed signal is 210µs long. (b) Time-reversed wave field observed at different times around point A on a 15 mm × 15 mm square

reversal in wave systems with chaotic ray dynamics. Paradoxically, in the case of one-channel time reversal, chaotic dynamics is not only harmless but also even useful, as it guarantees ergodicity. Secondly, using a source of vanishing aperture, we obtain an almost perfect focusing quality. The procedure approaches the performance of a closed TRC, which has an aperture of 360°. Hence, a one-point time reversal in a chaotic cavity produces better results than a TRM in an open system. Using reflections at the edge, focusing quality is not aperture limited, and in addition, the time-reversed collapsing wavefront approaches the focal spot from all directions.

As for the multiple-scattering medium, focusing properties of the time-reversed wave can be calculated using the spatial correlator approach. Taking into account the modal decomposition of the impulse response $h_{AB}(t)$ on the eigenmodes $\psi_n(\boldsymbol{x})$ of the cavity with eigenfrequency ω_n, we obtain

$$h_{AB}(t) = \sum_n \psi_n(A)\psi_n(B)\frac{\sin(\omega_n t)}{\omega_n}, \quad (t > 0). \tag{15}$$

If $h_{A'B}(t)$ is the propagation impulse response from point B to an observation point A' (with coordinates $\boldsymbol{r}_1$), which is different from the source location A, the time-reversed signal recreated at A' at time $t_1 = 0$ can be written as follows:

$$\phi^{\mathrm{tr}}(\boldsymbol{r}_1, 0) = \int h_{AB}(t)\, h_{A'B}(t)\, \mathrm{d}t. \tag{16}$$

Thus the directivity pattern of the time-reversed wave field is given by the cross correlation of the Green's functions that can be developed on the eigenmodes of the cavity:

$$\phi^{\mathrm{tr}}(\boldsymbol{r}_1, 0) = \sum_n \frac{1}{\omega_n^2}\psi_n(A)\psi_n(\boldsymbol{r}_1)\psi_n^2(B). \tag{17}$$

Note that in a real experiment one has to take into account the limited bandwidth of the transducers, so a spectral function $F(\omega)$ centered on frequency ω_c, with bandwidth $\Delta\omega$, must be introduced and we can write (17) in the form

$$\phi^{\mathrm{tr}}(\boldsymbol{r}_1, 0) = \sum_n \frac{1}{\omega_n^2}\psi_n(A)\psi_n(\boldsymbol{r}_1)\psi_n^2(B)F(\omega_n). \tag{18}$$

Thus the summation is limited to a finite number of modes, which is typically in our experiment of the order of a few hundred. As we do not know the exact eigenmode distribution for each chaotic cavity, we cannot evaluate this expression directly. However, one may use a statistical approach and consider the average over different realizations, which consists of summing over different cavity realizations. So we replace in (18) the eigenmodes' product by their expectation values $\langle\ldots\rangle$. We also use the qualitative argument proposed by *Berry* [22,23,24] to characterize irregular modes in a chaotic system. If chaotic rays support an irregular mode, it can be considered as a superposition of a large number of plane waves with random direction and phase. This implies that the amplitude of an eigenmode has a Gaussian distribution with $\langle\psi_n^2\rangle = \sigma^2$ and a short-range isotropic correlation function given by a Bessel function that can be written as follows:

$$\langle\psi_n(A)\psi_n(\boldsymbol{r}_1)\rangle = J_0(2\pi\,|\boldsymbol{r}_1 - \boldsymbol{r}_0|/\lambda_n), \tag{19}$$

where λ_n is the wavelength corresponding to ω_n. If A and A' are sufficiently far apart from B not to be correlated, then

$$\langle\psi_n(A)\psi_n(\boldsymbol{r}_1)\psi_n^2(B)\rangle = \langle\psi_n(A)\psi_n(\boldsymbol{r}_1)\rangle\langle\psi_n^2(B)\rangle. \tag{20}$$

One obtains finally

$$\langle \phi^{\text{tr}}(\boldsymbol{r}_1, 0) \rangle = \sum_n \frac{1}{\omega_n^2} J_0(2\pi \left| \boldsymbol{r}_1 - \boldsymbol{r}_0 \right| / \lambda_n) \sigma^2 F(\omega_n). \quad (21)$$

The experimental results obtained in Fig. 10 agree with this prediction and show that in a chaotic cavity the spatial resolution is independent of the TRM aperture. Indeed, with a one-channel TRM, the directivity patterns at $t = 0$ are closed to the Bessel function $J_0(2\pi \left| \boldsymbol{r}_1 - \boldsymbol{r}_0 \right| / \lambda_c)$ corresponding to the central frequency of the transducers.

One can also observe that a very good estimate of the eigenmode correlation function is experimentally obtained with only one realization. A one-channel omnidirectional transducer is able to refocus a wave in a chaotic cavity, and we have not averaged the data on different cavities or on different positions of the transducer B.

3.3.1 Phase Conjugation Versus Time Reversal

This interesting result emphasizes how interesting time-reversal experiments are compared to phase-conjugated experiments. In phase conjugation, one only works with monochromatic waves and not with broadband pulses. For example, if one works only at frequency ω_n, so that there is only one term in (18), one cannot refocus a wave on point A. An omnidirectional transducer, located at any position B, working in monochromatic mode, sends a diverging wave in the cavity that has no reason to refocus on point A. The refocusing process works only with broadband pulses, with a large number of eigenmodes in the transducer bandwidth. Here, the averaging process that gives a good estimate of the spatial correlation function is not obtained by summing over different realizations of the cavity, as in (18), but by a sum over "pseudo-realizations" which correspond to the different modes in the same cavity. This come from the fact that in a chaotic cavity we may assume a statistical decorrelation of the different eigenmodes. As the number of eigenmodes available in the transducer bandwidth increases, the refocusing quality becomes better and the focal spot pattern becomes closed to the ideal Bessel function. Hence, the signal-to-noise level should increase as the square-root of the number of modes in the transducer bandwidth.

A similar result has also been observed in the time-reversal experiment conducted in a multiple-scattering medium. A clear refocusing has been obtained with only a single array element (Fig. 4). The focusing process works with broadband pulses (the transducer center frequency is 3.5 MHz with a 50% bandwidth at −6 dB). For each individual frequency there is no focusing, and the estimate of the spatial correlation is very noisy. However, for a large bandwidth, if we have statistical decorrelation of the wave fields for different frequencies, the time-reversed field is self-averaging.

4 Applications of Time-Reversal Mirrors

The most promising area for the application of TRMs is *pulse–echo* detection. In this domain, one is interested in the detection, imaging and sometimes destruction of passive reflecting targets. The low velocities of ultrasonic waves allows separation of reflecting targets at different depths. A piezoelectric transducer first sends a short impulse and then detects the various echoes from the targets. In nondestructive evaluation (NDE), cracks and defects can be found within materials of various shapes. In medical imaging, one looks for organ walls, calcification, tumors, kidney or gallbladder stones. In underwater acoustics, one looks for mines, submarines, or objects buried under sediments. In all of these cases, the acoustic detection quality depends on the availability of the sharpest possible ultrasonic beams to scan the medium of interest. The presence of an aberrating medium between the targets and the transducers can drastically change the beam profiles. In medical imaging, a fat layer of varying thickness, bone tissues, or some muscular tissues may greatly degrade focusing. In the human body, ultrasonic velocity variations from $1440\,\mathrm{m/s}$ in fat to $1675\,\mathrm{m/s}$ in collagen defocus and deflect acoustic beams. In NDE, the samples to be evaluated are usually immersed in a pool; the interface shape between the samples and the coupling liquid currently limits the detectability of small defects. In underwater acoustics, refraction due to oceanic structure ranging in scale from centimeters to tens of kilometers are important sources of distortions. For all these applications, a TRM array can be controlled according to a three-step sequence [25,26]. One part of the array generates a brief pulse to illuminate the region of interest through an aberrating medium. If the region contains a point reflector, the reflected wavefront is selected using a temporal window and the information acquired is time reversed and reemitted. The re-emitted wavefront refocuses on the target and through the aberrating medium. It compensates also for unknown array deformation. In terms of signal theory, time-reversal processing makes the spatio-temporally matched filter [27] to the propagation transfer function between the array and the target. Although this self-focusing technique is highly effective, it requires the presence of a reflecting target in the medium. When this medium contains several targets, the problem is more complicated and iteration of the time-reversal operation may be used to select one target. Indeed, if the medium contains two targets of different reflectivity, the time reversal of the echoes reflected from these targets generates two wavefronts focused on each target. The mirror produces the real acoustic images of the two reflectors on themselves. The highest amplitude wavefront illuminates the most reflective target, while the weakest wavefront illuminates the second target. In this case, the time reversal process can be iterated. After the first time-reversed illumination, the weakest target is illuminated more weakly and reflects a fainter wavefront than the one coming from the strongest target. After some iterations, the process converges and produces a wavefront focused on the most reflective target. It converges if the target separation is sufficient to avoid the illumination of one target by the real acoustic image of the other one. *Prada* and *Fink* and *Prada* et al. [28,29] studied theoretically the convergence of time-reversal iterations in a multitarget medium and have determined the cases in which the target which reflects the most is selected among several targets. In some cases it is also interesting to learn how to focus on the other reflectors. The theoretical analysis of the iterative time-reversal process led to a very elegant solution to this problem (the DORT method), which is presented in Chap. 5. This analysis consists of determining the possible transmitted waveforms that are invariant under the time-reversal process. For these waveforms an iteration of the time-reversal operation gives stationary results. Such waveforms can be determined through the calculation of the eigenvectors of the so-called time-reversal operator.

Another interesting application of pulse–echo-mode TRMs is to put an elastic target in resonance. For example, if you illuminate an extended solid target with a short pulse, the backscattered field results in several contributions. A first reflected wave, "the specular echo," is determined by the target geometry. It is followed by a series of waves, "the resonant echo," which correspond to the propagation of surface and volume waves around and inside the scatterer. These waves are generated at particular points on the target. They propagate at the surface or in the solid, and they radiate into the fluid from different mode-conversion points on the scatterer, which behave as secondary sources. *Thomas* et al. [30] studied the case of a hollow target where the elastic part is mainly due to circumferential waves (dispersive Lamb waves): the first symmetrical and antisymmetrical Lamb waves, S_0 and A_0. As these two waves have different velocities, they can be separated experimentally by time windowing, and their generation points on the target are located at different positions in accordance with Snell's laws. Selecting and time reversing each wavefront separately, the time-reversed wave energy only concentrates at the generation points of the selected wave. This process enhances the generation of each specific Lamb wave compared to the other reflected waves. Iterating this process, we can build a waveform that is spatio-temporally matched to the vibration mode of the target. *Prada* et al. have extended the DORT technique [31] to this type of target, and they have demonstrated that each Lamb wave give rises to a set of eigenvectors of the time-reversal operator. These vectors can be calculated for a specific target and can be used to build optimal excitations of the array to put the target in resonance.

A first medical application of pulse–echo-mode TRMs is the destruction of kidney and gallbladder stones in the human body. Although the stones may be accurately located using X-ray imaging or ultrasonic scanners, it is difficult to focus precisely the ultrasonic waves in order to destroy the stones through inhomogeneous tissues. Furthermore, the stones move as much as several centimeters during breathing. Several thousand shots are required to destroy a stone and it is not currently possible to track the stone movements with a mechanical system. Consequently, it is estimated that, with current piezoelectric devices, only 30% of the ultrasonic shots reach the stone. Ultrasonic time-reversal techniques can solve these problems. To locate a reflecting target, such as a kidney stone in its environment of other stones and organ walls, the zone of interest is illuminated using a few elements of the transducer matrix. The reflected signals are recorded with the whole matrix and time-reversed. When the process is iterated several times, the ultrasonic beam converges towards the area of the stone that reflects the most. The time-reversal iteration selects one of the spots. Once the spot has been reliably located, intermittent amplified pulses can be applied to shatter the stone. As the stone moves, the process is repeated in order to locate it in real time. With *Thomas* and *Wu* [32], we have developed a 64-channel TRM 20 cm in diameter.

Another major promise of self-focusing TRM arrays that has not yet been fulfilled is ultrasonic medical hyperthermia. In this technique, high-intensity ultrasound produces thermal effects. A part of the ultrasound energy is absorbed by the tissue and converted to heat, resulting in an increase in local temperature. If a temperature of 60–70 °C is reached, irreversible and deleterious effects will occur within several seconds. Focused ultrasound surgery pioneered in the 1950s by Fry at the University of Illinois did not gain general acceptance until recently [33,34]. Focal probes consisting of annular phased arrays are now marketed for the treatment of prostate cancer. These techniques are limited to the production of necrosis in tissues that are not moving; however, applications to abdominal and cardiac surgery are limited by the tissue motion induced both by the cardiac cycle and by breathing. At the University of Michigan, Ebbini and his group are developing self-focusing arrays to solve this problem. In our group, we are working on a TRM application for brain hyperthermia. The challenge of this application is to focus through the skull bone, which induces severe refractions and scattering of the ultrasonic beam. With Thomas and Tanter, we have shown that the porosity of the skull bone produces a strong dissipation, which breaks the time-reversal symmetry of the wave equation. We have shown [35,36] that time-reversal focusing is no longer appropriate to compensate for the skull properties, and we have developed a new focusing technique which combines a correction for the dissipative effects with classical time-reversal focusing. This technique allows us to focus and steer, through the skull, an ultrasonic beam which converges on a 1.5 mm diameter spot with very low side lobes.

Another important application of TRMs is flaw detection in solids. The detection of small defects is difficult when the inspected object consists of heterogeneous or anisotropic material and when the sample has a complex geometry. Usually, the solid and the ultrasonic transducers are immersed in water, and the transducers are moved to scan the zone of interest. Due to refraction, the ultrasonic beams can be altered by the water–solid interface. In addition, the longitudinally polarized ultrasound traveling in water may produce waves of different polarizations and velocities in the solid (longitudinal, transverse and surface waves). Beam focusing and steering of the ultrasonic array in the solid is a difficult task for which self-focusing techniques have been proposed to enhance the flexibility of the process [37]. We have shown that using TRMs is a very effective technique to solve these problems. TRMs automatically compensate for refraction, mode conversion and anisotropy. In a joint program with SNECMA (Société Nationale d'Etudes et de Construction de Moteurs d'Avion), we have developed a 128-element TRM to detect the presence of low-contrast defects within titanium alloys used in jet engines. It is a difficult problem because titanium has a highly heterogeneous microstructure which produces large amounts of scattering noise that can hide the echo from a defect. With *Chakroun* et al. [38], we have shown that the iterative pulse–echo mode allows us to autofocus and to detect defects as small as 0.4 mm in 250-mm-diameter titanium billets. Compared to other techniques, the signal-to-noise ratio is enhanced in all situations, and smaller defects can be detected in the billet core, where ultrasonic beams are strongly distorted.

5 Conclusion

Time reversal has exciting applications in the field of acoustics. Because acoustic time-reversal technology is now easily accessible to modern electronic technology, it is expected that applications in various areas will expand rapidly. Initial applications show promise in medical therapy as well as in nondestructive testing. In addition to solving practical problems, time-reversal mirrors are also unique research tools that may allow us to better understand problems related to wave propagation in disordered media and reverberant cavities.

References

1. M. Fink, C. Prada, F. Wu, D. Cassereau, Self focusing in inhomogeneous media with time reversal acoustic mirrors, IEEE Ultras. Symp. Proc. **1**, 681–686 (1989)
2. M. Fink, Time reversal of Ultrasonic field-part I: Basic Principles, IEEE Trans. Ultrason. Ferroelec. Freq. Contr. **39**, 555–566 (1992)
3. M. Fink, Time Reversed Acoustics, Phys. Today **50**, 34–40 (1997)
4. D. R. Jackson, D. R. Dowling, Phase Conjugation in Underwater Acoustics, J. Acoust. Soc. Am. **89**, 171 (1991)
5. D. Cassereau, M. Fink, Time-reversal of ultrasonic fields, IEEE Trans. Ultrason. Ferroelec. Freq. Contr. **39**, 579–592 (1992)
6. J. de Rosny, M. Fink, submitted to Phys. Rev. Lett.
7. D. Cassereau, M. Fink, Time-reversal focusing through a plane interface separating two fluids, J. Acoust. Soc. Am. **96**, 3145–3154 (1994)
8. G. S. Kino, *Acoustics Waves* (Prentice Hall, Englewood Cliffs 1987)

9. A. Derode, P. Roux, M. Fink, Robust acoustic time-reversal with high order multiple scattering, Phys. Rev. Lett. **75**, 4206–4209 (1995)
10. R. Snieder, J. Scales, Time-reversed imaging as a diagnostic of wave and particle chaos, Phys. Rev. E **58**, 5668–5675 (1998)
11. A. Derode, A. Tourin, M. Fink, Limits of time-reversal focusing through multiple scattering: Long-range correlation, J. Acoust. Soc. Am. **107**, 2987 (2000)
12. A. Derode, A. Tourin, M. Fink, Ultrasonic pulse compression with one-bit time reversal through multiple scattering, J. Appl. Phys. **85**, 6343–6352 (1999)
13. P. Roux, B. Roman, M. Fink, Time-reversal in an ultrasonic waveguide, Appl. Phys. Lett. **70**, 1811 (1997)
14. A. Parvulescu, Matched signal ('MESS') processing by the ocean, J. Acoust. Soc. Am. **98**, 943 (1995)
15. A. Parvulescu, C. S. Clay, Reproducibility of signal transmissions in the ocean, Radio Elect. Eng. **29**, 233 (1965)
16. C. Feuillade, C. S. Clay, Source imaging and sidelobe suppression using time-domain techniques in a shallow water waveguide, J. Acoust. Soc. Am. **92**, 2165 (1992)
17. W. A. Kupperman, W. S. Hodgkiss, H. C. Song, T. Akal, T. Ferla, D. Jackson, Phase conjugation in the ocean: experimental demonstration of an acoustic time-reversal mirror, J. Acoust. Soc. Am. **103**, 25 (1998)
18. W. Hodgkiss, H. Song, W. Kuperman, T. Akal, C. Ferla, D. Jackson, A long range and variable focus phase-conjugation experiment in shallow water, J. Acoust. Soc. Am. **105**, 1597 (1999)
19. C. Draeger, M. Fink, One channel time-reversal of elastic waves in a chaotic 2D-silicon cavity, Phys. Rev. Lett. **79**, 407 (1997)
20. C. Draeger, M. Fink, One-channel time-reversal in chaotic cavities: theoretical limits, J. Acoust. Soc. Am. **105**, 618 (1999)
21. C. Draeger, M. Fink, One-channel time-reversal in chaotic cavities: experimental results, J. Acoust. Soc. Am. **105**, 611 (1999)
22. M. V. Berry, in Les Houches 1981 - *Chaotic Behaviour of Deterministic Systems* (North Holland, Amsterdam, 1983), p. 171
23. S. W. McDonald, A. N. Kaufman, Wave chaos in the stadium: Statistical properties of short-wave solutions of the Helmholtz equation, Phys. Rev. A **37**, 3067 (1988)
24. R. Weaver, J. Burkhardt, Weak Anderson localisation and enhanced backscatter in reverberation rooms and quantum dots, J. Acoust. Soc. Am. **96**, 3186 (1994)
25. C. Prada, F. Wu, M. Fink, The iterative time reversal mirror, J. Acoust. Soc. Am. **90**, 1119 (1991)
26. F. Wu, J.-L. Thomas, M. Fink, Time Reversal of Ultrasonic Fields—Part I: Experimental Results, IEEE Trans. Ultrason. Ferroelec. Freq. Contr. **39**, 567 (1992)
27. C. Dorme, M. Fink, Focusing in transmit-receive mode through inhomogeneous media: the time reversal matched filter approach, J. Acoust. Soc. Am. **98**, 1155 (1995)
28. C. Prada, M. Fink, Eigenmodes of the time reversal operator: a solution to selective focusing on two scatterers, Wave Motion **20**, 151 (1994)
29. C. Prada, J.-L. Thomas, M. Fink, The iterative time-reversal process: analysis of the convergence, J. Acoust. Soc. Am. **97**, 62 (1995)

30. J.-L. Thomas, P. Roux, M. Fink, Inverse problem in wave scattering with an acoustic time reversal mirror, Phys. Rev. Lett. **72**, 637–640 (1994)
31. C. Prada, J.-L. Thomas, P. Roux, M. Fink, Acoustic time reversal and inverse scattering, in H. D. Bui, M. Tamaka (Ed.), *Inverse problems in engineering mechanics*, Proc. 2nd Int. Symp. Inv. Prob. at Paris, Nov 2–4, 1994 (Balkema, Rotterdam 1994) pp. 309–316
32. J.-L. Thomas, F. Wu, M. Fink, Time Reversal Mirror applied to litotripsy, Ultras. Imag. **18**, 106 (1996)
33. L. Crum, K. Hyninen, Sound Therapy, Phys. World, 28, August (1996)
34. H. Wang, E. S. Ebinni, M. O. Donell, C. A. Cain, Phase aberration correction and motion compensation for ultrasonic hyperthermia phased arrays: experimental results, IEEE Trans. Ultrason. Ferroelec. Freq. Contr. **41**, 34 (1994)
35. J.-L. Thomas, M. Fink, Ultrasonic beam focusing through tissue inhomogeneities with a time reversal mirror: Application to transkull therapy, IEEE Trans. Ultrason. Ferroelec. Freq. Contr. **43**, 1122 (1996)
36. M. Tanter, J.-L. Thomas, M. Fink, Focusing and steering through absorbing and heterogeneous medium: application to ultrasonic propagation through the skull, J. Acoust. Soc. Am. **103**, 2403 (1998)
37. B. Beardsley, M. Peterson, J. D. Achenbach, A simple scheme for self-focusing of an array, J. Nondestr. Eval. **14**, 169 (1995)
38. N. Chakroun, M. Fink, F. Wu, Time reversal processing in non destructive testing, IEEE Trans. Ultrason. Ferroelec. Freq. Contr. **42**, 1087 (1995)

Ocean Acoustics, Matched-Field Processing and Phase Conjugation

William A. Kuperman[1] and Darrel R. Jackson[2]

[1] Scripps Institution of Oceanography, Marine Physical Laboratory, University of California, San Diego, La Jolla, Calif. 92093-0701, USA
wkuperman@ucsd.edu

[2] Applied Physics Laboratory, College of Ocean and Fishery Sciences, University of Washington, Seattle, Wash. 98105-6698, USA
drj@apl.washington.edu

Abstract. This chapter treats ocean acoustics and various applications of signal processing, phase conjugation and tomography to ocean acoustics. Phase conjugation is related to the standard principles of passive signal processing and to the more recent demonstrations of time-reversal mirrors in the ocean. Here we will give an overview of the relevant physics of ocean acoustics, present a passive signal processing method called matched-field processing and then discuss the related concept and implementation of time-reversal methods in ocean acoustics.

1 Review of Ocean Acoustics

The ocean is a waveguide, bounded above by a pressure release surface and below by a visco-elastic medium. The physical oceanographic parameters, as ultimately represented by the ocean sound speed structure, make up the index of refraction of the water column waveguide. The combination of water column and bottom properties leads to a set of generic sound propagation paths descriptive of most propagation phenomena in the ocean.

1.1 Qualitative Description of Ocean Sound Propagation

We summarize aspects of oceanography that impact propagation paths before we go into more detail on the propagation itself. Figure 1 illustrates a typical set of sound speed profiles indicating greatest variability near the surface as a function of season and time of day. In a warmer season (or warmer part of the day), the temperature increases near the surface and hence the sound speed decreases with depth. In nonpolar regions, the oceanographic properties of the water near the surface result from mixing activity originating from the air–sea interface. This near-surface mixed layer has a constant temperature (except in calm, warm surface conditions as described above). In this isothermal mixed layer, the sound speed profile can increase with depth due to the pressure gradient effect. This is the "surface duct" region.

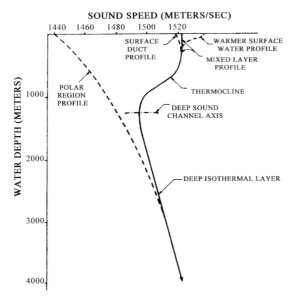

Fig. 1. Generic sound speed profiles

1.1.1 Sound Propagation Paths in the Ocean

Below the mixed layer is the thermocline, in which the temperature decreases with depth and the sound speed decreases with depth. Below the thermocline, the temperature is constant and the sound speed increases because of increasing pressure. Therefore, there exists a depth between the deep isothermal region and the mixed layer with a minimum in sound speed; this depth is often referred to as the axis of the deep sound channel.

However, in polar regions, the water is coldest near the surface and hence the minimum sound speed is at the ocean–air(or ice) interface, as indicated in Fig. 1. In continental shelf regions (shallow water) with water depths on the order of a few hundred meters, only the upper region of the sound speed profile in Fig. 1, which is dependent on season and time of day, affects sound propagation in the water column.

Figure 2 is a contour display of the sound speed structure of the North and South Atlantic [1], with the axis of the deep sound channel indicated by the heavy dashed line. Note the geographic (and climatic) variability of the upper ocean sound speed structure and the stability of this structure in the deep isothermal layer. For example, as explained above, the axis of the deep sound channel becomes shallower toward both poles, eventually going to the surface. Figure 3 is a schematic of the basic types of propagation in the ocean resulting from the sound speed profiles (indicated by the dashed lines) discussed in the last section. These sound paths can be understood from a simplified statement of Snell's Law: sound bends locally toward regions of low sound

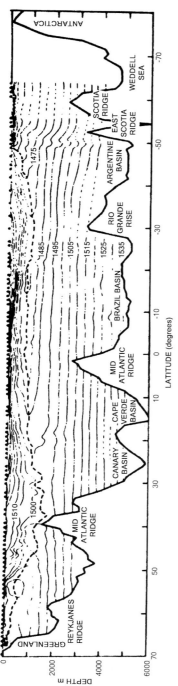

Fig. 2. Sound speed contours at 5 m/s intervals taken from the North and South Atlantic along 30.50° West. *Dashed line* indicates axis of deep sound channel. (From [1])

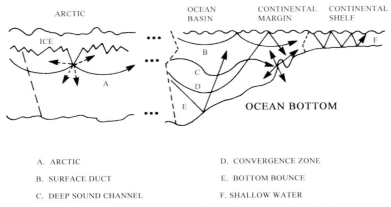

A. ARCTIC
B. SURFACE DUCT
C. DEEP SOUND CHANNEL
D. CONVERGENCE ZONE
E. BOTTOM BOUNCE
F. SHALLOW WATER

Fig. 3. Schematic of various types of sound propagation in the ocean

speed (or sound is "trapped" in regions of low sound speed). Paths A and B correspond to surface duct propagation, where the minimum sound speed is at the ocean surface (or at the bottom of the ice cover for the Arctic case). Path C, depicted by a ray leaving a deeper source at a shallow horizontal angle propagates in the deep sound channel whose axis is at the shown sound speed minimum. This local minimum tends to become more shallow towards polar latitudes, converging to the Arctic surface minimum. Hence for mid-latitudes, sound in the deep channel can propagate long distances without interacting with lossy boundaries; propagation via this path has been observed over distances of thousands of kilometers. Also, from the above description of the geographical variation of the acoustic environment combined with Snell's Law, we can expect that shallow sources coupling into the water column at polar latitudes will tend to propagate more horizontally around an axis which becomes deeper toward the mid-latitudes. Path D, which is at slightly steeper angles than those associated with path C, is convergence zone propagation, a spatially periodic ($\sim 35 - 65$ km) refocusing phenomenon producing zones of high intensity near the surface due to the upward refracting nature of the deep sound speed profile. Referring back to Fig. 1, there may be a depth in the deep isothermal layer at which the sound speed is the same as it is at the surface. This depth is called the "critical depth" and, in effect, is the lower limit of the deep sound channel. A "positive critical depth" specifies that the environment supports long distance propagation without bottom interaction, whereas "negative" implies that the bottom ocean boundary *is* the lower boundary of the deep sound channel. The bottom bounce path, E, which interacts with the ocean bottom is also a periodic phenomenon but with a shorter cycle distance and a shorter total propagation distance because of losses when sound is reflected from the ocean bottom. Finally, the right hand side of Fig. 3 depicts propagation in a shallow water region such as a continental shelf. Here sound is channeled in a waveguide bounded above by the ocean surface and below by the ocean bottom.

The modeling of sound propagation in the ocean is further complicated because the environment varies laterally (range dependence), and all environmental effects on sound propagation are dependent on acoustic frequency in a rather complicated way, which often makes the ray-type schematic of Fig. 3 misleading, particularly at low frequencies. Finally, a quantitative understanding of acoustic loss mechanisms in the ocean is required for modeling sound propagation. These losses are, aside from geometric spreading: volume attenuation, bottom loss (i.e., a smooth water–bottom interface is not a perfect reflector), and surface, volume (including fish) and bottom scattering loss. Here we will only review those aspects of bottom loss which impact propagation structure.

1.1.2 Bottom Loss

Ocean bottom sediments are often modeled as fluids, since the rigidity (and hence the shear speed) of the sediment is usually considerably less than that of a solid, such as rock. In the latter case, which applies to the "ocean basement" or the case where there is no sediment overlying the basement, the medium must be modeled as an elastic solid, which means it supports both compressional and shear waves.

Reflectivity, the amplitude ratio of reflected and incident plane waves at an interface separating two media, is an important measure of the effect of the bottom on sound propagation. For an interface between two fluid semi-infinite halfspaces with density ρ_i and sound speed $c_i (i = 1, 2)$, as shown in Fig. 4a [assuming a harmonic time dependence of $\exp(-i\omega t)$], the reflectivity is given by

$$\mathcal{R}(\theta) = \frac{\rho_2 k_{1z} - \rho_1 k_{2z}}{\rho_2 k_{1z} + \rho_1 k_{2z}}, \tag{1}$$

with

$$k_{iz} = (\omega/c_i) \sin \theta_i \equiv k_i \sin \theta_i \, , \; i = 1, 2. \tag{2}$$

The incident and transmitted grazing angles are related by Snell's Law,

$$k_\perp = k_1 \cos \theta_1 = k_2 \cos \theta_2, \tag{3}$$

where the incident grazing angle θ_1 is also equal to the angle of the reflected plane wave, θ_R. $\mathcal{R}(\theta)$ is also referred to as the Rayleigh reflection coefficient and has unit magnitude (total internal reflection) when the numerator and denominator of (1) are complex conjugates. This occurs when k_{2z} is purely imaginary. Using Snell's Law to determine θ_2 in terms of the incident grazing angle, we obtain the *critical grazing angle*, below which there is perfect reflection,

$$\cos \theta_c = c_1/c_2; \tag{4}$$

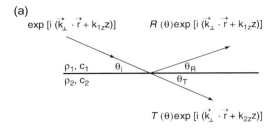

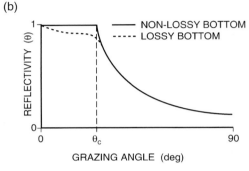

Fig. 4a,b. The reflection and transmission process. Grazing angles are defined relative to the horizontal

thus, a critical angle can exist only when the speed in the second medium is higher than that of the first.

Using (2), (1) can be rewritten as

$$\mathcal{R}(\theta) = \frac{\rho_2 c_2/\sin\theta_2 - \rho_1 c_1/\sin\theta_1}{\rho_2 c_2/\sin\theta_2 + \rho_1 c_1/\sin\theta_1} \equiv \frac{Z_2 - Z_1}{Z_2 + Z_1}, \tag{5}$$

where the right hand side is in the form of *impedances*, $Z_i(\theta_i) = \rho_i c_i/\sin\theta_i$, which are the ratios of the pressure to the vertical particle velocity at the interface in the ith medium. Written in this form, the reflectivity for more complicated boundaries follows in a straightforward manner [2].

In lossy media, attenuation can be included in the reflectivity formula by taking the sound speed to be complex, so that the wavenumbers are subsequently also complex, $k_i \to k_i + \alpha_i$.

Figure 4b depicts a simple bottom *loss* curve derived from the Rayleigh reflection coefficient formula where both the densities and sound speed of the second medium are larger than those in the first medium with unit reflectivity, indicating perfect reflection. For loss in dB, 0 dB is perfect reflection, 6 dB loss is an amplitude factor of one-half, 12 dB loss is an amplitude factor of one-fourth, etc. For a lossless bottom, severe loss occurs above the critical angle in the water column due to transmission into the bottom. For the lossy (more realistic) bottoms, only partial reflection occurs at all angles. With paths involving many bottom bounces (shallow water propagation), bottom losses as small as a few-tenths of a dB per bounce accumulate and become significant because the propagation path may involve many tens of bounces.

Path E in Fig. 3, the bottom bounce path, often involves paths which correspond to angles near or above the critical angle; therefore, after a few bounces, the sound level is highly attenuated. On the other hand, for shallow angles, many bounces are possible. Hence, in shallow water, path F, most of the energy that propagates is close to the horizontal and this type of propagation is most analogous to waveguide propagation. In fact, as shown in Fig. 5, there exists a small cone from which energy propagates long distances (θ_c is typically $10° - 20°$). Energy outside the cone is referred to as the near field (or continuous spectrum), which eventually escapes the waveguide. The trapped field originating from within the cone is referred to as the normal-mode field (or discrete spectrum) because there are a set of angles corresponding to discrete paths which constructively interfere and make up the normal (natural) modes of the shallow water environment.

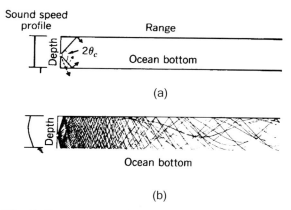

Fig. 5. Ocean waveguide propagation. (**a**) Long distance propagation occurs within a cone of $2\theta_c$. (**b**) Same as (**a**) but with nonisovelocity water column showing refraction

1.2 Sound Propagation Models

Sound propagation in the ocean is mathematically described by the wave equation, whose parameters and boundary conditions are descriptive of the ocean environment. There are essentially four types of models (computer solutions to the wave equation) to describe sound propagation in the sea: ray theory, the spectral method or fast field program (FFP), normal mode (NM) and parabolic equation (PE) [2]. All of these models allow for the fact that the ocean environment varies with depth. A model that also takes into account horizontal variations in the environment (i.e., sloping bottom or spatially variable oceanography) is termed range dependent. For high frequencies (a few kilohertz or above), ray theory is the most practical. The other three model types are more applicable and usable at lower frequencies (below a kilohertz). Here, we will confine the discussion to the NM model including its mild adiabatic range-dependent extension and the PE (see Appendix A).

1.2.1 The Wave Equation and Boundary Conditions

The wave equation for pressure in cylindrical coordinates with the range coordinates denoted by $\boldsymbol{r} = (x, y)$ and the depth coordinate denoted by z (taken positive downward) for a source-free region of constant density ρ is

$$\nabla^2 p(\boldsymbol{r}, z, t) - \frac{1}{c^2}\frac{\partial^2 p(\boldsymbol{r}, z, t)}{\partial t^2} = 0, \tag{6}$$

where c is the sound speed in the wave propagating medium. The wave equation is most often solved in the frequency domain; that is, a frequency dependence of $\exp(-i\omega t)$ is assumed to obtain the Helmholtz equation ($K \equiv \omega/c$),

$$\nabla^2 p(\boldsymbol{r}, z) + K^2 p(\boldsymbol{r}, z) = 0. \tag{7}$$

The most common plane interface boundary conditions encountered in underwater acoustics are as follows: For the ocean surface there is the pressure release condition where the pressure vanishes,

$$p = 0. \tag{8}$$

The interface between the water column (layer 1) and an ocean bottom sediment (layer 2) is often characterized as a fluid–fluid interface. The continuity of pressure and vertical particle velocity at the interface yields the following boundary conditions in terms of pressure:

$$p_1 = p_2, \quad \frac{1}{\rho_1}\frac{\partial p_1}{\partial z} = \frac{1}{\rho_2}\frac{\partial p_2}{\partial z}. \tag{9}$$

These boundary conditions applied to the plane wave fields in Fig. 4a yield the Rayleigh reflection coefficient given by (1).

The Helmholtz equation for an acoustic field from a point source with angular frequency ω is

$$\nabla^2 G(\boldsymbol{r}, z) + K^2(\boldsymbol{r}, z) G(\boldsymbol{r}, z) = -\delta^2(\boldsymbol{r} - \boldsymbol{r}_{\mathrm{s}})\delta(z - z_{\mathrm{s}});$$
$$K^2(\boldsymbol{r}, z) = \frac{\omega^2}{c^2(\boldsymbol{r}, z)}, \tag{10}$$

where the subscript "s" denotes the source coordinates. The range-dependent environment manifests itself as the coefficient $K^2(\boldsymbol{r}, z)$ of the partial differential equation for the appropriate sound speed profile. The range-dependent bottom type and topography appear as boundary conditions. The acoustic field from a point source, $G(r, z)$, is either obtained by solving the boundary value problem of (10) (spectral method or NM model) or by approximating (10) by an initial value problem (ray theory, PE).

1.2.2 Normal-Mode Model

For a range-independent environment, we assume a pressure field solution of (10) of the form

$$G(\boldsymbol{r}, z) = \frac{1}{2\pi} \int_{-\infty}^{\infty} d^2\boldsymbol{k} \, g(\boldsymbol{k}, z, z_s) \exp[i\boldsymbol{k}\cdot(\boldsymbol{r} - \boldsymbol{r}_s)], \tag{11}$$

which then leads to the equation for the depth-dependent Green's function, $g(\boldsymbol{k}, z, z_s)$,

$$\frac{d^2 g}{dz^2} + \left[K^2(z) - k^2\right] g = -\frac{1}{2\pi}\delta(z - z_s). \tag{12}$$

We can solve (12) for g using a NM expansion of the form

$$g(\boldsymbol{k}, z, z_s) = \sum a_n(\boldsymbol{k}) u_n(z), \tag{13}$$

where the quantities u_n are eigenfunctions of the following eigenvalue problem:

$$\frac{d^2 u_n}{dz^2} + \left[K^2(z) - k_n^2\right] u_n(z) = 0. \tag{14}$$

The eigenfunctions, u_n, are zero at $z = 0$, satisfy the local boundary conditions descriptive of the ocean bottom properties and satisfy a radiation condition for $z \to \infty$. They form an orthonormal set in a Hilbert space with weighting function $\rho(z)$, the local density. The range of discrete eigenvalues of (14) is given by the condition

$$\min\left[K(z)\right] < k_n < \max\left[K(z)\right]. \tag{15}$$

These discrete eigenvalues correspond to discrete angles within the critical angle cone in Fig. 5 such that specific waves constructively interfere. The mode functions form a complete set (for simplicity we omit discussion of the continuous spectrum, though a good approximation is to use a set of discrete mode functions obtained from a waveguide extended in depth and terminated by a pressure release or rigid boundary)

$$\sum_{\text{all modes}} \frac{u_n(z) u_n(z_s)}{\rho(z_s)} = \delta(z - z_s) \tag{16}$$

and satisfy the orthonormality condition

$$\int_0^\infty \frac{u_m(z) u_n(z)}{\rho(z)} dz = \delta_{nm}, \tag{17}$$

where δ_{nm} is the Kronecker delta symbol. The eigenvalues k_n typically also have a small imaginary part, α_n, which serves as the modal attenuation representative of all the losses in the ocean environment (see [2] for the formulation of NM attenuation coefficients).

Solving (10) using the NM expansion given by (13) yields (for the source at the origin).

$$G(r,z) = \frac{i}{4\rho(z_s)} \sum_n u_n(z_s) u_n(z) H_0^1(k_n r). \qquad (18)$$

The asymptotic form of the Hankel function can be used in (18) to obtain the well-known NM representation of a cylindrical (axis is depth) waveguide:

$$G(r,z) = \frac{i}{\rho(z_s)(8\pi r)^{1/2}} \exp(-i\pi/4) \sum_n \frac{u_n(z_s)u_n(z)}{k_n^{1/2}} \exp(ik_n r). \qquad (19)$$

Equation (19) is a far-field solution of the wave equation and neglects the continuous spectrum $\{k_n < \min[K(z)]$ of (15)$\}$ of modes. For purposes of illustrating the various portions of the acoustic field, we note that k_n is a horizontal wavenumber so that a "ray angle" associated with a mode with respect to the horizontal can be taken to be $\theta = \cos^{-1}[k_n/K(z)]$. For a simple waveguide the maximum sound speed is the bottom sound speed corresponding to $\min[K(z)]$. At this value of $K(z)$, we have from Snell's Law $\theta = \theta_c$ the bottom critical angle. In effect, if we look at a ray picture of the modes, the continuous portion of the mode spectrum corresponds to rays with grazing angles greater than the bottom critical angle of Fig. 4b and therefore outside the cone of Fig. 5. This portion undergoes severe loss. Hence, we note that the continuous spectrum is the near (vertical) field and the discrete spectrum is the (more horizontal, profile dependent) far field (falling within the cone in Fig. 5).

The advantages of the NM procedure are as follows: the solution is available for all source and receiver configurations once the eigenvalue problem is solved; it is easily extended to moderately range-dependent conditions using the adiabatic approximation; it can be applied (with more effort) to extremely range-dependent environments using coupled-mode theory. However, it does not include a full representation of the near field.

1.2.3 Adiabatic-Mode Theory

All of the range-independent NM "machinery" developed for environmental ocean acoustic modeling applications can be adapted to mildly range-dependent conditions using adiabatic-mode theory. The underlying assumption is that individual propagating normal modes adapt (but do not scatter or "couple" into each other) to the local environment. The coefficients of the mode expansion, a_n in (13), now become mild functions of range, i.e., $a_n(\boldsymbol{k}) \to a_n(\boldsymbol{k}, \boldsymbol{r})$. This modifies (19) as follows:

$$G(\boldsymbol{r}, z) = \frac{\mathrm{i}}{\rho(z_\mathrm{s})(8\pi r)^{1/2}} \exp(-\mathrm{i}\pi/4) \sum_n \frac{u_n(z_\mathrm{s}) v_n(z)}{\overline{k_n}^{1/2}} \exp(\mathrm{i}\overline{k_n} r). \tag{20}$$

where the range-averaged wavenumber (eigenvalue) is

$$\overline{k_n} = \frac{1}{r} \int_0^r k_n(r') \, \mathrm{d}r', \tag{21}$$

and the $k_n(r')$ are obtained at each range segment from the eigenvalue problem (14) evaluated at the environment for that particular range along the path. The quantities u_n and v_n are the sets of modes at the source and the field positions, respectively.

Simply stated, the adiabatic-mode theory leads to a description of sound propagation such that the acoustic field is a function of the modal structure at both the source and the receiver and some average propagation conditions between the two. Thus, for example, when sound emanates from a shallow region where only two discrete modes exist and propagates into a deeper region with the same bottom (same critical angle), the two modes from the shallow region adapt to the form of the first two modes in the deep region. However, the deep region can support many more modes; intuitively, we therefore expect the resulting two modes in the deep region will take up a smaller more horizontal part of the cone of Fig. 5 than they take up in the shallow region. This means that sound rays going from shallow to deep tend to become more horizontal, which is consistent with a ray picture of downslope propagation. Finally, fully coupled-mode theory for range-dependent environments has been developed [3] but requires extremely intensive computation.

1.3 Quantitative Description of Propagation

All of the models described above attempt to describe reality and to solve in one way or another the Helmholtz equation. They therefore should be consistent, and there is much insight to be gained from understanding this consistency. The models ultimately compute propagation loss, which is taken as the decibel ratio (see Appendix B) of the pressure at the field point to a reference pressure, typically 1 m from the source.

Figure 6 shows convergence-zone-type propagation for a simplified profile. The ray trace in Fig. 6b shows the cyclic focusing discussed in Sect. 1.1. The same profile is used to calculate the normal modes shown in Fig. 6c, which, when summed according to (19) yield the same cyclic pattern as the ray picture. Figure 6d shows both the NM (wave theory) and ray theory results. Ray theory exhibits sharply bounded shadow regions as expected, whereas the NM theory, which includes diffraction, shows that the acoustic field does exist in the shadow regions and the convergence zones have structure.

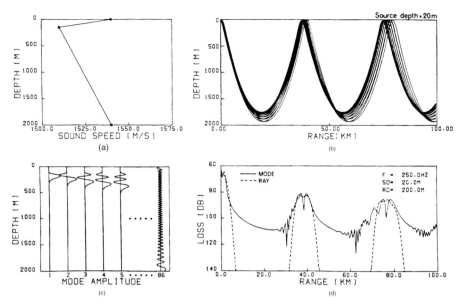

Fig. 6. Ray and normal mode theory. (**a**) Sound speed profile, (**b**) ray trace, (**c**) normal modes, and (**d**) propagation calculations

NM models sum the discrete modes, which roughly correspond to angles of propagation within the cone of Fig. 5. The spectral (FFP) method can include the full field, discrete plus continuous, the latter corresponding to larger angles. The spectral model involves solving (12) numerically for a continuum of wavenumbers, k, without the NM expression of (13) (see Fig. 7a). Therefore, the consistency we expect between the NM model and the spectral method and the physics of Fig. 5 is that the continuous portion of the spectral solution decays rapidly with range, so that there should be complete agreement at long ranges between NM and spectral solutions. The Lloyd's mirror effect, a near-field effect, should also be exhibited in the spectral solution but not the NM solution. These aspects are apparent in Fig. 7b. The PE solution appears in Fig. 7b and is in good agreement with the other solutions but with some phase error associated with the average wavenumber that must be chosen in the split step method. The PE solution, which contains part of the continuous spectrum including the Lloyd mirror beams, is more accurate than the NM solution at short range; more recent PE results [27] can be made arbitrarily accurate in the forward direction.

Range-dependent results [2] are shown in Fig. 8. A ray trace, a ray-trace field result, a PE result and data are plotted together for a range-dependent sound speed profile environment. The models agree with the data in general, with the exception that the ray results predict too sharp a leading edge of the convergence zone.

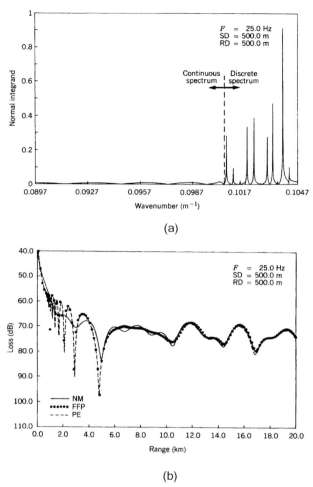

Fig. 7. Relationship between FFP, NM and PE computations. (a) FFP Green's function from (12). (b) NM, FFP and PE propagation results showing some agreement in near field and complete agreement in far field

Upslope propagation is modeled with the PE in Fig. 9. As the field propagates upslope, sound is dumped into the bottom in what appear to be discrete beams [4]. The flat region has three modes and each is cut off successively as sound propagates into shallower water. The ray picture also has a consistent explanation of this phenomenon. The rays for each mode become steeper as they propagate upslope. When the ray angle exceeds the critical angle the sound is significantly transmitted into the bottom. The locations where this takes place for each of the modes is identified by the three arrows.

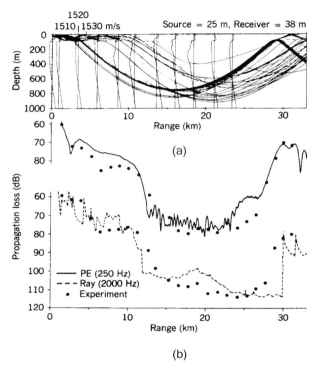

Fig. 8. Model and data comparison for a range-dependent case. (**a**) Profiles and ray trace for a case of a surface duct disappearing. (**b**) 250-Hz PE and 2-kHz ray trace comparisons with data

2 Matched-Field Processing

Spatial sampling of a sound field is usually performed by an array of transducers, although the synthetic aperture array, in which a sensor is moved through space to obtain measurements in both the time and space domains, is an important exception. Spatial sampling is analogous to temporal sampling with the sampling interval replaced by the sensor spacing vectors. The Nyquist criterion requires that the sensor spacing be at least twice the spatial wavenumber of the measured sound field. Aliasing and signal reconstruction follow the same principles as for temporal data, except that ambiguity can arise in the reconstructed wavenumber components due to symmetries in the array geometry. The simplest example of array processing is phase shading in the frequency domain (or time delay in the time domain) to search in bearing for a plane wave signal and is referred to as plane wave beamforming, or delay and sum beamforming in the time domain. This form of beamforming is discussed elsewhere, and we will move on to a process referred to as matched-field processing (MFP).

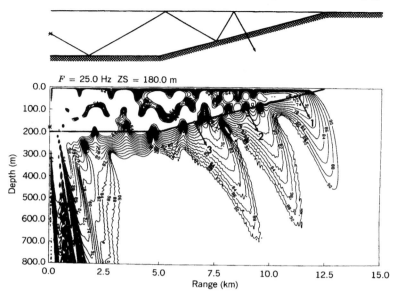

Fig. 9. Relation between upslope propagation (from PE calculation) showing individual mode cutoff and energy dumping in the bottom, and a corresponding ray schematic. *Arrows*: ray angle exceeds critical angle and sound is significantly transmitted into the bottom

MFP [5] is the 3-dimensional generalization of the conventional lower-dimensional plane wave beamformer. The one or two dimensions in the latter case are bearing and/or elevation and the matching is done to plane waves. Localization in this case refers to determining a direction. The generalized matched-field beamformer matches the measured field at the array with replicas of the expected field for all source locations. These replicas are derived from propagation models as discussed above or in more detail in [2]. The unique spatial structure of the field permits localization in range, depth and azimuth depending on the array geometry and complexity of the ocean environment. The complex interference pattern of the acoustic field is a function of the source location and this pattern can be "matched." In terms of rays we can say that the refractive properties of the waveguide generate a pattern of arrival angles that can also be matched. The process is shown schematically in Fig. 10. The process consists of systematically placing a test point source at each point of a search grid, computing the acoustic field (replicas) at all the elements of the array and then correlating this modeled field with the data from the real point source whose location is unknown. When the test point source is collocated with the true point source, the correlation will be a maximum. The two main factors which limit performance of MFP are noise, natural and man made, and the ability to accurately model the ocean acoustic environment.

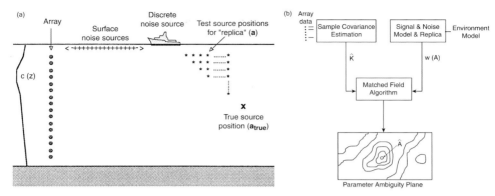

Fig. 10. Matched-field processing. (a) Schematic of the procedure, and (b) block diagram of the processor

Now construct the replica field, $w(a)$ on the array for each candidate position, a, from one of the numerical solutions to the acoustic wave equation, (10). The replica vector is normalized, i.e., it has unit length, so that the output of the processor is the source strength,

$$w(a) = \frac{G(a)}{|G(a)|}, \qquad (22)$$

where $G(a)$ is the vector whose ith element is the solution to the waveguide equation, e.g., (19), at the field (hydrophone) position $(r, z) = (r_i, z_i)$ for a source at a.

Rather than find just the direction of the source from the array, we can search for its actual location by matching the received data, $d(a_{\text{true}})$ from the true source location against solutions of the wave equation. The matching process is then similar to plane-wave beamforming with the plane waves being replaced by solutions of the wave equation. The output of this "Bartlett" matched field process, denoted $S(a)$, at each point in space a is given by

$$S(a) = |w^\dagger(a)d(a_{\text{true}})|^2. \qquad (23)$$

This equation can be rewritten in terms of the cross-spectral density matrix of the array data, $\mathbf{K}(a_{\text{true}}) = d(a_{\text{true}})d^\dagger(a_{\text{true}})$,

$$S(a) = w^\dagger(a)\mathbf{K}(a_{\text{true}})w(a). \qquad (24)$$

There are an assortment of interesting issues concerning the actual measurement of cross-spectral density matrix, particularly when adaptive processing is utilized.

The peak of the output of the beamformer, $S(a)$, is at a_{true}. $S(a)$ is also referred to as the ambiguity function (or surface) of the matched-field processor because it also contains ambiguous peaks which are analogous to the sidelobes of a conventional plane-wave beamformer. Sidelobe suppression can often be accomplished by adaptive processing. Here we write down the form of one of the adaptive processors: the minimum variance distortionless processor

$$S_{\text{MVDP}}(\boldsymbol{a}) = [\boldsymbol{w}^\dagger(\boldsymbol{a})\mathbf{K}^{-1}(\boldsymbol{a}_{\text{true}})\boldsymbol{w}(\boldsymbol{a})]^{-1}, \qquad (25)$$

where again we note from above that there are measurement issues associated with estimating an invertible cross spectral density matrix.

A vertical receiver array simulation example of the Bartlett and MVDP matched-field processors for an ocean acoustic waveguide with a high signal-to-noise ratio is shown in Fig. 11. In MFP, mismatch between replicas, which are derived from a usually incomplete knowledge of the environment, and observed data is a significant problem that has received much attention. Matched-field tomography (MFT) searches for the environmental parameters controlling the propagation (for example, the index of refraction, which may be a spatially dependent coefficient of the wave equation) rather than source location. As with the case of plane-wave beamforming, there are an assortment of processors that can be used for MFP/MFT [5]

There have been two approaches to deal with environmental mismatch in MFP. The first has been to construct processors which are tolerant to environmental uncertainty [6,7,8]. These processors trade resolution for lack of precise knowledge of the environment. The second approach is focalization [9], which uses nonlinear optimization methods to simultaneously search for the environment and the source position. All of these methods have limitations, but under various conditions they have been successfully applied to experimental data.

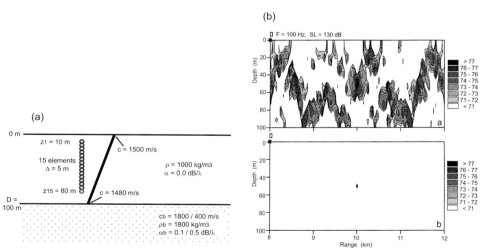

Fig. 11. Matched-field processing. (**a**) Environment and geometry for shallow-water example. (**b**) Matched-field ambiguity surfaces for a 130 dB source ($r_s = 10$ km, $z_s = 50$ m). Bartlett (*top*) and MVDP (*bottom*)

3 Phase Conjugation in the Ocean

Phase conjugation (PC) [10], and its time domain equivalent, referred to as a time-reversal mirror (TRM), is a process that was first demonstrated in nonlinear optics. The TRM in ultrasonics [11,12] is the subject of Chap. 2. Aspects of PC as applied to underwater acoustics also have been explored recently [13,14,15,16]. The Fourier conjugate of PC is time reversal; implementation of such a process over a finite spatial aperture results in a TRM. This section will begin with a description of the basic theory and implementation in the ocean based on the ocean acoustics discussion above. This will be followed by a discussion of some of the more complex issues associated with the real ocean, such as fluctuations. Some of the material given below will overlap with and supplement Chap. 2.

3.1 Basic Properties of Phase Conjugation

PC takes advantage of reciprocity, which is a property of wave propagation in a static medium and is a consequence of the invariance of the linear lossless wave equation to time reversal. In the frequency domain, time reversal corresponds to conjugation invariance of the Helmholtz equation. The property of reciprocity allows one to retransmit a time-reversed version of a multipath dispersed probe pulse back to its origin, arriving there time reversed, with the multipath structure having been undone [17,18]. This process is equivalent to using the ocean as a matched filter, since the probe pulse arrival has embedded in it the transfer function of the medium. This process can be extended further by receiving and retransmitting the probe signal with a source–receiver array (SRA). Depending on the spatial extent of the array, the above process results in some degree of spatial focusing of the signal at the origin of the probe signal.

A TRM can therefore be realized with an SRA. The incident signal is received, time-reversed and transmitted from sources contiguous with the receiving hydrophones. The time reversal can be accomplished in a straightforward way, for example, by using the rewind output of an analog tape recorder or by a simple program that reverses a digitized segment of a received signal.

PC or the implementation of a TRM in the ocean will be shown to be related to MFP, which requires detailed knowledge of the environment. PC is an environmentally self-adaptive process which may therefore have significant applications to localization and communications in complicated ocean environments. Though the "effective" ocean environment must remain static over the turnaround time of the PC process, ocean variability on time scales shorter than the turnaround time might be compensated for with feedback algorithms. However, an understanding of relevant ocean time scales vis-à-vis the stability of the PC process will be required.

As noted earlier, PC in the ocean shares many features in common with ultrasonic time reversal. These general aspects will be discussed first and then

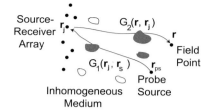

Fig. 12. Schematic of the generic PC focusing experiment

those aspects that demand treatment of the ocean waveguide will be covered. The generic PC focusing experiment is depicted in Fig. 12. A point acoustic source, denoted the "probe source" transmits a signal which is received by the elements of an array capable of both reception and transmission (the SRA). The signal received by each element is time-reversed and retransmitted on the *same* element (or perhaps on a nearby source element if the receiving elements are unsuitable for acoustic sources). The field produced by the array approximates a time-reversed version of the probe source field, consequently it focuses at the location of the probe source.

3.1.1 Idealized Arrays

In order to assess the degradation in focusing due to environmental effects, it is helpful to have a reference point for comparison. Thus, we will first consider ideal arrays capable of providing the sharpest possible focus. While one might think of the ideal case as the production of a simple time-reversed version of the field due to the probe source, this ideal is unreachable, even in principle. To see this, consider a point probe source, which produces a field that approaches infinity as one approaches the source location. This infinite field does not satisfy the source-free wave equation at the source location; therefore it is impossible to produce such a field using an array whose elements are at a distance from the source location.

These considerations lead to the following question: How sharp can the focus of the time-reversed field be? It is useful to first consider the ideal case, in which a continuous SRA *surrounds* a probe source, as shown in Fig. 13.

The probe source, situated at the location r_{ps} emits a time-harmonic field of angular frequency ω. We will take the source to have unit strength, obeying the inhomogeneous Helmholtz equation

$$\nabla^2 G_\omega(\mathbf{r}, \mathbf{r}_{\mathrm{ps}}) + k(\mathbf{r})^2 G_\omega(\mathbf{r}, \mathbf{r}_{\mathrm{ps}}) = -\delta(\mathbf{r} - \mathbf{r}_{\mathrm{ps}}). \tag{26}$$

Note that this Green's function incorporates all the complexity of the problem at hand. That is, it satisfies appropriate boundary conditions such as those at the sea surface and bottom and includes all the effects of inhomogeneity in the sound speed in the position-dependent wavenumber $k(\mathbf{r}) = \omega/c(\mathbf{r})$. At a general point $\mathbf{r}'$ on its surface, the ideal SRA receives the field $G_\omega(\mathbf{r}', \mathbf{r}_{\mathrm{ps}})$

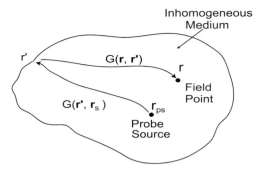

Fig. 13. Schematic of the ideal, closed SRA

of the probe source, conjugates it, and retransmits to a point r in the interior according to Huygen's principle as embodied in the Helmholtz–Kirchhoff integral [30]. The resulting field at r is

$$P_{\text{pc}}(r;\omega) = \int [G_\omega(r,r')G^*_\omega(r',r_{\text{ps}}) + \nabla G_\omega(r,r') \cdot \nabla G^*_\omega(r',r_{\text{ps}})]\, d\mathbf{S}'. \quad (27)$$

As *Porter* [30] has shown, in the lossless case [that is, when $k(r)$ is real], the field produced by the ideal array can be expressed exactly as

$$P_{\text{pc}}(r;\omega) = 2i\Im\{G_\omega(r,r_{\text{ps}})\}. \quad (28)$$

By taking the imaginary part of (26) it can be seen that $P_{\text{pc}}(r, r_{\text{ps}};\omega)$ is a solution of the sourceless (homogeneous) Helmholtz equation. Thus, it is seen that the ideal SRA produces a field which is similar to the original field, but is source free in the focus region, as it must be. The most remarkable aspect of this ideal case is that the focus field is independent of the shape and location of the closed array, as long as it surrounds the original source. We will see that this ideal behavior has implications in more practical situations.

How good is the ideal focus? In the case where the medium is homogeneous in the vicinity of the original source and in which backscattering due to boundaries and inhomogeneities produces a negligible field near the original source, the original field can be approximated by the free-space Green's function,

$$G_\omega(r, r_{\text{ps}}) \approx \frac{\exp(ik|r - r_{\text{ps}}|)}{4\pi|r - r_{\text{ps}}|}. \quad (29)$$

While this field is singular at $r = r_{\text{ps}}$, its imaginary part is not, as shown in Fig. 14. It can be seen from the figure that the ideal focus has a width comparable to the wavelength. This ideal is not generally realized in more practical situations. Note that the focus field is oscillatory. It is, in fact, a standing wave created by the interference of waves launched from all directions, since the ideal SRA surrounds the focus.

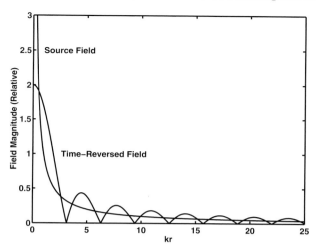

Fig. 14. Comparison of source and ideal focus fields

3.1.2 The Open Array

In more realistic cases, the array will not be closed and will not surround the original source. The ideal array provides insight into such cases, as one can, as a thought experiment, divide the closed SRA into two parts, as in Fig. 15, the smaller of these parts representing an actual 2-dimensional SRA. Consider deforming this array, while leaving the other portion undeformed. We know from the previous result that the focus field will not change as a result of this deformation. The focus field in this case is the sum of the fields from the two parts, and, as one part of the array is fixed, its contribution to the field must be fixed as well. It follows that the field of the smaller part must also be unchanging. Thus, an ideal open two-dimensional SRA should display an interesting invariance property: Its focus field will be independent of deformations of the array provided the outer edge of the array is held fixed. Of course, this array is rather special; it is continuous and employs both the incident field and its normal derivative. Nevertheless, we shall see that this invariance survives at least as an approximation in realistic situations.

It is possible to extract an additional useful piece of information from this example. Consider an infinite, planar SRA in a homogeneous acoustic medium. It can be considered as one-half of a closed array formed from two planes placed symmetrically with respect to the probe source. In the plane parallel to the arrays and passing through the source, symmetry dictates that each array produces the same field; thus the infinite planar SRA must produce a $\sin(kR)/kR$ field in the plane through the source, where R is the distance from the source.

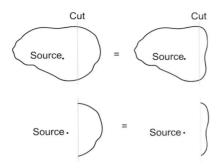

Fig. 15. Demonstration of shape invariance of ideal open arrays

3.1.3 The Discrete Array

In practice, creating an array to approximate even an open version of the array represented by (13) is very difficult, as this array must be capable of sensing both the incident field and its normal derivative. Furthermore, it must include both monopole sources [to produce $G_\omega(\mathbf{r},\mathbf{r}')$] and dipole sources [to produce $\nabla G_\omega(\mathbf{r},\mathbf{r}')$]. The general problem of using arrays with various combinations of monopole and dipole sources has been discussed by *Cassereau* and *Fink* [31]. In low-frequency underwater acoustics, arrays are usually composed of discrete elements that are omnidirectional or nearly so. Such arrays are more accurately represented by a set of discrete monopole receivers/sources, as illustrated in Fig. 12.

In the harmonic case, such an array produces the following field in response to a point probe source of unit strength:

$$P_{\mathrm{pc}}(\mathbf{r};\omega) = \sum_{j=1}^{J} G_{2,\omega}(\mathbf{r},\mathbf{r}_j) G_{1,\omega}^{*}(\mathbf{r}_j,\mathbf{r}_{\mathrm{ps}}) \,. \tag{30}$$

The sum is over the J elements of the SRA, whose position vectors are denoted $\mathbf{r}_j$. Propagation from the probe source to the SRA is described by the Green's function $G_{1,\omega}(\mathbf{r}_j,\mathbf{r}_{\mathrm{ps}})$, while propagation from the SRA to the field point is described by $G_{2,\omega}(\mathbf{r},\mathbf{r}_n)$. The subscripts 1 and 2 allow for the possibility that time variation of the ocean might cause changes in the Green's function between the probe and PC transmission cycles. During either propagation cycle, the ocean is assumed to be "frozen" in the sense that it behaves as a time-invariant linear system. In this view, the Green's function is the frequency-dependent system transfer function for acoustic propagation between any two points in the ocean. It is instructive to consider the focus field at the location of the probe source. By suppressing the subscripts 1 and 2 in (30), the field at $\mathbf{r} = \mathbf{r}_{\mathrm{ps}}$ may be written as follows:

$$P_{\mathrm{pc}}(\mathbf{r}_{\mathrm{ps}};\omega) = \sum_{j=1}^{J} |G_\omega(\mathbf{r}_{\mathrm{ps}},\mathbf{r}_j)|^2 \,. \tag{31}$$

In (31), reciprocity has been used in the form $G_\omega(\mathbf{r}_{\text{ps}}, \mathbf{r}_j) = G_\omega(\mathbf{r}_j, \mathbf{r}_{\text{ps}})$, which neglects gradients in density. The field at the focus is a sum of positive terms, that is, the transmissions from each element of the SRA arrive exactly in phase at the probe source location. Even though each term in this sum is likely to have complicated dependence on frequency, ω, owing to modal interference (equivalently, owing to multipath propagation), the sum will tend to have less severe frequency dependence than the individual terms. This is because each element of the SRA is subject to different modal interference, with destructive interference tending to occur at different frequencies for each element. This lack of frequency dependence at the source location implies that a time-reversed pulse will suffer little multipath distortion at the focus. See Sect. 3.2.2 for a discussion of this matter in the time domain. Note that the field at the focus is proportional to the sum over the SRA of the squared probe source pressure at each element. If the SRA samples nearly the entire water depth, this sum will tend to be stable and independent of details of propagation in the intervening medium, as the sum is a measure of the total power propagating through the waveguide. Thus, if one imagines perturbing the environment by altering the seafloor roughness or by introducing slight random changes in the sound speed of the medium, one would not expect the field value at the focus to change appreciably. This is another indication of the robust nature of PC focusing.

3.2 Background Theory and Simulation for Phase Conjugation/Time-Reversal Mirror in the Ocean

The theory of PC *vis-à-vis* ocean acoustics has already been presented [19,20]. Here we briefly review salient issues using the basic geometry of an ocean TRM experiment [19,21], shown schematically in Fig. 16. The figure also indicates the types of environmental measurements that were made. The TRM was implemented by a 77 m SRA in 125 m deep water. The SRA consisted of 20 hydrophones with 20 contiguously located sources with a nominal resonance frequency of 445 Hz. The sources were operated at a mean nominal 165 dB source level. The received signals from a probe source (PS) were digitized, time reversed and, after being converted back to analog form, retransmitted. A vertical 46-element receiver array (VRA) spanning 90 m was located initially 6.3 km from the SRA at the PS range.

3.2.1 Harmonic Point Source

For a unit harmonic source of angular frequency ω located at $(\mathbf{r}_{\text{ps}}, z_{\text{ps}})$, the pressure field, $G_\omega(R; z_j, z_{\text{ps}})$, at the jth receiver element of the SRA from the PS in Fig. 16 is determined from (19):

$$G_\omega(r; z, z_{\text{ps}}) = \frac{i}{\rho(z_{\text{ps}})(8\pi r)^{1/2}} \exp(-i\pi/4) \sum_n \frac{u_n(z_{\text{ps}})u_n(z)}{k_n^{1/2}} \exp(ik_n r). \quad (32)$$

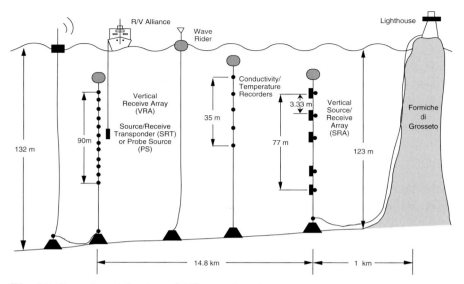

Fig. 16. Experimental setup of PC experiment

For a vertical line of discrete sources, (30) (with the subscripts 1 and 2 dropped, because time dependence of the medium is neglected) takes the form

$$P_{\mathrm{pc}}(r,z;\omega) = \sum_{j=1}^{J} G_\omega(r;z,z_j) G_\omega^*(R;z_j,z_{\mathrm{ps}}), \tag{33}$$

where R is the horizontal distance of the SRA from the PS and r is the horizontal distance from the SRA to a field point. We can implement (33) with propagation models other than the normal-mode example we are using. For mildly range-dependent environments, we can use the adiabatic-mode theory for $G_\omega(r;z,z_j)$ described in Sect. 1.2.3. For more highly range-dependent environments, the parabolic equation model is not only more appropriate, it is intuitively suitable in the sense that it uses a range-marching algorithm; hence, it is a direct implementation of back propagation. This model is described in Appendix A, and we note here that it has recently been applied to some interesting medical ultrasonics problems [22].

Note that the magnitude squared of the right hand side (r.h.s.) of (33) is the unnormalized (22) ambiguity function of the matched field processor of (23) where the data, d, are given by $G_\omega(R;z_j,z_{\mathrm{ps}})$ and the unnormalized replica field by $G_\omega(r;z,z_j)$. In effect, the process of PC is an implementation of matched-field processing where the ocean itself is used to construct the replica field. Or, alternatively, matched-field processing simulates the experimental implementation of PC in which a SRA is used. To demonstrate that $P_{\mathrm{pc}}(r,z)$ focuses at the position of the PS, (R, z_{ps}), we simply substitute (32) into (33), which specifies that we sum over all modes and array sources,

$$P_{\text{pc}}(r,z;\omega) \approx \sum_m \sum_n \sum_j \frac{u_m(z)u_m(z_j)u_n(z_j)u_n(z_{\text{ps}})}{\rho(z_j)\rho(z_{\text{ps}})\sqrt{k_m k_n r R}} \exp\left[\text{i}(k_m r - k_n R)\right]. \quad (34)$$

For an array which substantially spans the water column and adequately samples most of the modes, we may approximate the sum of sources as an integral and invoke orthonormality as specified by (17). Then the sum over j selects out modes $m = n$ and (34) becomes

$$P_{\text{pc}}(r,z;\omega) \approx \sum_m \frac{u_m(z)u_m(z_{\text{ps}})}{\rho(z_{\text{ps}})k_m\sqrt{rR}} \exp\left[\text{i}k_m(r - R)\right]. \quad (35)$$

The individual terms change sign rapidly with mode number. However, for the field at the PS, $r = R$, the closure relation of (16) can be applied approximately (we assume that k_n is nearly constant over the interval of the contributing modes) with the result that $P_{\text{pc}}(r,z) \approx \delta(z - z_{\text{ps}})$. Figure 17b is a simulation of the PC process using (33) for a PS at 40 m depth and at a range of 6.3 km from a 20-element SRA as specified in Fig. 16, verifying the above discussion. The measured range-dependent bathymetry, bottom properties and sound speed profile were used as the input to an adiabatic-mode model (Sect. 1.2.3). Notice that the focusing in the vertical is indicative of the closure property of the modes. As a matter of fact, for an SRA with substantially fewer elements, we see that the focusing still is relatively good. For example, Fig. 17c shows a result for the bottom 10 elements of the SRA, which are below the thermocline.

3.2.2 Pulse Excitation

In the actual experiments, a 50-ms pure-tone pulse with center frequency 445 Hz was used for the probe transmission. We can Fourier synthesize the above results to examine PC for pulse excitation. Here, in the context of this experiment, we remind the reader that PC in the frequency domain is equivalent to time reversal in the time domain. The jth element of the SRA receives the following time-domain signal, given by Fourier synthesis of the solution of (28),

$$P(R,z_j;t) = \int G_\omega(R;z_j,z_{\text{ps}})S(\omega)\text{e}^{-\text{i}\omega t}\text{d}\omega, \quad (36)$$

where $S(\omega)$ is the Fourier transform of the PS pulse. This expression incorporates all waveguide effects, including time elongation due to multipath propagation. For convenience, take the time origin such that $P(R,z_j;t) = 0$ outside the time interval $(0,\tau)$. Then the time-reversed signal that will be

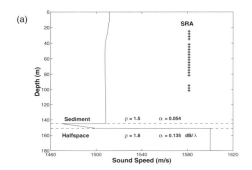

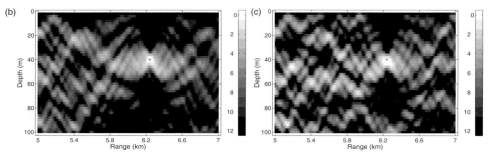

Fig. 17a–c. Single-frequency simulation of PC for the geometry of Fig. 16 for a PS located at a depth of 40 m and a range of 6.3 km. (**a**) Sound speed profile. The density, ρ, and attenuation, α (in dB/wavelength), of the bottom two layers are also given. (**b**) Simulation for a 20-element SRA. Note the sharp focus in depth. (**c**) Simulation for only the bottom 10 elements of the SRA

used to excite the jth transmitting element of the SRA is $P(R, z_j; T - t)$ such that $T > 2\tau$. This condition is imposed by causality; the signal has to be completely received before it can be time reversed. Then

$$P(R, z_j; T - t) = \int G_\omega(R; z_j, z_{\text{ps}}) S(\omega) e^{-i\omega(T-t)} d\omega$$
$$= \int \left[G_\omega^*(R; z_j, z_{\text{ps}}) e^{i\omega T} S^*(\omega) \right] e^{-i\omega t} d\omega, \qquad (37)$$

where the sign of the integration variable, ω, has been reversed and the conjugate symmetry of the frequency-domain Green's function and probe pulse has been used. The quantity in brackets in (37) is the Fourier transform of the signal received by the jth SRA receiver element after time reversal and time delay. Hence, there is an equivalence of time reversal and phase conjugation in their respective time and frequency domains.

Noting that the bracketed quantity in (37) is the frequency-domain representation of the signal retransmitted by the jth element of the SRA, Fourier synthesis can be used to obtain the time-domain representation of the field produced by the TRM. Using (33),

$$P_{\rm pc}(r,z;t) = \sum_{j=1}^{J} \int G_\omega(r,z,z_j) G_\omega^*(R,z_j;z_{\rm ps}) e^{i\omega T} S^*(\omega) e^{-i\omega t} d\omega. \tag{38}$$

This expression can be used to show that the TRM produces focusing in time as well as in space. Focusing in time occurs because a form of matched filtering occurs. To understand this, examine the TRM field at the focus point [that is, take $r = R$, $z = z_{\rm ps}$ in (38)]. Neglecting density gradients, reciprocity allows the interchange $G_\omega(R, z_{\rm ps}, z_j) = G_\omega(R, z_j, z_{\rm ps})$. Then the time-domain equivalent of (38) is

$$P_{\rm pc}(r,z;t) = \frac{1}{(2\pi)^2} \int \sum_{j=1}^{J} \left(\int G_{t'+t''}(R, z_j, z_{\rm ps}) G_{t'}(R, z_j, z_{\rm ps}) dt' \right)$$
$$\times S(t'' - t + T) dt'', \tag{39}$$

where the time-domain representations of the Green's function and probe pulse are used. Note that the Green's function is correlated with itself. This operation is matched filtering, with the filter matched to the impulse response for propagation from the PS to the jth SRA element. This operation gives focusing in the time domain, that is, it reduces the time elongation due to multipath propagation [17]. The sum over array elements is a form of spatial matched filtering, analogous to that employed in the matched-field processor. In addition, this sum further improves temporal focusing, as the temporal sidelobes of the matched filters for each channel tend to average to zero, which is also analogous to broadband matched-field processing results [23]. Finally, note that the integral over t'' in (39) is a convolution of each matched-filtered channel impulse response with the time-reversed and delayed probe pulse. As a consequence, this pulse is *not* matched filtered, for example, a linear FM up-sweep will appear as a down-sweep at the focus and will not be compressed.

Figure 18a shows a simulation for a 50-ms rectangular pulse with center frequency 445 Hz for the same geometry used in Fig. 17a as received at the SRA, and Fig. 18b shows the pulse as transmitted to a plane at a range of 6.3 km, the range of the PS. Four sources were excluded from the simulation because these phones were not used in the experiment. Note the temporal focusing, that is, the 50-ms pulse disperses to about 75 ms at the SRA but the time-reversed pulse received at the VRA is compressed (focused) to 50 ms as opposed to exhibiting even further time dispersion. On the other hand Fig. 18c shows a pulse 500 m outbound from the PS (i.e., the VRA is at the same location, but the PS is 500 m closer to the SRA). The pulse is not spatially focused, and it is temporally more diffuse than the result for the focal spot.

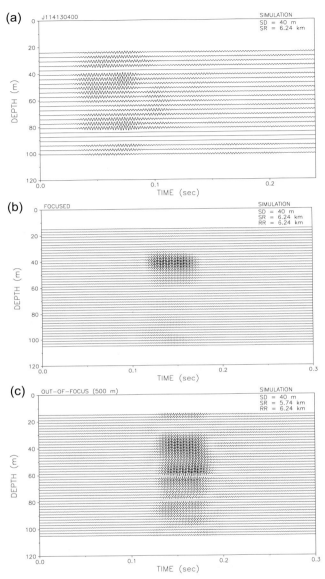

Fig. 18a–c. Simulation of a 445-Hz, 50-ms transmitted pulse for the geometry in Fig. 16 for a PS located at a depth of 40 m. (**a**) Pulse received on the SRA at a range of 6.3 km from the PS. There is a temporal dispersion of about 75 ms and significant energy throughout the water column. (**b**) The focus of the time-reversed pulse at the VRA. There is pulse compression back to the original transmitted 50-ms duration as well as spatial focusing in depth. (**c**) Vertical and temporal distribution for a pulse 500 m outbound from the PS (the VRA is at the same location, but the PS is 500 m closer to the SRA)

3.2.3 Properties of the Focal Region

The TRM focus is robust provided the SRA adequately samples the field in the water column. First, the focus tends to depend primarily on the properties of the ocean near the focus and tends to be independent of the (possibly range-dependent) properties of the medium between the SRA and the focus. Temporal changes in the medium due to, for example, surface waves and internal waves degrade the focus, but this degradation will be tolerable if the average (or coherent) Green's function is not severely reduced by these time variations. Generally, the shape of the focus is approximated by the field that a PS placed at the focus generates after nonpropagating modes are subtracted. Thus, if absorption or scattering tends to eliminate high-order modes, the focus will be comprised of the remaining lower-order modes and will be relatively broader. Very roughly, the vertical width of the focus will be equal to the water depth (or depth of the duct) divided by the number of contributing modes if the sound speed (in the duct) is not strongly dependent upon depth.

The TRM focus is also robust with respect to array shape [13] provided the shape does not change between the probe reception and time-reversed transmission. This property makes it unnecessary to know the exact shape of the TRM array and offers a considerable advantage over conventional beamforming. The focal properties are discussed in detail in Sect. 4.

3.2.4 Variable-Range Focusing

The TRM technique has been extended to refocus at ranges other than that of the PS [20]. The technique involves retransmitting the data at a shifted frequency according to the desired change in focal range, such that

$$(\Delta \omega/\omega) = \beta \left(\Delta R/R \right), \tag{40}$$

where the invariant β is determined by the properties of the medium. Typically, β is approximately equal to 1 in a large class of acoustic waveguides, which means that a 5% increase in frequency results in a 5% increase in focal range. On the other hand, $\beta \approx -3$ for a surface sound channel or in a waveguide with a deeply submerged channel axis. In this case, the retransmitting frequency should be decreased to increase the focal range. The frequency shift can be implemented easily in near real-time by an FFT bin shift prior to retransmission. Recently, variable focusing in range has been applied to medical ultrasonics in free space [22].

3.3 Implementation of a Time-Reversal Mirror in the Ocean

Two experiments [19,21] were conducted by the SACLANT Undersea Research Centre (Dr. T. Akal was the Chief Scientist of the experiments) and the Marine Physical Laboratory of Scripps Oceanographic Institution. An assortment of runs was made to examine the structure of the focal point region and the temporal stability of the process. The TRM process was successfully demonstrated out to a range of 30 km in these experiments. Here we will be reporting on some of the results. (Note that range refers to the distance from the SRA.)

3.3.1 Demonstration of a Time-Reversal Mirror in the Ocean

In the first experiment the VRA was deployed at a range, determined by DGPS, of 6.24 km from the SRA and the PS was deployed at two different depths, 40 m and 75 m. Figure 19 shows the pulse as received on the SRA and VRA for both source depths. The data at the SRA is a combination of signal and noise. A 233-ms window was digitized and time reversed for transmission to the VRA. When the VRA and PS are at the same range (experimentally within 40 m by a DGPS measurement) from the SRA, we see the focusing as predicted in Sect. 3.1 for a probe source at 40 m depth and similar results for a probe source at 75 m depth. Clearly, we have implemented a TRM focusing at the range and depth of the PS.

In the second experiment, the PS was deployed out to a range of 30 km. Figure 20 shows the results for the PS at 81 m depth for five different ranges from the SRA: 4.5, 7.7, 15, 20, and 30 km. As expected, the temporal focus remains compact, while the spatial focus broadens with range due to mode stripping. The latter can be seen more quantitatively in Fig. 21a. Simulations, shown in Fig. 21b, using the measured environment confirm that the focal structure is consistent with the above theory.

To summarize, the TRM is still quite effective at 30 km, particularly for a deep PS in the stable part of the water column. The focal region has a vertical extent of about 25 m (3 dB down from the peak) at a range of 30 km, which corresponds to 20% of the waveguide thickness at a range of about 250 waveguide depths. Although we did not map out the radial extent of the focus, a simulation using the measured environment which has successfully described this process indicates that the radial size of the focus was on the order of 800 m at 30 km range and 300 m at 5 km range.

3.3.2 Stability Measurements

Here we only show examples of stability measurements from the first experiment. Two stability data collection periods for the PS depths of 40 m and 75 m were made for 1 h and 2 h, respectively (the lengths of the runs were dictated by experimental circumstance). Figure 22 shows the results of these runs. These plots indicate that the focus was considerably more stable for the deep PS versus the shallower PS and that the focus is broader for the shallower PS. This is consistent with the measured sound speed structure. A discussion on the stability of the TRM is given in Sect. 5.

Ocean Acoustics, Matched-Field Processing and Phase Conjugation

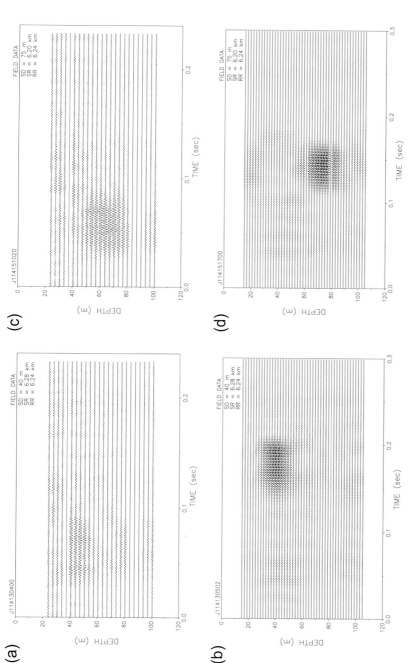

Fig. 19a–d. Experimental results for the PS and VRA at same range. (**a**) The pulse data received on the SRA for PS at depth of 40 m. (**b**) The pulse data received on the SRA for the PS at a depth of 75 m. (**c**) The data received on the VRA from the time-reversed transmission of pulses shown in (**a**). The VRA is 40 m inbound from the focus as determined by DGPS. (**d**) The data received on the VRA from the time-reversed transmission of the pulse shown in (**b**). The VRA is 40 m outbound from the focus as determined by DGPS

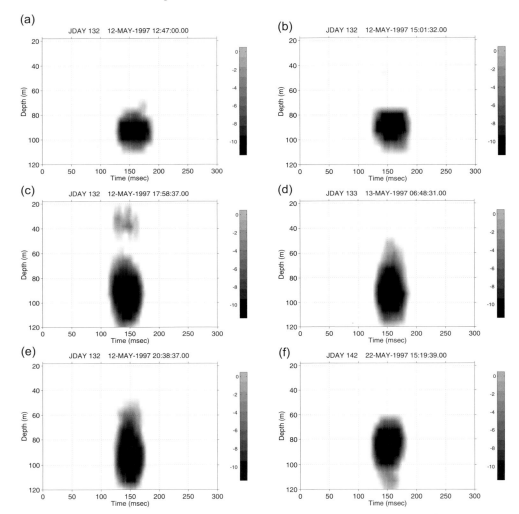

Fig. 20a–f. Experimental results for a 50-ms, 445-Hz center-frequency PS) at 81 m depth and various ranges, R, between the PS and the SRA. (**a**) $R = 4.5$ km, (**b**) $R = 7.7$ km, (**c**) $R = 15$ km, (**d**) $R = 20$ km, (**e**) $R = 30$ km, and (**f**) $R = 15$ km. The VRA is within 60 m of the PS. Both the VRA and PS were suspended from the ALLIANCE, except for (**f**) which was from a rf-telemetered VRA at 15 km range

3.3.3 A Time-Reversal Mirror with Variable-Range Focusing

Here we present results which experimentally confirm a technique to change the range focus of a TRM as described in Sect. 3.2 [20]. During the experiment, the frequency shift was implemented in near real-time by a simple FFT bin shift of the PS data received by the SRA prior to retransmission.

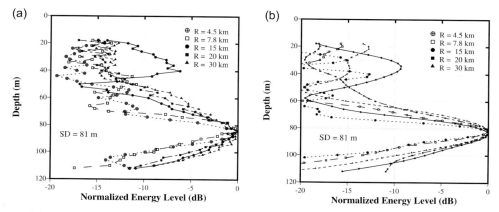

Fig. 21. The energy over a 0.3 s time window of the pulse received on VRA as a function of depth for various ranges. The depth of the probe source was 81 m. (**a**) Experimental results. (**b**) Simulation results. Note the sidelobe at around 40 m depth and 15 km range in both cases

Figure 23a shows the considerable defocusing for the PS at a depth of 68 m when the VRA was 600 m inbound of the PS. Figure 23b–d show frequency shifts of 20 Hz, 25 Hz and 32 Hz, with the best focus resulting from the 32 Hz shift. It is interesting to note that this latter result corresponds to $\beta = 1.4$, and in essence, this procedure is a way of determining β.

To summarize, it is possible to shift the focal range on the order of 10% of the nominal range of the PS. The theory on which this shift is dependent is valid only over a frequency range in which the mode shapes do not change significantly. Frequency shifts of greater than about 10% violate this condition. A practical limitation also comes from the transducer characteristics of the SRA, the resonance of which around 445 Hz with a 3 dB bandwidth of approximately 35 Hz as shown in Fig. B1 in [19]. Therefore, it is difficult to excite the pulse at a carrier frequency more than 10% offset from the original resonance frequency.

3.4 Summary of Time-Reversal Mirror Experiments

The description above has been limited mostly to the implementation of a TRM in the ocean. The experiments also studied other aspects of the TRM process such as stability, fluctuations, range dependence, and out-of-plane defocusing. These topics will be discussed elsewhere in this chapter.

4 The Range-Dependent Ocean Waveguide

The factors that control phase-conjugate focusing in severely range-dependent environments will be examined by considering a general nonuniform, nonadiabatic waveguide. The conditions for "ideal" phase-conjugate

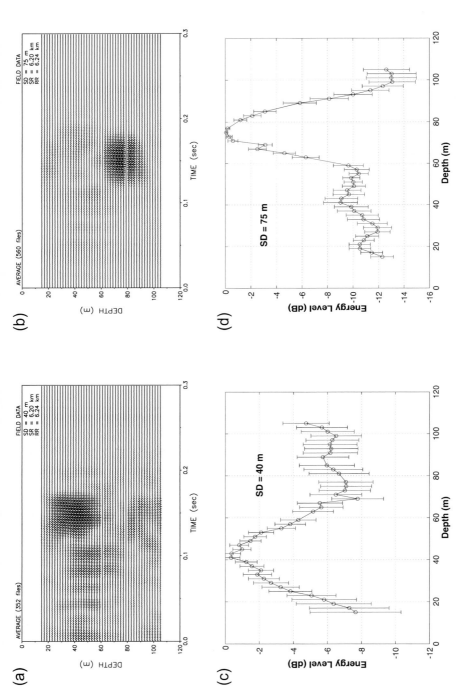

Fig. 22a–d. Results on stability of the focal region. (a) Pulse arrival structure at the VRA for PS at 40 m depth averaged over 1 h. (b) Pulse arrival structure at the VRA for PS at 75 m depth averaged over 2 h. (c) Mean and standard deviation of energy in a 0.3 s window for a 40 m PS. (d) Mean and standard deviation of energy in a 0.3 s window for a 75 m PS

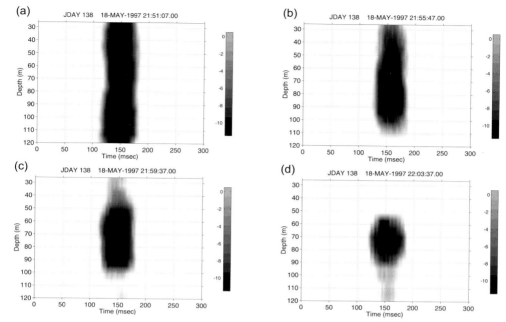

Fig. 23a–d. Experimental results for the PS at a depth of 68 m. (**a**) Out-of-focus results on the VRA when the VRA is 600 m inbound of the PS. (**b**) -20 Hz frequency shift. (**c**) -25 Hz frequency shift. (**d**) -32 Hz frequency shift. The -32 Hz shift in (**d**) shows the best focus, which corresponds to $\beta = 1.4$

focusing in such a waveguide will be derived, and this will implicitly identify the factors that degrade focusing. To simplify the discussion, only vertical SRAs will be considered. The main objective is to generalize (35) to the range-dependent case, using the approach given by *Siderius* et al. [32] in connection with the "guide source" concept. In this approach, small regions near the PS and SRA are assumed to be range independent, but the larger region between is allowed to have arbitrary range dependence in bathymetry and sound speed. Losses are neglected and will be discussed later in qualitative terms.

The Green's function for the probe field near the PS is approximated using range-independent normal modes:

$$G_\omega(r, z; R, z_{\rm ps}) = \sum_n \frac{a_n(z_{\rm ps})u_n(R, z)}{\sqrt{k_n(R)|r-R|}} e^{ik_n(R)|r-R|} . \qquad (41)$$

Similarly, the Green's function for the probe field at the SRA is written in the form

$$G_\omega(0, z_j; R, z_{\rm ps}) = \sum_n \frac{b_n(z_{\rm ps})u_n(0, z_j)}{\sqrt{k_n(0)R}} e^{ik_n(0)R} . \qquad (42)$$

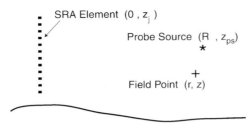

Fig. 24. Coordinate definitions for source-receiver array (SRA), probe source, and field point in range-dependent ocean

The modal eigenfunctions in the vicinity of the PS and SRA are denoted $u_n(R, z)$ and $u_n(0, z)$, respectively. The corresponding eigenvalues are $k_n(R)$ and $k_n(0)$. These Green's functions do not bear the superscripts 1 and 2 introduced earlier because a time-invariant environment is being considered. The subscript ω is used here in the same sense as in the earlier text. The mode amplitudes for the near-source Green's function are

$$a_n(z_{\mathrm{ps}}) = \frac{\mathrm{i}e^{-\mathrm{i}\pi/4}}{\sqrt{8\pi\rho(z_{\mathrm{ps}})}} u_n(R, z_{\mathrm{ps}}), \tag{43}$$

and the mode amplitudes for the Green's function near the SRA are given by the linear transformation

$$b_m(z) = \sum_n U_{mn} a_n(z). \tag{44}$$

For convenience, it is assumed that there are the same number of modes near the source and near the SRA, so that U_{mn} is a square matrix. Cases for which these numbers are similar but not equal can be treated by discarding high-order modes. The matrix U_{mn} includes any mode coupling that is due to the range dependence of the ocean and is defined in such a way as to be independent of source depth. Furthermore, to the extent that absorption loss in the water column and seafloor can be neglected, U_{mn} is unitary.

The field produced by the SRA is

$$P_{\mathrm{pc}}(r, z; \omega) = \sum_{j=1}^{J} G_\omega(r, z; 0, z_j) G_\omega^*(0, z_j; R, z_{\mathrm{ps}}). \tag{45}$$

The Green's function for propagation from the jth array element to the field point (r, z) can be expressed in terms of the Green's function for propagation in the opposite direction by using reciprocity:

$$G_\omega(r, z; 0, z_j) = \frac{\rho(z)}{\rho(z_j)} G_\omega(0, z_j; r, z). \tag{46}$$

In terms of mode amplitudes,

$$G_\omega(r, z; 0, z_j) = \frac{\rho(z)}{\rho(z_j)} \sum_n \frac{c_n(z) u_n(0, z_j)}{\sqrt{k_n(0) r}} e^{i k_n(0) R}, \quad (47)$$

where the mode amplitudes, $c_n(z)$, are

$$c_m(z) = \sum_{n'} U_{mn} a_n(z) e^{i k_n(R)(r-R)}. \quad (48)$$

Note that the mode amplitudes, $c_n(z)$, are essentially the same as $b_n(z)$, but with the source range coordinate shifted by $r - R$.

Equations (42) and (47) can be inserted in (45) to obtain an expression for the phase-conjugate field in a range-dependent waveguide:

$$P_{\rm pc}(r, z; \omega) = \frac{\rho(z)}{\sqrt{Rr}} \sum_{m,n} \frac{c_m(z) \Delta_{mn} b_n^*(z_{\rm ps})}{\sqrt{k_m(0) k_n(0)}}, \quad (49)$$

where

$$\Delta_{mn} = \sum_{j=1}^{J} \frac{u_m(0, z_j) u_n(0, z_j)}{\rho(z_j)}. \quad (50)$$

In the ideal case, the SRA spans the entire water column with elements having uniform spacing, $d_{\rm a}$, and the modal eigenfunctions have negligible amplitude in the bottom. In this case, the sum over array elements in (50) approximates the orthonormality integral for modal eigenfunctions (17), and $\Delta_{mn} d_{\rm a}$ can be taken equal to δ_{mn}. This ideal can be approached quite closely in the environment of the Mediterranean experiments. Using the environmental parameters defined in Fig. 17, and considering only the first 12 modes, an array with 36 elements with spacing $d_a = 3.33$ m and with the shallowest element 4.44 m below the surface gives diagonal elements in $\Delta_{mn} d_{\rm a}$ that are within 3% of unity and off-diagonal elements that are of the order of 0.03 or less.

Returning to the derivation of the conditions for ideal phase-conjugate focusing, take $\Delta_{mn} = \delta_{mn}/d_{\rm a}$ in (49) to obtain

$$P_{\rm pc}(r, z; \omega) = \frac{\rho(z)}{d_{\rm a}\sqrt{Rr}} \sum_{m,n} Q_{mn} a_m(z) a_n^*(z_{\rm ps}) e^{i k_n(R)(r-R)}, \quad (51)$$

where

$$Q_{mn} = \sum_l \frac{U_{lm} U_{ln}^*}{k_l(0)}. \quad (52)$$

Losses due to absorption and scattering are detrimental to phase-conjugate focusing, as they cause attenuation of higher-order modes, yielding a blurrier focus than would be possible with lower loss. Furthermore, this blurring will increase as the range between the source and the array increases

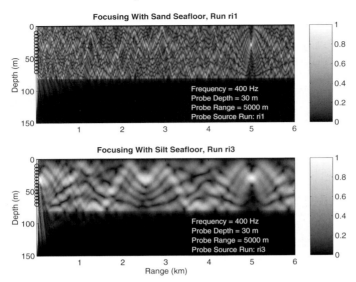

Fig. 25. Comparison of focusing for sand and silt seafloors. The sand seafloor has a sediment/water density ratio of 2.0 and a sound speed ratio of 1.13, with an absorption of 0.135 dB/wavelength. The corresponding numbers for the silt seafloor are 1.5, 1.027 and 0.1 dB/wavelength. The gray scale is linear in the relative field magnitude

owing to the strong range- and mode-number dependence of attenuation. Figure 25 shows the loss of focus resolution in a high-loss silty environment as compared to a lower-loss sandy environment. As explained in Sect. 1.1, losses in the ocean waveguide are primarily due to refraction of energy into the seafloor, so that seafloors with relatively high sound speed (sand) are less lossy than seafloors with relatively low sound speed (silt). As a result, the sandy environment supports a greater number of acoustic modes, providing a sharper focus.

In developing the ideal PC case, losses are set to zero and the mode-coupling matrix, U_{mn}, is taken to be unitary. If the mode dependence of $k_l(0)$ in (52) is neglected,

$$Q_{mn} = \frac{\delta_{mn}}{k_m(0)}. \tag{53}$$

and the phase-conjugate field for an ideal array in a lossless environment can be approximated as

$$P_{\text{pc}}(r, z; \omega) = \sum_m \frac{u_m(R, z) u_m(R, z_{\text{ps}}) e^{ik_m(R)(r-R)}}{8\pi\rho(z_{\text{ps}}) k_m(0) d_a \sqrt{Rr}}. \tag{54}$$

Apart from inessential factors, this expression is the same as (35), which was derived for the range-independent case. Even though (54) represents the ideal

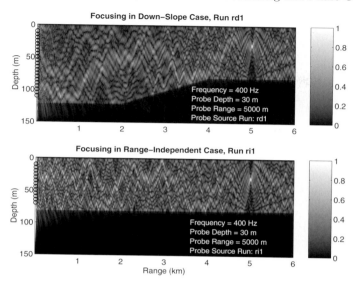

Fig. 26. Comparison of focusing for range-independent and range-dependent (downslope propagation of PS) environments. The sediment acoustic properties are the same as those of the sand seafloor in Fig. 25. The gray scale is linear in the relative field magnitude

case, it illustrates properties that actual phase-conjugate arrays may possess provided they are not too far from ideal. One such property is independence of the focus pattern upon the distance between the PS and the array (when absorption can be neglected and apart from the cylindrical spreading factor $1/\sqrt{Rr}$). Even more strikingly, the focus field is independent of the (possibly range-dependent) environment between the focus and the array (see examples presented in [32]). That is, the focus depends only on the local properties of the water column and sea floor, and it is not affected by bathymetry or range-dependent water-column properties in the region between the array and the focus provided they do not change appreciably during the two propagation cycles. This means that, in the ideal case, PC is not affected by time-invariant forward scattering due to bathymetry, fronts, etc. It also implies that, in simulations of phase-conjugate focusing, it is important to accurately model the ocean in the vicinity of the focus, but less accuracy is required for the more distant parts of the propagation path. One important reservation must be added at this point. The derivation above is essentially two-dimensional in that cross-range spatial variation of the ocean is neglected.

Figure 26 shows the results of a simulation which illustrates the invariance of the focus. In this figure, range-dependent and range-independent environments having the same seafloor type (sand) yield very similar focal fields, even though the bathymetry between the PS and SRA is greatly different in the two cases. This is a non-ideal case with realistic parameters, including loss.

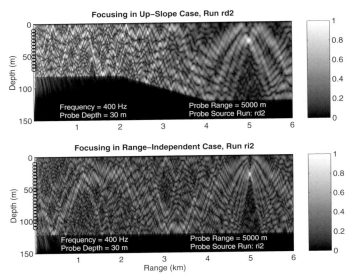

Fig. 27. Comparison of focusing for range-independent and range-dependent (upslope propagation of PS) environments. The sediment acoustic properties and gray scale are the same as those in Fig. 25

Figure 27 shows a case in which losses destroy focus invariance. In this case, the PS field propagates upslope to reach the SRA. Modes are lost during upslope propagation, leading to a broader focus than in the range-independent situation. In contrast, modes are not lost in the downslope propagation of Fig. 26.

The invariance seen in the ideal ocean waveguide case is similar to that predicted for the ideal, closed array, except that the time-reversed field is a standing wave in the latter case. In the present case, the PS field (including only propagating modes) is given by (41) which can be put in the form

$$G_\omega(r, z; R, z_\mathrm{ps}) = \frac{\mathrm{i} e^{-\mathrm{i}\pi/4}}{\rho(z_\mathrm{ps})\sqrt{8\pi|r-R|}} \\ \times \sum_n \frac{u_n(R, z) u_n(R, z_\mathrm{ps}) e^{\mathrm{i} k_n(R)|r-R|}}{\sqrt{k_n(R)}}. \quad (55)$$

Apart from a difference in spreading loss and an overall phase difference, (54) and (55) are quite similar. There is a slight term-by-term difference owing to differing factors involving modal eigenvalues, but the primary difference is in the propagation phase factor. The source field propagates *away* from the source location, while the phase-conjugate field propagates *past* the source location in the direction away from the SRA.

5 The Effect of Ocean Fluctuations on Phase Conjugation

The ocean is random in several respects. The surface and bottom boundaries are randomly rough, and the seawater is inhomogeneous owing to stratification, turbulence, and internal waves. In addition, wave-generated bubbles introduce inhomogeneity in a layer near the surface, and biological, hydrodynamic, and geophysical processes make the ocean sediment inhomogeneous. Sound scattering due to randomness of the seafloor is time independent and is not expected to have deleterious effects on phase-conjugate focusing. In fact, the range-dependent environment discussed earlier can be considered a kind of scattering example, and it was seen that PC adapts nicely to the complexities of this case. Generally, PC can be expected to operate robustly in the presence of time-independent scattering. Problems arise when the scattering is time dependent, since the medium changes between the probe and retransmission cycles. This is the case with scattering by surface waves and internal waves, for which one would like to know the characteristic stability time over which a given probe will yield satisfactory focusing.

5.1 Time-Independent Volume Scattering

As noted above, volume inhomogeneities in the ocean are inherently time dependent, owing to the presence of internal waves. These waves have rather long periods (minutes to hours), however, so volume scattering can be approximated as time independent if the time between probing and retransmission is sufficiently short. In this case one does not expect scattering to degrade the focus, in fact, it will be seen that the focus is improved by scattering. In static environments, the subscripts 1 and 2 can be dropped (see Fig. 12) because there is no change in the environment between the probe transmission and the time-reversed retransmission. It is useful to consider the formal average of the focus field, imagining an ensemble of time-reversal measurements made using a hypothetical ensemble of random environments. Often in underwater acoustics the average of a field over an ensemble of realizations is not comparable in magnitude to the field seen in single realizations, because scattering often causes fluctuations that are comparable to or greater than the mean. This is not the situation with PC if the SRA spans an appreciable portion of the water column, as the arguments made in connection with (54) show that the focus field is not very strongly dependent on perturbations of the environment. Thus, it is useful to examine the average focus field, taking it as an approximation to the field seen in any given realization.

$$\langle P_{\rm pc}(\boldsymbol{r};\omega)\rangle = \sum_{j=1}^{J} K(\boldsymbol{r},\boldsymbol{r}_{\rm ps},\boldsymbol{r}_j)\,, \tag{56}$$

where

$$K(\boldsymbol{r}, \boldsymbol{r}_{\mathrm{ps}}, \boldsymbol{r}_j) = \langle G_\omega(\boldsymbol{r}, \boldsymbol{r}_j) G_\omega^*(\boldsymbol{r}_{\mathrm{ps}}, \boldsymbol{r}_j) \rangle. \qquad (57)$$

The first moment of the focus field depends on the second moment of the Green's function. In some simple cases, the second moment of the Green's function is of the form (see [33] for an example and references and [34] for dynamic case)

$$\langle G_\omega(r, z, 0, z_j) G_\omega^*(R, z_{\mathrm{ps}}, 0, z_j) \rangle = G_{0,\omega}(r, z, 0, z_j) G_{0,\omega}^*(R, z_{\mathrm{ps}}, 0, z_j) e^{-H}, \qquad (58)$$

where

$$H = \frac{\Phi^2}{R_{\mathrm{s}}} \int_0^{R_{\mathrm{s}}} \{1 - \rho[\Delta z(R')]\} \mathrm{d}R', \qquad (59)$$

and where $G_{0,\omega}(r, z, 0, z_j)$ is the Green's function for the average medium, $\rho(z)$ is the vertical correlation function for refractive index ($\rho(0) = 1$), Φ^2 is the phase variance for a path of length R, and $\Delta z(R')$ is vertical separation of ray paths (Fig. 28).

Using this result, the mean phase-conjugate field is

$$\langle P_{\mathrm{pc}}(R, z; \omega) \rangle = P_{\mathrm{pc0}}(R, z; \omega) e^{-H}, \qquad (60)$$

where

$$P_{\mathrm{pc0}}(R, z; \omega) = \sum_{j=1}^{J} G_{0,\omega}(r, z, 0, z_j) G_{0,\omega}^*(R, z_{\mathrm{ps}}, 0, z_j) \qquad (61)$$

is the phase-conjugate field for the average medium. Following *Uscinski* and *Reeve* [35], expand the correlation function for refractive-index fluctuations in a power series:

$$\rho(z) = 1 - \frac{z^2}{l_v^2} + \dots. \qquad (62)$$

Assuming that the average medium is infinite and nonrefractive, the ray paths are straight lines and the integral for H can be performed:

$$H = \frac{\Phi^2 (z - z_s)^2}{3 l_v^2}. \qquad (63)$$

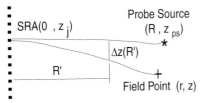

Fig. 28. Coordinate definitions for static volume scattering environment

The factor e^{-H} has a Gaussian shape in z centered on the PS with $1/\mathrm{e}$ full width

$$W_v = \frac{2\sqrt{3}l_v}{\Phi}. \tag{64}$$

In some cases, this width can be less than the focal spot width for the average medium [14,36]. This sharpening of the focus has been observed experimentally but in a stronger scattering regime than the present formalism can describe [37].

5.2 Time-Dependent Scattering by Surface Waves

Time-dependent forward scattering due to surface waves causes degradation of phase-conjugate focusing. Following [19], the Green's function will be decomposed into coherent (unscattered) and incoherent (scattered) parts:

$$G_\alpha(\boldsymbol{r}, \boldsymbol{r}') = \overline{G}(\boldsymbol{r}, \boldsymbol{r}') + \delta G_\alpha(\boldsymbol{r}, \boldsymbol{r}'). \tag{65}$$

The subscript α takes on the values 1 and 2 for the probe and conjugate transmission cycles, respectively. The coherent, or mean, Green's function, $\overline{G}(\boldsymbol{r}, \boldsymbol{r}')$ is not assigned a subscript because the random time variations are assumed to be stationary in the statistical sense. It will be assumed that sufficient time has elapsed between the probe and conjugate transmission cycles that variations in the two Green's functions are uncorrelated:

$$\langle \delta G_2(\boldsymbol{r}_\mathrm{d}, \boldsymbol{r}_\mathrm{c}) \delta G_1^*(\boldsymbol{r}_\mathrm{b}, \boldsymbol{r}_\mathrm{a}) \rangle = \langle \delta G_2(\boldsymbol{r}_\mathrm{d}, \boldsymbol{r}_\mathrm{c}) \delta G_1(\boldsymbol{r}_\mathrm{b}, \boldsymbol{r}_\mathrm{a}) \rangle = 0. \tag{66}$$

This condition was very likely satisfied in the Mediterranean experiments with respect to scattering by surface waves, which have correlation time scales on the order of seconds, while the time between transmission cycles was measured in minutes and hours.

Combining (30), (65), and (66), the mean phase-conjugate field is

$$\overline{P}_\mathrm{pc}(r, z; \omega) = \sum_{j=1}^{J} \overline{G}(\boldsymbol{r}, \boldsymbol{r}_j) \overline{G}^*(\boldsymbol{r}_j, \boldsymbol{r}_\mathrm{ps}), \tag{67}$$

and the variance of the field is

$$\overline{|P_\mathrm{pc}(r, z; \omega)|^2} - |\overline{P}_\mathrm{pc}(r, z; \omega)|^2 = \sum_{j=1}^{J} \sum_{j'=1}^{J} [\overline{G}(\boldsymbol{r}, \boldsymbol{r}_j) \overline{G}^*(\boldsymbol{r}, \boldsymbol{r}_{j'}) K_{jj'}(\boldsymbol{r}_\mathrm{ps})$$
$$+ \overline{G}(\boldsymbol{r}_\mathrm{ps}, \boldsymbol{r}_j) \overline{G}^*(\boldsymbol{r}_\mathrm{ps}, \boldsymbol{r}_{j'}) K_{jj'}(\boldsymbol{r}) + K_{jj'}(\boldsymbol{r}) K_{jj'}(\boldsymbol{r}_\mathrm{ps})], \tag{68}$$

where

$$K_{jj'}(\boldsymbol{r}) = \langle \delta G_\alpha(\boldsymbol{r}_j, \boldsymbol{r}) \delta G_\alpha^*(\boldsymbol{r}_{j'}, \boldsymbol{r}) \rangle. \tag{69}$$

The covariance, $K_{jj'}(\boldsymbol{r})$, is proportional to the correlation between the incoherent field at elements j and j' of the array, with a unit PS situated at $\boldsymbol{r}$.

In deriving (68), free use was made of reciprocity (which allows interchange of the two arguments of the Green's function) and stationarity (which means that δG_1 and δG_2 have identical statistics).

Equations (67) and (68) are general and include 3-dimensional scattering (i.e., in-plane and out-of-plane scattering). They lead to two general conclusions regarding focusing in ocean experiments for those cases in which sufficient time elapsed between the two transmission cycles. First, the mean focus field, that is, the focus field averaged over many independent probe-conjugate-transmission cycles, is obtained by using the coherent Green's function in place of the actual (random) Green's function. Second, and most important, the field near the focus does not fluctuate appreciably, that is, it is well approximated by the mean focus field. As demonstrated in [19], careful inspection of (68) shows that the variance of the phase-conjugate field is not localized near the focus, but is spread diffusely in range and depth. Thus, near the focus, the mean field dominates, unless scattering is strong enough to diminish the mean Green's function to such a degree that focusing is essentially destroyed. It is rather straightforward to compute the mean field in the presence of surface-wave scattering [38]. Figure 29 is an example of such a calculation using the modal propagation code KRAKEN [39] with parameters appropriate to the 1996 Mediterranean experiment. For root-mean-square wave heights within the range measured during the experiment, the degradation of the mean focus is predicted to be slight. This agrees with the experimental fact that the measured focus was stable on the short time scales characteristic of surface wave motion.

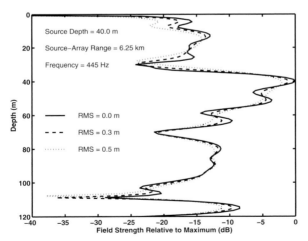

Fig. 29. Vertical slice through mean focus for selected root-mean-square surface-wave heights

5.3 Time-Dependent Scattering by Internal Waves

Internal waves are ubiquitous in the ocean and arise because of density stratification. Like surface waves, internal waves involve an interplay of kinetic energy and gravitational potential energy. Internal waves cause a slow orbital motion of the water which distorts the sound speed profile in both the vertical and horizontal, leading to random inhomogeneity in the acoustic index of refraction. This inhomogeneity, in turn, causes time-dependent fluctuations in sound propagation, which affect the stability of PC.

In this section, a brief summary of relevant internal wave properties is given, concentrating on the statistical description needed for modeling of propagation fluctuations. For more complete accounts of internal wave dynamics and statistics, see [40,41]. For treatments of the problem of acoustic scattering by internal waves, see [42,43,44].

The vertical displacement due to internal waves, $\zeta(\boldsymbol{r},t)$, can be written as a sum over normal modes

$$\zeta(\boldsymbol{r},t) = \sum_{\boldsymbol{K}} \sum_{j=1}^{J} a_j(\boldsymbol{K}) W_j(z) e^{i[2\pi f_j(K)t - \boldsymbol{K}\cdot\boldsymbol{R}]}, \tag{70}$$

where $\boldsymbol{r} = (\boldsymbol{R}, z)$, $\boldsymbol{R} = (x,y)$, $\boldsymbol{K} = (K_x, K_y)$. The frequencies, $f_j(K)$, and eigenfunctions, $W_j(z)$, are obtained by solving the eigenvalue equation

$$(f_j^2 - f_I^2) W_j'' + K^2 \left[N^2(z) - f_j^2 \right] W_j = 0 \tag{71}$$

subject to the conditions that the displacement field vanishes at the surface ($z = 0$) and bottom ($z = h$). The parameter f_I is the inertial frequency:

$$f_I = 2\sin(\theta_L)/T. \tag{72}$$

Here, θ_L is the latitude and T is the rotation period of the earth.

The function $N(z)$ is the depth-dependent buoyancy frequency, determined by the vertical gradient of density:

$$N^2(z) = (2\pi)^{-2} \frac{g}{\rho(z)} \frac{\partial}{\partial z} [\rho(z) - \rho_a(z)], \tag{73}$$

where $\rho_a(z)$ is the density profile in an adiabatically mixed ocean and g is the acceleration of gravity.

The eigenfunctions obey the following orthonormality relation:

$$\int_0^h \left[N^2(z) - f_I^2 \right] W_j(z) W_{j'}(z) dz = \delta_{jj'}. \tag{74}$$

Figure 30 shows a buoyancy frequency profile obtained from oceanographic data in the 1997 Mediterranean experiment [45]. Several of the resulting eigenfunctions computed for a particular wavenumber are shown in the figure, and modal frequencies are shown in Fig. 31.

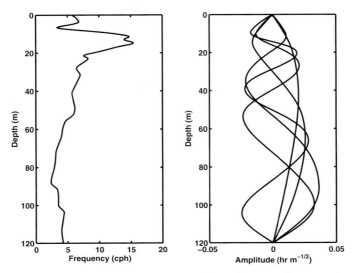

Fig. 30. Buoyancy frequency profile and internal wave eigenfunctions for the 1997 Mediterranean experiment. The eigenfunctions are computed for horizontal wavenumber $K = 0.001\,\mathrm{m}^{-1}$

The space–time covariance of the displacement field is defined as the following average:

$$\langle \zeta(\boldsymbol{R}_0 + \boldsymbol{R}, z_1, t+\tau)\zeta(\boldsymbol{R}_0, z_2, t)\rangle = K(R, z_1, z_2, \tau), \tag{75}$$

where it is assumed that the internal wave field is statistically isotropic and stationary in the transverse directions and stationary with respect to time so that the covariance depends on $R = \sqrt{x^2 + y^2}$ and τ, where (x, y) is the horizontal vector separation of the two points for which the covariance is measured. The displacement field is not stationary in the vertical; hence the covariance depends on both z_1 and z_2. A useful general expression for the covariance can be obtained by squaring and averaging (70):

$$K(R, z_1, z_2, \tau) = \sum_j \int_{f_1}^{\infty} F(f, j) W_j(z_1) W_j(z_2) J_0(KR) \cos(2\pi f \tau) \mathrm{d}f. \tag{76}$$

The eigenfunctions, $W_j(z)$, and the horizontal wavenumber, K, depend upon both the mode number, j, and the frequency, f.

Henyey et al. [42] provide a shallow-water form of the Garrett–Munk internal-wave statistics as follows:

$$F(f, j) = H(j) S(f), \tag{77}$$

$$H(j) = \frac{n_\mathrm{s}}{j^2 + j_*^2}, \tag{78}$$

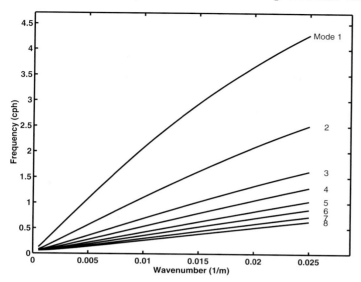

Fig. 31. Internal wave frequency vs. wavenumber for the 1997 Mediterranean experiment

$$S(f) = \frac{2f_{\rm I}\sqrt{f^2 - f_{\rm I}^2}\, bE_{\rm GM}}{\pi f^3} \left(\int_0^h N(z)\mathrm{d}z \right)^2. \tag{79}$$

The two parameters b and $E_{\rm GM}$ appear as a product, and this product (having units of length) is treated as a single parameter determining the energy of the internal wave field. Typically, $bE_{\rm GM}$ is of order 1 m. The parameter $n_{\rm s}$ is chosen so that the sum of $H(j)$ over all modes is unity. The parameter j_* determines which modes are most important. In deep water, $j_* = 3$ is typical, but smaller values are expected in shallow water [42].

Simulated internal-wave displacement fields can be generated by assigning random values to the amplitudes $a_j(\boldsymbol{K})$ in (70). The resulting random sound speed field follows from knowledge of the sound speed profile at zero displacement, which may be estimated using an average over several measured profiles. Figure 32 shows simulations of focusing using a probe sent at $t = 0$. These simulations, for an acoustic frequency of 400 Hz, employed internal wave parameters $j_* = 1$ and $bE_{\rm GM} = 2.0$ m. This value of $bE_{\rm GM}$, the internal-wave energy parameter, is much larger than the value suggested by oceanographic data gathered during the 1997 Mediterranean experiment, $bE_{\rm GM} = 0.135$ m [45]. The smaller, more realistic value produced negligible internal-wave effects, consistent with the observed stability of the focus. With the stronger internal-wave field, Fig. 32 shows that the initial sharp focus degrades substantially on time scales of several minutes.

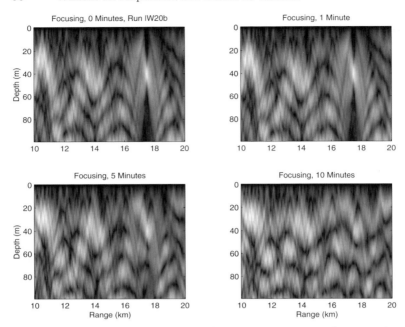

Fig. 32. Instability of focus in simulated environment with strong internal-wave activity. The gray scale is the same as that in Fig. 25

6 Conclusions

The ocean is an acoustic waveguide whose fundamental properties are well understood. Prediction of acoustic phenomena requires extensive ocean acoustic environmental knowledge, which is often difficult or impossible to obtain. Matched-field processing, a passive localization technique is dependent on accurate acoustic information to construct "replicas" and therefore is extremely sensitive to mismatch. Phase conjugation and time-reversal mirrors have been shown to be the active implementation of a matched-field processor where an array of sources is used to propagate the measured data on an array back to its origin. In essence, the ocean is used to construct the replicas, eliminating the possibility of mismatch in a reasonably stable ocean environment. Probably the most amazing aspect of implementing a TRM in the ocean over distances as great as 30 km is that the performance of the ocean TRM is almost exactly the same as that of a scaled (MHz regime) ultrasonic TRM implemented in a totally stable laboratory environment out to distances of tens of centimeters [24]. The TRM process is extremely robust.

Acknowledgement

This research was sponsored by the U.S. Office of Naval Research.

7 Appendix A: Parabolic Equation (PE) Model

The PE method was introduced into ocean acoustics and made viable with the development of the "split-step algorithm" which utilized fast Fourier transforms (FFTs) at each range step [25]. Subsequent numerical developments greatly expanded the applicability of the PE.

7.1 Standard Parabolic Equation Split-Step Algorithm

The PE method is presently the most practical and encompassing wave-theoretic range-dependent propagation model. In its simplest form, it is a far-field, narrow-angle ($\sim \pm 20°$ with respect to the horizontal — adequate for most underwater propagation problems) approximation to the wave equation. Assuming azimuthal symmetry about a source, we express the solution of (10) in cylindrical coordinates in a source-free region in the form

$$G(r,z) = \psi(r,z) \cdot J(r), \tag{80}$$

and we define $K^2(r,z) \equiv K_0^2 n^2$, n therefore being an "index of refraction" c_0/c, where c_0 is a reference sound speed. Substituting (80) into (10) in a source-free region and taking K_0^2 as the separation constant, J and ψ satisfy the following two equations:

$$\frac{d^2 J}{dr^2} + \frac{1}{r}\frac{dJ}{dr} + K_0^2 J = 0 \tag{81}$$

$$\frac{\partial^2 \psi}{\partial r^2} + \frac{\partial^2 \psi}{\partial z^2} + \left(\frac{1}{r} + \frac{2}{J}\frac{\partial J}{\partial r}\right)\frac{\partial \psi}{\partial r} + K_0^2 n^2 \psi - K_0^2 \psi = 0. \tag{82}$$

Equation (81) is a Bessel equation, and we take the outgoing solution, a Hankel function, $H_0^1(K_0 r)$, in its asymptotic form and substitute it into (82), together with the "paraxial" (narrow angle) approximation,

$$\frac{\partial^2 \psi}{\partial r^2} \ll 2K_0 \frac{\partial \psi}{\partial r}, \tag{83}$$

to obtain the PE (in r),

$$\frac{\partial^2 \psi}{\partial z^2} + 2iK_0 \frac{\partial \psi}{\partial r} + K_0^2(n^2 - 1)\psi = 0, \tag{84}$$

where we note that n is a function of range and depth. We use a marching solution to solve the PE. There has been an assortment of numerical solutions, but the one that still remains the standard is the so-called split-step algorithm [25].

We take n to be a constant; the error this introduces can be made arbitrarily small by the appropriate numerical gridding. The Fourier transform of ψ can then be written as

$$\chi(r,s) = \frac{1}{2\pi} \int_{-\infty}^{\infty} \psi(r,z) e^{-isz} \, dz, \tag{85}$$

which together with (84) gives

$$-s^2\chi + 2\mathrm{i}K_0\frac{\partial \chi}{\partial r} + K_0^2(n^2 - 1)\chi = 0. \tag{86}$$

The solution of (86) is simply

$$\chi(r, s) = \chi(r_0, s)\exp\left(-\frac{K_0^2(n^2 - 1) - s^2}{2\mathrm{i}K_0}(r - r_0)\right), \tag{87}$$

with specified initial condition at r_0. The inverse transform gives the field as a function of depth,

$$\psi(r,z) = \int_{-\infty}^{\infty} \chi(r_0, s)\exp\left(\frac{\mathrm{i}K_0}{2}(n^2 - 1)\Delta r\right)\exp\left(-\frac{\mathrm{i}\Delta r}{2K_0}s^2\right)\exp(\mathrm{i}sz)\,\mathrm{d}s, \tag{88}$$

where $\Delta r = r - r_0$. Introducing the symbol $\mathcal{F}$ for the Fourier transform operation from the z-domain [as performed in (85)] and $\mathcal{F}^{-1}$ for the inverse transform, (88) can be summarized by the range-stepping algorithm

$$\psi(r + \Delta r, z) = \exp\left(\frac{\mathrm{i}K_0}{2}(n^2 - 1)\Delta r\right)$$
$$\times \mathcal{F}^{-1}\left\{\left[\exp\left(-\frac{\mathrm{i}\Delta r}{2K_0}s^2\right)\right]\cdot\mathcal{F}[\psi(r,z)]\right\}, \tag{89}$$

which is often referred to as the split-step marching solution to the PE. The Fourier transforms are performed using FFTs. Equation (89) is the solution for constant n, but the error introduced when n (profile or bathymetry) varies with range and depth can be made arbitrarily small by increasing the transform size and decreasing the range-step size. It is possible to modify the split-step algorithm to increase its accuracy with respect to higher-angle propagation [26].

7.2 Generalized or Higher-Order Parabolic Equation Methods

Methods of solving the PE, including extensions to higher-angle propagation, elastic media, and direct time-domain solutions including nonlinear effects have recently appeared (see [27,28] for additional references). In particular, accurate high-angle solutions are important when the environment supports acoustic paths that become more vertical, such as when the bottom has a very high speed and, hence, a large critical angle with respect to the horizontal. In addition, for elastic propagation, the compressional and shear waves span a wide-angle interval. Finally, Fourier synthesis for pulse modeling requires high accuracy in phase, and the high-angle PEs are more accurate in phase, even at the low angles.

Equation (82) with the second-order range derivative which was neglected because of (83) can be written in operator notation as

$$[P^2 + 2iK_0 P + K_0^2(Q^2 - 1)]\psi = 0, \tag{90}$$

where

$$P \equiv \frac{\partial}{\partial r}, \quad Q \equiv \sqrt{n^2 + \frac{1}{K_0^2}\frac{\partial^2}{\partial z^2}}. \tag{91}$$

Factoring (90) assuming weak range dependence and retaining only the factor associated with outgoing propagation yields a one-way equation:

$$P\psi = iK_0(Q-1)\psi, \tag{92}$$

which is a generalization of the PE beyond the narrow-angle approximation associated with (83). If we define $Q = \sqrt{1+q}$ and expand Q in a Taylor series as a function of q, the standard PE method is recovered by $Q \approx 1 + 0.5q$. The wide-angle PE to arbitrary accuracy in angle, phase, etc., can be obtained from a Padé series representation of the Q operator [27]:

$$Q \equiv \sqrt{1+q} = 1 + \sum_{j=1}^{n} \frac{a_{j,n} q}{1 + b_{j,n} q} + \mathcal{O}(q^{2n+1}), \tag{93}$$

where n is the number of terms in the Padé expansion and

$$a_{j,n} = \frac{2}{2n+1} \sin^2\left(\frac{j\pi}{2n+1}\right), \quad b_{j,n} = \cos^2\left(\frac{j\pi}{2n+1}\right). \tag{94}$$

The solution of (92) using (49) has been implemented using finite-difference techniques for fluid and elastic media [27]. A split-step Padé algorithm [29] has recently been developed which greatly enhances the numerical efficiency of this method.

8 Appendix B: Units

The decibel (dB) is the dominant unit in underwater acoustics and denotes a ratio of intensities (not pressures) expressed in terms of a logarithmic (base 10) scale. Two intensities, I_1 and I_2 have a ratio, I_1/I_2, in decibels of $10\log(I_1/I_2)$ dB. Absolute intensities can therefore be expressed by using a reference intensity. The presently accepted reference intensity is based on a reference pressure of one micropascal (µPa): the intensity of a plane wave having an root-mean-square (rms) pressure equal to 10^{-5} dynes/cm. Therefore, taking 1 µPa as I_2, a sound wave having an intensity, of, say, one million times that of a plane wave of rms pressure 1 µPa has a level of $10\log(10^6/1)$ $\equiv$ 60 dB re 1 µPa. Pressure (p) ratios are expressed in dB re 1 µPa by taking

$20\log(p_1/p_2)$, where it is understood that the reference originates from the intensity of a plane wave of pressure equal to $1\,\mu\text{Pa}$.

The average intensity, I, of a plane wave with rms pressure p in a medium of density ρ and sound speed c is $I = p^2/\rho c$. In seawater, ρc is $1.5 \times 10^5 \text{g}\,\text{cm}^{-2}\,\text{s}^{-1}$ so that a plane wave of rms pressure $1\,\text{dyne}/\text{cm}^2$ has an intensity of $0.67 \times 10^{-12}\,\text{W}/\text{cm}^2$. Substituting the value of a micropascal for the rms pressure in the plane wave intensity expression, we find that a plane wave pressure of $1\,\mu\text{Pa}$ corresponds to an intensity of $0.67 \times 10^{-22}\,\text{W}/\text{cm}^2$ (i.e., $0\,\text{dB}$ re $1\,\mu\text{Pa}$).

References

1. J. Northrup, J. G. Colborn, Sofar Channel Axial Sound Speed and Depth in the Atlantic Ocean, J. Geophys. Res., **79**, 5633 (1974)
2. F. B. Jensen, W. A. Kuperman, M. B. Porter, H. Schmidt, *Computational Ocean Acoustics* (AIP, Woodbury, N.Y. 1994)
3. R. B. Evans, A coupled mode solution for acoustic propagation in a waveguide with stepwise depth variations of a penetrable bottom, J. Acoust. Soc. Am. **74**, 188 (1983)
4. F. B. Jensen, W. A. Kuperman, Sound propagation in a wedge shaped ocean with a penetrable bottom, J. Acoust. Soc. Am. **67**, 1564 (1980)
5. A. B. Baggeroer, W. A. Kuperman, P. N. Mikhalevsky, An Overview of Matched Field Methods in Ocean Acoustics, IEEE J. Ocean Eng. **18**, 401–424 (1993)
6. H. Schmidt, A. B. Baggeroer, W. A. Kuperman, E. K. Scheer, Environmentally tolerant beamforming for high-resolution matched field processing: Deterministic mismatch, J. Acoust. Soc. Am. **88**, 1851–1862 (1990)
7. A. M. Richardson, L. W. Nolte, *A posteriori* probability source localization in an uncertain sound speed, deep ocean environment, J. Acoust. Soc. Am. **89**, 2280–2284 (1991)
8. J. L. Krolik, Matched field minimum variance beamforming in a random ocean channel, J. Acoust. Soc. Am. **92**, 1408–1419 (1992)
9. M. D. Collins, W. A. Kuperman, Focalization: Environmental focusing and source localization, J. Acoust. Soc. Am. **90**, 1410–1422 (1991)
10. B. Y. Zel'dovich, N. F. Pilipetsky, V. V. Shkunov, *Principles of Phase Conjugation* (Springer, Berlin 1985)
11. M. Fink, C. Prada, F. Wu, D. Cassereau, Self focusing with time reversal mirror in inhomogeneous media, Proc. IEEE Ultrason. Symp. **2**, 681–686 (1989)
12. M. Fink, Time Reversal Mirrors, In Acoust. Imaging **21**, ed. by J. P. Jones (Plenum, New York 1995) p. 1–15
13. D. R. Jackson, D. R. Dowling, Phase conjugation in underwater acoustics, J. Acoust. Soc. Am. **89**, 171–181 (1991)
14. D. R. Dowling, D. R. Jackson, Narrow-band performance of phase-conjugate arrays in dynamic random media, J. Acoust. Soc. Am. **91**, 3257–3277 (1992)
15. D. R. Dowling, Phase-conjugate array focusing in a moving medium, J. Acoust. Soc. Am. **94**, 1716–1718 (1993)
16. D. R. Dowling, Acoustic pulse compression using passive phase-conjugate processing, J. Acoust. Soc. Am. **95**, 1450–1458 (1994)

17. A. Parvulescu, C. S. Clay, Reproducibility of signal transmissions in the ocean Radio Electr. Eng. **29**, 223–228 (1965)
18. A. Parvulescu, Matched-signal (Mess) processing by the ocean, J. Acoust. Soc. Am. **98**, 943–960 (1995)
19. W. A. Kuperman, William S. Hodgkiss, Hee Chun Song, T. Akal, C. Ferla, Darell Jackson, Phase Conjugation in the ocean: Experimental demonstration of an acoustic time-reversal mirror, J. Acoust. Soc. Am. **103**, 25–40 (1998)
20. H. C. Song, W. A. Kuperman, W. S. Hodgkiss, A time-reversal mirror with variable range focusing, J. Acoust. Soc. Am. **103**, 3234–3240 (1998)
21. W. S. Hodgkiss, H. C. Song, W. A. Kuperman, T. Akal, C. Ferla, D. Jackson, A long-range and variable focus phase-conjugation experiment in shallow water, J. Acoust. Soc. Am. **105**, 1597–1604 (1999)
22. P. Roux, H. C. Song, M. B. Porter, W. A. Kuperman, Application of parabolic equation method to medical ultrasonics, Wave Motion **31**, 181–196 (2000)
23. R. K. Brienzo, W. S. Hodgkiss, Broadband matched field processing, J. Acoust. Soc. Am. **94**, 2821–2831 (1993)
24. P. Roux, B. Roman, M. Fink, Time-reversal in an ultrasonic waveguide, Appl. Phys. Lett. **70**, 1811–1813 (1997)
25. F. D. Tappert, The Parabolic Approximation Method, In *Wave Propagation and Underwater Acoustics*, ed. by J. B. Keller, J. S. Papadakis (Springer, Berlin, Heidelberg 1977)
26. D. J. Thomson, N. R. Chapman, A wide-angle split-step algorithm for the parabolic equation, J. Acoust. Soc. Am. **74**, 1848 (1983)
27. M. D. Collins, Higher-order Padé approximations for accurate and stable elastic parabolic equations with applications to interface wave propagation, J. Acoust. Soc. Am. **89**, 1050 (1991)
28. B. E. McDonald, W. A. Kuperman, Time domain formulation for pulse propagation including nonlinear behavior at a caustic, J. Acoust. Soc. Am. **81**, 1406 (1987)
29. M. D. Collins, A split-step Padé solution for the parabolic equation method, J. Acoust. Soc. Am. **93**, 1736 (1993)
30. R. P. Porter, Generalized holography as a framework for solving inverse scattering and inverse source problems, Prog. Opt. XXVII (Elsevier, New York 1989)
31. D. Cassereau, M. Fink, Focusing with plane time-reversal mirrors: An efficient alternative to closed cavities, J. Acoust. Soc. Am. **94**, 2373–2386 (1992)
32. M. Siderius, D. R. Jackson, D. Rouseff, R. P. Porter, Multipath compensation in range dependent shallow water environments using a virtual receiver, J. Acoust. Soc. Am. **102**, 3439–3449 (1997)
33. D. R. Jackson, T. E. Ewart, The effect of internal waves on matched-field processing, J. Acoust. Soc. Am. **96**, 2945–2955 (1994)
34. S. R. Khosla, D. R. Dowling, Time-reversing array retrofocusing in simple dynamic underwater environments, J. Acoust. Soc. Am. **104**, 3339–3350 (1998)
35. B. J. Uscinski, D. E. Reeve, The effect of ocean inhomogeneities on array output, J. Acoust. Soc. Am. **87**, 2527–2534 (1990)
36. A. I. Saichev, Effect of compensating distortions due to scattering in an inhomogeneous medium by use of a reflector turning the front, Radio Eng. Electr. **27**, 23–30 (1982)
37. A. Derode, P. Roux, M. Fink, Robust time reversal with high-order multiple scattering, Phys. Rev. Lett. **75**, 4206–4209 (1995)

38. W. A. Kuperman, F. Ingenito, Attenuation of the coherent component of sound propagating in shallow water with rough boundaries, J. Acoust. Soc. Am. **61**, 1178–1187 (1977)
39. M. B. Porter, The KRAKEN normal mode program, SACLANTCEN Memorandum **SM-245**, La Spezia (1991)
40. W. Munk, Internal waves and small-scale processes, In *Evolution of Physical Oceanography*, ed. by B. A. Warren, C. Wunsch (MIT Press, Cambridge, MA 1981) pp. 264–291
41. R. Dashen, W. H. Munk, K. M. Watson, F. Zachariasen, In *Sound Transmission Through a Fluctuating Ocean*, ed. by S. M. Flatte (Cambridge Univ. Press, Cambridge 1979) pp. 44–61
42. F. S. Henyey, D. Rouseff, J. M. Grochocinski, S. A. Reynolds, K. L. Williams, T. E. Ewart, Effects of internal waves and turbulence on a horizontal aperture sonar, IEEE J. Ocean. Eng. **22**, 270–280 (1997)
43. D. Tielbürger, S. Finette, S. Wolf, Acoustic propagation through an internal wave field in a shallow water waveguide, J. Acoust. Soc. Am. **101**, 789–808 (1997)
44. K. B. Winters, E. A. D'Asaro, Direct simulation of internal wave energy transfer, J. Phys. Oceanogr. **27**, 1937–1945 (1997)
45. S. A. Reynolds (private communication) Analysis of oceanographic data obtained during the 1997 Mediterranean experiment

Time Reversal, Focusing and Exact Inverse Scattering

James H. Rose[1,2]

[1] Laboratoire Ondes et Acoustique, Université de Paris VI,
Paris 75005, France
jhrose@iastate.edu
[2] Department of Physics and Astronomy and Ames Laboratory
Iowa State University, Ames, Iowa 50011, USA

Abstract. Focusing combined with time reversal is argued to be the physical basis of exact inverse scattering theory. These ideas yield the Newton–Marchenko equation, a foundational equation of inverse scattering theory, in a simple and physically convincing way. Since most applied imaging techniques are based on focusing and back-propagation, time reversal and focusing provide an important conceptual bridge between applied imaging and exact inverse scattering theory.

1 Introduction

The exact equations [1,2,3] of inverse scattering theory have long remained nonintuitive, despite their critical importance for imaging and for wave propagation in nonlinear systems [4,5]. The initial lack of physical insight was not surprising, since mathematicians, who were interested in questions of mathematical rigor, solved the problem. Burridge [6] took an important step towards establishing the physical content of exact inverse scattering theory. He derived the 1-dimensional (1D) Marchenko equation using a "time-domain" picture that exposed the role of causality in a clear and convincing way. Later Rose et al. [7] derived the 3-dimensional (3D) Newton–Marchenko equation (NME) using the same "time domain" and explicitly noted the critical role played by time reversal. Nonetheless important aspects of the exact inverse scattering problem remain nonintuitive. For example, Newton, in deriving the 3D NME, referred to one of his key results as the "miracle" [8].

I claim that focusing is the physical basis of exact inverse scattering theory. Since focusing is the fundamental paradigm for imaging, this claim argues a strong connection between inverse scattering theory and more approximate imaging methods. In the past, the gulf between the mathematical theory of inverse scattering and standard imaging techniques has been bridged by considering various approximate *linearizations* of the inverse scattering problem, e. g., the inverse Born approximation, synthetic aperture radar, and microscopy. These linearizations can be viewed as answering the question "How do we focus the beam?" Linearization's answer is only approximate: "Ignore the unknown aspects of the target and focus using *a priori* data." For example, the inverse Born approximation ignores the target altogether and focuses the beam as if in free space.

Focusing is introduced to exact inverse scattering theory and combined with time reversal to provide a bridge between the exact mathematical theory and applied imaging science. This work is carried out in the time domain in terms of pulses and delta functions. The definition of various terms will be given since the time domain differs substantially from the more usual frequency domain. "Focus" denotes the concentration of the beam to a very small region of space at a certain time; conventionally and in the limit, an exactly focused beam becomes $\delta(x)$ at $t = 0$. "Time-reversal focus" denotes a focus accomplished using time-reversal mirrors, i.e., processing equipment that digitally records the reflected waves, truncates and reverses them in time (first-in, last-out), and rebroadcasts them.

The 1D NME will be shown to arise trivially as soon as we demand that the beam be focused at some point x inside the potential. This physically obvious and trivial derivation shows the power of introducing focusing into exact inverse scattering theory. The derivation's physical essence is as follows: Consider those incident wave fields required to focus the beam inside the potential at x. Given a focus, the scattered waves can be determined from the incident waves in two ways. First, since the beam focuses to a point in space and time, the incident waves at early times determine the scattered waves at late times by time reversal. Second, the scattered waves are also determined from the incident waves by the impulse response function – the data for the inverse scattering problem. The NME is the consistency requirement for these two relations.

The NME answers the question, "How do we exactly focus the beam to a point x at time $t = 0$?" The answer: "Measure the impulse response function and solve the NME." The input field that focuses the beam is just that solution of the NME. Thus, the NME is seen to be closely related to a myriad of approximate imaging methods. However, the Newton–Marchenko method is exact.

This chapter has four sections. Section 2 reviews the scattering and inverse scattering problems for the 1D plasma wave equation. The main result, a simple physical derivation of the 1D NME, follows in Sect. 3. Finally, the chapter is concluded with a discussion and summary.

2 Direct and Inverse Scattering Problems

The direct and inverse scattering problems for the 1D plasma wave equation are reviewed. The forward scattering problem and the definition of the scattering data form Sect. 2.1. The inverse scattering problem is defined in Sect. 2.2.

2.1 The Forward Problem

The plasma wave equation occurs in the analysis of many linear, nondissipative scattering problems in 1D. For example, it is closely related to single-particle scattering in quantum mechanics, electromagnetic propagation in plasmas, and acoustic scattering in 1D [9]. The plasma wave equation is

$$\frac{\partial^2 u(t,x)}{\partial x^2} - \frac{\partial^2 u(t,x)}{\partial t^2} - v(x)u(t,x) = 0. \tag{1}$$

For simplicity, we assume that the potential $v(x)$ is a bounded, smooth, positive real function with compact support centered about zero. The coordinate x and the time t span plus and minus infinity. The wave field is denoted by u.

The direct scattering problem is defined as follows: given the initial conditions, the potential and the wave equation, determine the scattered wave field (i.e., the final conditions). The hypothetical experiment is shown in Fig. 1. Imagine a causal, rightward propagating delta-function pulse incident from the left at early times. This incident pulse propagates to the potential and interacts with it in a complicated way. Consequently, waves scatter to both the left and the right. The corresponding solution to the wave equation will be denoted by u^+, which is determined from the potential and the following initial conditions

$$u^+(t,e,x) = \delta(t - ex) \tag{2}$$

and

$$\frac{\partial u^+(t,e,x)}{\partial t} = \frac{\partial \delta(t - ex)}{\partial t}. \tag{3}$$

Here, $e = 1$ denotes propagation towards the right, while $e = -1$ denotes propagation towards the left.

The scattered wave field (the final conditions) will be described by the impulse response function. First, define the scattered wave field to be the difference between the total and incident wave fields

$$u^{+\text{sc}}(t,e,x) \equiv u^+(t,e,x) - \delta(t - ex). \tag{4}$$

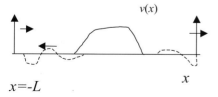

$x=-L$

Fig. 1. The basic scattering experiment. Incident from the left at early times, a delta function pulse is shown by the arrow at $x = -L$. At later times waves scatter to the left and right due to the interaction of the incident pulse with the potential. The *dashed lines* show the scattered waves, which are measured at $x = \pm L$

Next, record the transmitted and reflected scattered fields u^{+sc} at two points $x = \pm L$ situated to either side of the potential. Finally, describe the scattered wave compactly by the impulse response function,

$$R(t, e', e) \equiv u^{+sc}(t + L, e, e'L). \tag{5}$$

Here, e' denotes the direction of propagation of the scattered wave, while e denotes the direction of incidence. The time argument of u^+ is shifted by L in order to account for the travel time from the origin to the points of measurement at $x = \pm L$. The impulse response function is defined so that it is independent of L as long as the measurement points lie outside the support of the potential.

2.2 Inverse Scattering Problem

The inverse scattering problem is: determine the potential given the initial conditions and the final conditions. Within the context of the plasma wave equation, an important inverse scattering problem is to determine the potential $v(x)$ given the impulse response function. Both the Marchenko equation and the NME solve this problem in 1D. Marchenko's equation is economical in that it only requires reflected waves. The NME is less economical, since it takes both reflected and transmitted waves as data. However, the NME has one important advantage: it can be generalized to 2D and 3D. The 1D NME is given for $t > ex$ by

$$u^{+sc}(t, e, x) = \sum_{e'=-1,1} R(\tau + e'x, -e', e)$$

$$+ \sum_{e'=-1,1} \int_{-\infty}^{\infty} d\tau R(\tau + \tau', -e', e) u^{+sc}(\tau, e', x), \tag{6}$$

and for $t < ex$ by

$$u^{+sc}(t, e, x) = 0. \tag{7}$$

Note that this is a coupled set of equations for the wave incident from the left $u^+(t, 1, x)$ and the wave incident from the right $u^+(t, -1, x)$.

3 Physics of the Newton–Marchenko Equation

The NME is an integral equation that determines the wave field $u^+(t, e, x)$ at any point x given the impulse response function $R(t)$. The potential can be found trivially from the wave equation once u^+ is known. I will show that the NME for a point x follows immediately from the requirement that the beam focuses there. More precisely, consider the question "What input field focuses the beam to $\delta(x)$ at $t = 0$?" The NME is just the mathematical formulation of this question, and its solution $u^+(t, e, x)$ is the answer.

Some terminology is needed. This work uses *input pulses* (incident pulses) and the correspondingly *output pulses* (scattered pulses). The input pulses are transmitted towards the potential from two measurement points lying to either side of the potential's support (conventionally at $\pm L$). These input pulses consist of a square integrable part for $t < -L$ and a delta function $1/2\delta(t+L)$ and are zero for $t > -L$. They propagate to the potential, interact in a complicated way and then scatter out to infinity. The "output fields" are just the outgoing wave fields measured at $\pm L$ for $t > 0$. See Fig. 2 for the measurement geometry. The meaning of the word focus needs to be elaborated slightly for a hyperbolic equation. Specifically, as t approaches zero, the total wave field approaches $\delta(x)$ as $1/2\left[\delta(t+x) + \delta(t-x)\right]$.

The 1D NME follows in three steps from the requirement that the beam focus at $t = 0$ to a point x. First, the input and output fields are related by a simple reversal of time due to the point-wise focus in space and time. In fact, the total wave field for $t < 0$ is just the time reversal of the total wave field for $t > 0$. Second, the output field can also be calculated from the input field given the impulse response function – the assumed data for the inverse problem. Since there are two equations relating the input and output fields, we can eliminate one field. Upon eliminating the output field and after some algebra, one finds the Newton–Marchenko integral equation.

I will now derive the NME for the point $x = 0$ – the result for a general point x follows trivially. There are three steps. First, the input and output fields are shown to determine each other by time reversal if the beam is focused. Second, the output field is shown to follow from the input field and the impulse response. Third, the integral equation for the input field that focuses the beam is found.

A focus implies that the input and output fields are simply related by time reversal. In a crude analogy time reversal says that you can "run the movie backwards". Lets consider the properties of the proposed focused wave field. Focusing implies that the wave field approaches $1/2\left[\delta(t-x) + \delta(t+x)\right]$ for

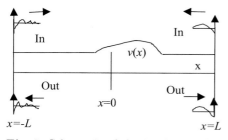

Fig. 2. Schematic of the incident and scattered fields for a field focused at $x = 0$. Waves are both input and output at $x = \pm L$. As shown, the output fields are the time-reversed input fields

small t approaching zero. Thus focusing defines the wave field and its time derivative over all space. This data determines both the outgoing scattered waves and the incoming incident waves. Measure the resulting outgoing waves at $\pm L$ for all times $t > 0$ and call these measured fields $\phi_{\text{out}}(t, e)$. Time-reversal invariance ("running the movie backwards") states that the delta-function impulse will be recovered if we time-reverse these output fields and transmit them back into the potential. That is, $\phi_{\text{out}}(-t, -e)$ is just the *input field* $\phi_{\text{in}}(t, e)$ needed to obtain a focus. Explicitly,

$$\phi_{\text{in}}(t, e) = \phi_{\text{out}}(-t, -e) \tag{8}$$

Figure 3 shows a space–time diagram for the input and output pulses (at $x = \pm L$) for a field that is focused at $x = 0$ and $t = 0$. Input pulses are nonzero for $t < -L$, and output pulses are nonzero for $t > L$. The dark solid circles show the delta functions. The content of (8) is that the input field at $x = -L$, read starting from the delta function at $t = -L$ and moving downward, is the same as the output field at $x = L$, read starting from the delta function and moving upwards.

The impulse response function is the given data for the inverse scattering problem. It determines the output wave field due to an arbitrary input wave field. In particular,

$$\phi_{\text{out}}(t, e) = \phi_{\text{in}}(t - 2L, e) + \sum_{e'=-1,1} \int_{-\infty}^{\infty} d\tau R(t - \tau - 2L, e, e') \phi_{\text{in}}(\tau, e'). \tag{9}$$

This equation is just the book-keeping of scattering theory and can be understood as follows: Consider, the output field measured at $x = L$, which consists of outgoing waves propagating to the right ($e = 1$). The first term on the right-hand side denotes the incident wave launched from the left at $x = -L$. The second term consists of the two scattered waves: transmitted

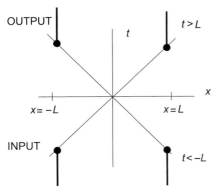

Fig. 3. Space-time diagram for the input and output pulses (at $x = \pm L$) for a field that is focused at $x = 0$ and $t = 0$. *Thick solid lines*: input pulses are nonzero for $t < -L$ and output pulses are nonzero for $t > L$. *Dark solid circles*: delta functions

and propagating to the right ($e' = 1$, $e = 1$) and reflected and propagating to the right ($e' = -1, e = 1$).

The NME is obtained by solving (8) and (9) for ϕ_{in}. Upon eliminating ϕ_{out}, we obtain

$$\phi_{\text{in}}(-t, -e) = \phi_{\text{in}}(t - 2L, e)$$
$$+ \sum_{e'=-1,1} \int_{-\infty}^{\infty} d\tau R(t - \tau - 2L, e, e') \phi_{\text{in}}(t, e'). \qquad (10)$$

By construction, $\phi_{\text{in}}(t, e) = 0$ for $t > -L$. Equation (10) is our main result. We see immediately that a solution of the NME, $\phi_{\text{in}}(t, e)$, focuses the beam to $x = 0$ at $t = 0$.

The form of (10) is nonstandard. For example, the dependence on x has been dropped to concentrate only on the point at $x = 0$. The rest of this section is an epilogue devoted to showing that (10) reduces to the standard form of the NME after some algebraic manipulation and the invocation of reciprocity. First, change variables for the integration and sum ($\tau \to -\tau, e' \to -e'$) and use the reciprocity of the impulse response function $R(t, e, e') = R(t, -e', -e)$ to obtain

$$\phi_{\text{in}}(-t, -e) = \phi_{\text{in}}(t - 2L, e)$$
$$+ \sum_{e'=-1,1} \int_{-\infty}^{\infty} d\tau R(t + \tau - 2L, -e', e) \phi_{\text{in}}(-\tau, -e'). \qquad (11)$$

Next, shift the definition of time ($t \to t+L$) so that the input wave commences at $t = 0$ rather than at $t = -L$. The result is

$$\phi_{\text{in}}(-t - L, -e) = \phi_{\text{in}}(t - L, e)$$
$$+ \sum_{e'=-1,1} \int_{-\infty}^{\infty} d\tau R(t + \tau, -e', e) \phi_{\text{in}}(-\tau - L, -e'). \qquad (12)$$

Next, redefine the input field to take this time shift into account:

$$\psi_{\text{in}}(t, e) \equiv \phi_{\text{in}}(t - L, e). \qquad (13)$$

The effect is that $\psi_{\text{in}}(t, e)$ is zero for $t > 0$, and (13) becomes

$$\psi_{\text{in}}(-t, -e) = \psi_{\text{in}}(t, e) + \sum_{e'=-1,1} \int_{-\infty}^{\infty} d\tau R(t + \tau, -e', e) \psi_{\text{in}}(-\tau, -e'). \qquad (14)$$

Finally, define

$$w(t, e) \equiv \psi(-t, -e), \qquad (15)$$

and insert in (14). The result is the standard form of the NME specialized for $x = 0$:

$$w(t, e) = w(-t, -e) + \sum_{e'=-1,1} \int_{-\infty}^{\infty} d\tau' R(t + \tau', -e', e) w(\tau', e'). \qquad (16)$$

For $t < 0$, $w(t, e) = 0$.

Finally, what is the NME for points x other than $x = 0$? The arguments given above remain valid if we change the measurement (and transmission) points from $-L$ and L to $-L + x$ and $L + x$. Equation (16) then refers to a focus at x. We can indicate this by extending the argument list for w to include the position: $w(t, e) \to w(\tau', e', x)$. Upon rewriting (16) using this new notation, we find for $t > x$

$$w(t, e, x) = w(-t, -e, x) + \sum_{e'=-1,1} \int_{-\infty}^{\infty} d\tau' R(t + \tau', -e', e) w(\tau', e, x), \quad (17)$$

and for $t < x$, $w(t, e, x) = 0$. The form of the equation is now identical with (6) – the NME.

4 Discussion and Summary

Two experimental results that arose in the recent development of time-reversal cavities and mirrors by *Fink* and collaborators [10,11,12,13] were particularly important for this work. First, Fink et al. showed that one can obtain sharp foci in 3D acoustic scatterers even though they exhibit large variations in the speed of sound. Second, they showed that repeated use of time-reversal mirrors focuses the beam on the "strongest" scatterer in the field of view. Furthermore, they developed a systematic procedure for finding the second and third "strongest scatterer" etc. – thus partially solving the inverse scattering problem [13].

This paper addresses the desire to focus the beam to any pre-specified point in the sample and furthermore to recover the potential at that point – completely solving the inverse scattering problem. The Newton–Marchenko equation is the equation that governs focusing for the plasma wave equation. The Marchenko equation can be derived in an entirely analogous way; a result that is reserved for separate publication. The 1D NME has been rederived by combining the ideas of scattering, focusing and time reversal. The solution of the NME for a point x furnishes those input fields that create a focus at x. Thus, the idea of focusing seems to be central. Why? Qualitatively, it seems that something like a focus is inevitable. An exact solution implies an arbitrarily fine spatial resolution, which is the ability to distinguish two arbitrarily close points from each other. Consequently, we must somehow reconstruct from the scattered data structures that overlap one of the points but not the other.

Acknowledgement

I would like to thank Mathias Fink for his encouragement and for providing a lively research environment that greatly contributed to this work.

References

1. I. M. Gel'fand, B. M. Levitan, On the determination of a differential equation from its spectral function, Am. Math. Soc. Transl. **1**, 253 (1951)
2. V. A. Marchenko, On the reconstruction of the potential energy from the phases of the scattered waves, Dokl. Akad. Nauk. SSSR **104**, 695 (1955)
3. Z. S. Agranovich, V. A. Marchenko, *The Inverse Problem of Scattering Theory* (Gordon Breach, New York 1963)
4. G. L. Lamb, Jr., *Elements of Soliton Theory* (Wiley, New York 1980)
5. A. C. Newell, *Solitons in Mathematics and Physics* (SIAM, Philadelphia 1985)
6. R. Burridge, The Gelfand–Levitan, the Marchenko and the Gopinath–Sondi integral equations of inverse scattering theory, regarded in the context of inverse impulse-response theory, Wave Motion **2**, 305 (1980)
7. J. H. Rose, M. Cheney and B. DeFacio, The connection between time- and frequency- domain three-dimensional inverse scattering methods, J. Math. Phys. **25**, 2995 (1984)
8. R. G. Newton, *Inverse Schrödinger Scattering in Three Dimensions* (Springer, Berlin, Heidelberg 1989)
   The term "miracle" was applied by Newton to the equation that extracts the potential from the reconstructed 3D wave field. The left-hand side of this equation (the potential) is a function of three variables, while the right-hand side is a function of five variables (time, direction of incidence and direction of scattering). For wave fields that correspond to actual potentials, the right-hand side is redundant in two of its dimensions. Newton referred to this fact as the "miracle"
9. G. M. L. Gladwell, *Inverse Problems in Scattering: An Introduction* (Kluwer Academic, Dordrecht 1993)
10. M. Fink, C. Prada, F. Wu, D. Cassereau, Self-focusing in inhomogeneous media with time-reversal acoustic mirrors, Proc. IEEE Ultrason. Symp. **681** (1989)
11. D. Cassereau, F. Wu, M. Fink, Limits of self-focusing using closed time-reversal cavities and mirrors – theory and experiment, Proc. IEEE Ultrason. Symp. **1613** (1990)
12. M. Fink, Time-reversed acoustics, Phys. Today **50**, 34 (1997)
13. C. Prada, M. Fink, Separation of interfering acoustic scattered signals using the invariants of the time-reversal operator. Application to Lamb wave characteristics, J. Acoust. Soc. Am. **104**, 801 (1998)

Detection and Imaging in Complex Media with the D.O.R.T. Method

Claire Prada

Laboratoire Ondes et Acoustique, ESPCI
10 rue Vauquelin, 75231 Paris, Cedex 05, France
claire.prada-julia@espci.fr

Abstract. Acoustic waves are used for detection, localization and sometimes destruction of passive targets. In most fields of acoustics, arrays of transmitters and arrays of receivers are available, and if not synthetic aperture techniques can be used. With such arrays, a great amount of data can be collected, and the general problem is to extract the relevant information from these data to detect (or to form an image of) a scattering object. This problem appears in applications ranging from medical imaging to underwater acoustics and even in seismology. The D.O.R.T. method is a new approach to active detection and focusing of acoustic waves using arrays of transmitters and receivers. This method was derived from the theoretical study of iterative time-reversal mirrors. It consists essentially of the construction of the invariants in the time-reversal process. After explaining the basic theory of the D.O.R.T. method, several experimental results are shown: (a) detection and selective focusing through an inhomogeneous medium; (b) detection and focusing in a water waveguide, where high resolution is achieved by taking advantage of the multiple paths in the guide; (c) an analysis of scattering by a thin hollow cylinder, where the various components of the elastic waves circumnavigating in the shell are separated; and (d) in some cases the eigenvectors obtained at different frequencies can be combined to obtain the time-domain Green's function of each scatterer.

1 Introduction

In various domains, such as medicine, nondestructive evaluation (NDE), underwater science and seismology, acoustic waves are used for detection, localization and sometimes destruction of passive targets. In most fields of acoustics, arrays of transmitters and arrays of receivers are available, and if not synthetic aperture techniques can be used.

The D.O.R.T. method provides a new approach to active detection and focusing of acoustic waves using an array of transmitters and an array of receivers. It is an analysis technique of pulse–echo measurements (or reflection data) that may be interesting for applications ranging from medical imaging, where arrays of transmit-receive transducers are currently used, to underwater acoustics and even seismology, where an ensemble of seisms can be considered as an array of transmitters and an ensemble of stations as an array of receivers.

With multiple sources and multiple receivers, a large amount of scattering data can be collected which correspond to different acoustic paths in the medium under study. The general problem is to extract the relevant information from these data to detect (or to form an image of) a scattering object in the medium.

For typical length scales, the low velocity of acoustic waves allows information coming from different parts of the studied medium to be separated. However, the detection sensitivity as well as the quality of the image of scattering objects depend on the ability to focus energy in the medium either in transmission or in reception. This focusing is also crucial for destruction techniques like lithotripsy. The presence of an aberrating medium between the object and the arrays can detrimentally alter the beam profiles.

In NDE, cracks and defects can be found within materials of various shapes. The samples to be evaluated are usually immersed in a pool, and the interface shape between the samples and the coupling liquid currently limits the detectability of small defects. In medical imaging, one looks for organ walls, calcification, tumors, kidney or gallbladder stones. A fat layer of varying thickness, bone tissue, or some muscular tissues may greatly degrade focusing. In underwater acoustics, one looks for mines, submarines, or objects buried under sediments. Refraction by the oceanic structure ranging in scale from centimeters to tens of kilometers is an important source of distortions. In seismology, detection of discontinuities such as local changes in reflectivity at the core–mantle interface can be achieved by inversion of reflection data. However, the poor knowledge of the acoustic properties of the medium renders this problem rather difficult.

During the past 10 years, *Fink* and his team have developed time-reversal techniques in order to achieve optimum focusing through distorting media (this volume) [1,2]. They have shown through several ultrasonic experiments that high-quality focusing can be obtained with acoustic time-reversal mirrors. In echographic mode, this self-focusing technique is effective in the presence of a single scatterer in the medium. When this medium contains several scattering centers, the time-reversal operation need to be iterated in order to select one of them. In general, after some iterations, the process converges and produces a wave front that focuses on the most reflective scatterer [3].

In some situations, it is interesting to learn how to focus on weaker scattering centers. The D.O.R.T. method provides a solution to this problem [4,5,6]. This method was derived from the theoretical study of iterative time-reversal mirrors and consists essentially of the construction of the signal patterns that are invariants under a time-reversal process. Those invariants appear as the eigenvectors of a matrix called the time-reversal operator which describes the time-reversal process. The eigenvectors are calculated offline after the measurement of the response function of the array in the presence of the scattering medium. It is not a real-time procedure; however, it can still be applied in many experimental situations. The D.O.R.T. method shares some of the principles of eigenvector decomposition techniques that are used in passive source detection [7,8,9]. However the latter assume statistically uncorrelated sources, while the D.O.R.T. method is active and deterministic; thus they should not be considered as competing techniques.

The basic theory of the D.O.R.T. method is explained in Sect. 2; experimental illustrations follow. An example of detection and selective focusing through an inhomogeneous medium is shown in Sect. 3. Examples of detection and focusing in a water waveguide are presented in Sect. 4. The method can also take advantage of the matched-filter property of the waveguide in order to separate the echoes from different scatterers with high resolution.

Sect. 5 is devoted to the analysis of the scattering by a thin hollow cylinder. It is shown how the D.O.R.T. method separates the various components of the elastic waves circumnavigating in the shell. In Sects. 3 to 5 the pulse–echo measurements are analysed frequency by frequency. In general, it is not possible to get back to the time domain; however, in Sect. 6, we show that in some cases the eigenvectors obtained in the whole frequency band of the transducers can be combined to obtain the time-domain Green's function of each scatterer.

2 Basic Principle of the D.O.R.T. Method

The D.O.R.T. method has been presented in several papers [4,5,6]. It consists of determining the transmitted waveforms that are invariant under the time-reversal process. We show that for a set of well-resolved scatterers the focusing on one of them is invariant in this manner. The analysis is based on a matrix formalism that describes the transmit–receive process. To begin, we introduce the transfer matrix of the system and the corresponding time-reversal operator.

2.1 The Transfer Matrix

An array of N transmitters (array No. 1) insonifying a scattering medium and an array of receivers (array No. 2) are considered. This system is assumed to be a linear and time-invariant system of N inputs and L outputs. Thus it is characterized by $N \times L$ inter-element impulse response functions. Let $h_{lm}(t)$ be the signal delivered by receiver l when a temporal delta function $\delta(t)$ is applied on the transmitter number m. These $N \times L$ functions provide a complete description of the transmit–receive process. Indeed, if $e_m(t), 1 < m < N$ are the transmitted signals, then the received signals are given by the equation

$$r_l(t) = \sum_{m=1}^{N} k_{lm}(t) \underset{t}{\otimes} e_m(t). \tag{1}$$

These L equations simplify in the frequency domain in a matrix form: any transmit–receive operation is described with a complex transfer $L \times N$ matrix $\mathbf{K}(\omega)$ by the equation:

$$\boldsymbol{R}(\omega) = \mathbf{K}(\omega)\boldsymbol{E}(\omega), \tag{2}$$

where ω is the frequency, $\boldsymbol{E}(\omega)$ is the transmitted vector of N components, $\boldsymbol{R}(\omega)$ is the received vector of L components and $\mathbf{K}(\omega)$ is the $L \times N$ transfer matrix of the system.

2.2 Invariants of the Time-Reversal Process and Decomposition of the Transfer Matrix

In most time-reversal experiments shown by Fink and his team, a single array is used in both transmit and receive modes. In the case where the transmitter and the receiver arrays are distinct, the time-reversal experiment can only be an hypothetical experiment: one can imagine a time-reversal operation between the two arrays, assuming that array No. 1 becomes the receiver and array No. 2 becomes the transmitter. Neglecting the impulse responses of the transmitter and receivers, the reciprocity principle guaranties that the response from element number l of array No. 2 to element number n of array No. 1 is equal to the response from element number n to element number m. Consequently, the transfer matrix from array No. 2 to array No. 1 is $^t\mathbf{K}(\omega)$.

A time-reversal operation $t \to -t$ is equivalent to a phase conjugation in the frequency domain. Thus, in a time-reversal process, if $\boldsymbol{E}_0(\omega)$ is the first transmitted signal (applied on array No. 1) of a time-reversal process, then the second transmitted signal (applied on array No. 2) is the phase conjugate of the received signal:

$$\boldsymbol{E}_2(\omega) = \mathbf{K}^*(\omega)\boldsymbol{E}_1^*(\omega). \tag{3}$$

After transmission of $E_2(\omega)$, the signal received on array No. 1 is

$$\boldsymbol{R}(\omega) = {^t}\mathbf{K}(\omega)\mathbf{K}^*(\omega)\boldsymbol{E}_1^*(\omega). \tag{4}$$

This signal is linked to the first transmitted signal through a phase conjugation, and the product by the matrix $^t\mathbf{K}^*(\omega)\mathbf{K}(\omega)$. Thus, this matrix allows any time-reversal process to be described and is called the time-reversal operator. Note that this approach is more general than the one presented in [4,5]. Indeed, in these papers, the same array acts as transmitter and receiver. In this case, the reciprocity principle insures that the transfer matrix $\mathbf{K}$ is symmetrical and the time reversal operator is simply $\mathbf{K}^*(\omega)\mathbf{K}(\omega)$. In the experiments that will be shown in the following sections, a single array is used. However, the theory presented here as well as in [6] is more general and broadens the field of application of the D.O.R.T. method. The important property of the time-reversal operator is that it is hermitic with positive eigenvalues.

Let the first transmitted signal be $\boldsymbol{V}(\omega)$, an eigenvector of ${}^t\mathbf{K}^*(\omega)\mathbf{K}(\omega)$ associated to the eigenvalue $\lambda(\omega)$, then after a time-reversal process the received signal is $\lambda(\omega)\boldsymbol{V}^*(\omega)$, which is proportional to the conjugate of $\boldsymbol{V}(\omega)$. Consequently one can say that the eigenvectors of ${}^t\mathbf{K}^*(\omega)\,\mathbf{K}(\omega)$ correspond to waveforms that are invariants of the time-reversal process.

In fact, from a mathematical point of view, the diagonalization of ${}^t\mathbf{K}^*(\omega)\mathbf{K}(\omega)$ is equivalent to the singular value decomposition (SVD) of the transfer matrix $\mathbf{K}(\omega)$. Indeed, the SVD is $\mathbf{K}(\omega) = \mathbf{U}(\omega)\boldsymbol{\Lambda}(\omega)\mathbf{V}^+(\omega)$, where $\boldsymbol{\Lambda}(\omega)$ is a real diagonal matrix of the singular values and $\mathbf{U}(\omega)$ and $\mathbf{V}(\omega)$ are unitary matrices. The eigenvalues of ${}^t\mathbf{K}^*(\omega)\mathbf{K}(\omega)$ are the squares of the singular values of $\mathbf{K}(\omega)$; its eigenvectors are the columns of $\mathbf{V}(\omega)$. We shall use this decomposition in the following. Note that when a column of $\mathbf{V}(\omega)$ is transmitted on the first array, the signal received on the second array is proportional to a column of $\mathbf{U}(\omega)$.

2.3 Transfer Matrix for Point-Like Scatterers

In the case of point-like scatterers the transfer matrix can be derived easily. We assume that the medium contains d point-like (Rayleigh) scatterers with complex frequency-dependent reflectivity coefficients $C_1(\omega), C_2(\omega), \ldots, C_d(\omega)$. Then the transfer matrix can be written as the product of three matrices: (a) a propagation matrix that describes the transmission and the propagation from the transducers to the scatterers, (b) a scattering matrix which is diagonal in the case of a single scattering process, and (c) the back propagation matrix.

Let $h_{1,il}(t)$ be the diffraction impulse response function of the transducer number l of array 1 to the scatterer number i with Fourier transform $H_{1,il}(\omega)$. Let $a_e(t)$ and $a_r(t)$ be the transducer acousto-electrical response in emission and in reception, with Fourier transforms $A_e(\omega)$ and $A_r(\omega)$. If the input signal at each element l is $e_l(t)$, then the pressure at the scatterer i is

$$p_i(t) = \sum_{l=1}^{N} h_{1,il}(t) \underset{t}{\otimes} a_e(t) \underset{t}{\otimes} e_l(t). \tag{5}$$

This equation is written in the frequency domain as

$$P_i = A_e \sum_{l=1}^{N} H_{1,il} E_l. \tag{6}$$

The expression is simplified using matrix notation:

$$\boldsymbol{P} = A_e \mathbf{H}_1 \boldsymbol{E}, \tag{7}$$

where $\boldsymbol{E}$ is the input vector signal, $\boldsymbol{P}$ is the vector representing the pressure received by the d scatterers and $\mathbf{H}_1$ is a matrix of dimensions $N \times d$ called the diffraction matrix.

In the case of single scattering, the pressure reflected by scatterer number i is $C_i P_i$. Therefore, the vector of reflected pressures is the matrix $\mathbf{CP}$ where $\mathbf{C}$ is a diagonal matrix of coefficients $C_{ij} = \delta_{ij} C_i$ for all i, j in $1, \ldots, d$.

According to the reciprocity principle, the propagation from the scatterer number i to the transducer number n of array 2 is $h_{2,in}(t)$, so that the back-propagation matrix is the transpose of the propagation matrix ${}^t\mathbf{H}_2$. Consequently, the output signal $\boldsymbol{R}$ at array number 2 is

$$\boldsymbol{R} = A_e A_r^t \mathbf{H}_2 \mathbf{C} \mathbf{H}_1 \boldsymbol{E} \tag{8}$$

The expression of the transfer matrix is then

$$\mathbf{K} = A_e A_r^t \mathbf{H}_2 \mathbf{C} \mathbf{H}_1 \,. \tag{9}$$

2.4 Decomposition of K for Well-Resolved Scatterers

The iterative time-reversal process allows selective focusing on the most reflective of a set of scatterers. In some cases it might be interesting to learn how to focus on the other scatterers. For example, this is the case in lithotripsy when several lithiases exist. The theoretical analysis of the iterative time-reversal process has provided a solution to this problem.

We now propose to compute the SVD of the transfer matrix in a scattering medium containing D point-like scatterers. According to Sect. 2.3, if $\mathbf{H}_1$ ($\mathbf{H}_2$) is the matrix of size $D \times N$ ($D \times L$) that describes the propagation from array No. 1 (array No. 2) to the D scatterers and $\mathbf{C}$ the diagonal matrix of size $D \times D$ that describes single scattering, and if the responses of the transmitter and receivers are assumed to be temporal delta functions, then

$$\mathbf{K} = {}^t\mathbf{H}_2 \mathbf{C} \mathbf{H}_1 \,. \tag{10}$$

We can say that the targets are ideally resolved if the time-reversal focusing on one of them does not produce energy on the others. From a mathematical point of view, it means that the columns of the propagation matrix $\mathbf{H}_i$ are orthogonal. If this property is satisfied for both arrays and if the "apparent reflectivities" (this phrase will be discussed further) of each target are different, the eigenvectors of the time-reversal operator can be analytically determined. We note that $\overline{\mathbf{H}_i} i = (1, 2)$ is the matrix whose rows are the normalized rows of $\mathbf{H}_i$. We can then write $\mathbf{H}_i = \Delta_i \overline{\mathbf{H}_i}$ where Δ_i is a diagonal matrix of coefficients the norms of the column of $\mathbf{H}_i$. The following equation is deduced:

$$\mathbf{K} = {}^t\overline{\mathbf{H}_2} \Delta_2 \mathbf{C} \Delta_1 \overline{\mathbf{H}_1} \tag{11}$$

This decomposition is the SVD of $\mathbf{K}$. Indeed, as $\overline{\mathbf{H}_i} i = (1, 2)$ are normalized, they satisfy the equations ${}^t\overline{\mathbf{H}_1} \overline{\mathbf{H}_1^*} = \mathbf{I}$ and ${}^t\overline{\mathbf{H}_2} \overline{\mathbf{H}_2^*} = \mathbf{I}$; they are unitary matrices.

The matrix $\Delta_2 C \Delta_1$ is real diagonal, with terms equal to

$$\lambda_i = C_i \sqrt{\sum_{n=1}^{N} |H_{1,in}|^2} \sqrt{\sum_{n=1}^{N} |H_{2,in}|^2} \quad \text{for} \quad 1 < i < D. \tag{12}$$

λ_i is precisely what we call the apparent reflectivity of scatterer number i. It is proportional to its reflectivity C_i and to a term that depends on the diffraction pattern of both the transmitter and receiver arrays.

Consequently, the eigenvectors of ${}^t\mathbf{K}^*\mathbf{K}$ are the complex conjugate of the first columns of $\overline{\mathbf{H}_1}$:

$$\frac{H_{1,i}^*}{\sqrt{\sum_{n=1}^{N} |H_{1,in}|^2}} \quad \text{for } 1 < i < D. \tag{13}$$

According to the reciprocity theorem, each of them is precisely the time-reversed form of the signal that would be received on array No. 1 if target number i was acting as a point-like acoustic source.

Finally, for well-resolved point-like scatterers the number of nonzero eigenvalues is equal to the number of scatterers. Furthermore, if the scatterers have different apparent reflectivities (λ_i), each eigenvector is associated with one scatterer; the phase and amplitude of the eigenvector should be applied to each transducer in the array in order to focus on that particular scatterer.

If the scatterers have equal apparent reflectivities, or are not resolved, the problem is more complicated and beyond the scope of this paper. However, some results in this regard can be found in [5].

2.5 The D.O.R.T. Method in Practice

The first step is the measurement of the inter-element impulse responses of the system. Since the reception system operates in parallel, this measurement requires N transmit–receive operations for an array of N transducers. The first transducer of the array is excited with a signal $e(t)$, and the signals received on the N channels are stored. This operation is repeated for all the transducers of the array with the same transmitted signal $e(t)$. The components of the transfer matrix $\mathbf{K}(\omega)$ are obtained by a Fourier transform of each signal. This measurement could also be done with any multiplexed system by N^2 transmit–receive operations.

The second step is the SVD of the transfer matrix $\mathbf{K}(\omega)$ at a chosen frequency. The singular value distribution contains valuable information as the number of secondary sources in the scattering medium.

The third step is to back-propagate each eigenvector. This can be done either numerically or experimentally. This is useful to focus through an aberrating medium selectively on each scatterer.

Detection consists of the following steps: First, the inter-element impulse response functions $k_{lm}(t)$ are measured. Second, the transfer matrix is calculated at one chosen frequency (more often the central frequency of the transducers). Last, the SVD of the transfer matrix is calculated: $\mathbf{K}(\omega) = \mathbf{U}(\omega)\,\mathbf{\Lambda}(\omega)\,\mathbf{V}^{+}(\omega)$, where $\mathbf{\Lambda}(\omega)$ is a real diagonal matrix of the singular values and $\mathbf{U}(\omega)$ and $\mathbf{V}(\omega)$ are unitary matrices.

3 Selective Focusing Through an Inhomogeneous Medium with the D.O.R.T. Method

The first application of this method is to learn how to focus selectively through an inhomogeneous layer. A simple example is now presented. A single linear array of 128 transducers is used for transmission and reception. The array pitch is 0.4 mm, and the central frequency is 3 MHz with 60% bandwidth. The sampling frequency of received signals is 20 MHz. The scatterers are two wires placed at 90 mm perpendicular to the array. A rubber layer of varying thickness is placed between the array and the wires (Fig. 1).

First, cylindrical time delay laws are applied to the transducers in order to focus on the wires as if there was no aberration. The pressure pattern is measured with a hydrophone in the plane of the wires: the layer induces severe defocusing (Fig. 2). Now the question is: how can we obtain a good focusing without any information on the acoustic properties of the layer.

The inter-element impulse response functions of all the elements of the array were measured. The responses of element 64 to the 128 elements of the array are shown in Fig. 3: the wavefronts corresponding to the echo of the two wires can easily be seen.

The transfer matrix $\mathbf{K}$ was computed at the frequency of 3 MHz. Note that the Fourier transforms of the signals displayed in Fig. 3 provide column 64 of $\mathbf{K}$.

The SVD of $\mathbf{K}$ reveals 2 significant singular values among 128 (Fig. 4), which indicates the presence of two scattering centers. The phase laws of the corresponding eigenvectors (Fig. 5) are unwrapped to form time-delay laws (Fig. 6). Each time delay law is well defined and corresponds to one of the wires. To confirm this, the time-delay laws are used to focus through the

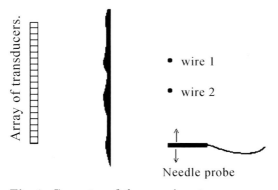

Fig. 1. Geometry of the experiment

Detection and Imaging in Complex Media with the D.O.R.T. Method 115

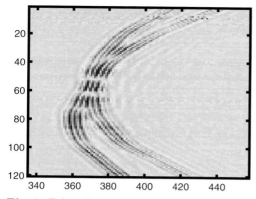

Fig. 2. Echo of the two wires after illumination by the center element of the array (horizontal axis: time in µs, vertical axis: array element)

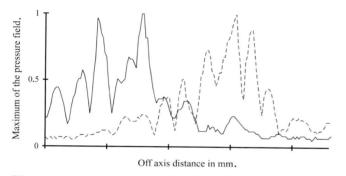

Fig. 3. Pressure pattern obtained by cylindrical focusing through the rubber layer

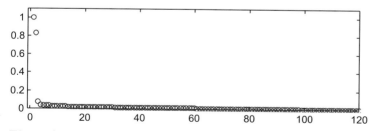

Fig. 4. Singular values of the transfer matrix

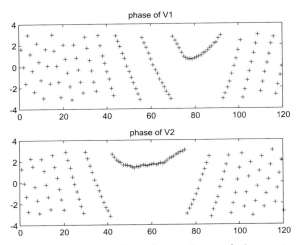

Fig. 5. Phase laws of the first and second eigenvectors

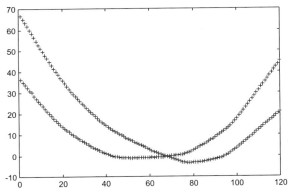

Fig. 6. Unwrapped phase laws of the first and second eigenvectors

rubber layer. The so-produced pressure field is measured in the plane of the wires (Fig. 7). Each eigenvector focuses on one of the wires and the accuracy of the focusing is similar to what can be obtained with this array in water only.

4 Highly Resolved Detection and Selective Focusing in a Waveguide

The problem of optimum signal transmission and source location in a waveguide has been the subject of many theoretical and experimental works. The propagation of an acoustic pulse inside a waveguide is a complex phenomenon. This complexity renders the detection and imaging process very difficult.

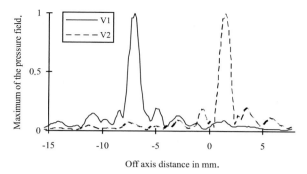

Fig. 7. Pressure patterns measured at the position of the wires after transmission of the first and second eigenvectors

Because of multiple-path effects, the Green's function that is used in matched-field processing is nontrivial, and its calculation requires accurate knowledge of the medium. However, several studies have shown how to take advantage of this complexity.

In waveguide transmission, the guide can be considered as a linear filter. *Parvulescu* et al. [10] reported a matched-filter experiment in the ocean between a source and a receiver. They recorded the reception of an impulsive transmission and replayed the time-reversed signal through the source. They obtained a high temporal compression, which was explained by the coherent recombination of the energy received over the different multiple paths. They also showed the high sensitivity to small displacements of the source, suggesting that this property should be used to locate the source. As proposed by *Clay* [11] and *Li* [12], the combination of array-matched filter and time-domain-matched signal techniques improve the accuracy in source localization. In these papers, the focusing is explained in terms of matched signal: the waveguide plays the role of a correlator.

The possibility of taking advantage of the invariance of the acoustic wave equation under time reversal in order to achieve highly resolved spatial and temporal focusing in a waveguide arose afterward. In 1991, *Jackson* et al. [13] provided a theoretical analysis of the time-reversal process in a water channel. Focusing experiments inside a water waveguide with a time-reversal mirror were made by *Roux* et al. [14]; then Kuperman and his team implemented a time-reversal mirror in the Mediterranean Sea [15,16,17]. They demonstrated how to refocus an incident acoustic field back to its origin and to achieve high temporal and spatial compression by time reversal of the wave field.

In the above-mentioned papers, only transmission from sources to receivers is considered. A natural question is how to use this super focusing property in echographic mode to detect and separate scatterers. In echographic mode, the signal reflected from a scatterer is extremely complex because it has undergone a double path through the guide. The D.O.R.T.

method was applied to this problem [18]: the separation and selective focusing on two scatterers and then the detection of a scatterer placed near an interface of the guide are shown in the following.

Finally, in a steel-water-air waveguide, the consequences of surface waves produced at the water-air interface on the performance of the D.O.R.T. method are studied.

4.1 Selective Highly Resolved Focusing in a Waveguide

The experiment is performed in a 2-dimensional water waveguide, delimited by two water–steel plane interfaces. The water layer is 35 mm thick. The array consists of 60 transducers with a central frequency of 1.5 MHz; it spans the whole height of the guide. The array pitch is equal to 0.58 mm. The scatterers are two wires of diameters 0.1 mm and 0.2 mm, spaced 2 mm, and placed perpendicular to the array axis at a distance of 400 mm (Fig. 8). As the average wavelength is 1 mm, both wires behave almost like point scatterers. For this range and this frequency, the free-space diffraction focal width is 12 mm, so the two wires are not resolved by the system.

The echographic signals recorded after a pulse is applied to one transducer of the array are very complex with a low signal-to-noise ratio. The interelement response $k_{2840}(t)$ is a typical example (Fig. 9). After approximately 5 reflections at the interfaces the signal can no longer be distinguished from noise. The echoes of the two wires are superimposed and cannot be separated in a simple manner.

The 60 × 60 impulse response functions are measured, and the transfer matrix is calculated at a frequency of 1.5 MHz. The decomposition reveals two singular values that are separated from the 58 "noise" singular values (Fig. 10). The "noise" singular values are partly explained by electronic and quantification noises. However, different second-order acoustical phenomena not taken into account in the model probably contribute to these singular values; these include the defects of the interfaces, the elastic responses of the wires, the multiple echoes between the wires, and also coupling between the transducers.

The eigenvectors V_1 and V_2 have a complicated phase and amplitude distribution, and it is impossible to tell to which scatterer each of them corresponds. These distributions are applied to the array of transducers. Namely, if $\mathbf{V}_1 = (A_1 e^{i\varphi_1}, A_2 e^{i\varphi_2}, \ldots, A_n e^{i\varphi_n})$ is the first eigenvector, then the signal $s_p(t) = A_p \cos(\omega t - \varphi_p)$ is applied to transducer number p. The

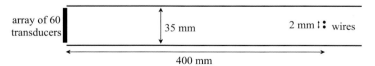

Fig. 8. Geometry of the experiment

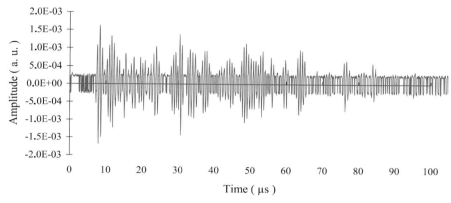

Fig. 9. Typical echo of the wires: inter-element impulse responses $k_{2840}(t)$

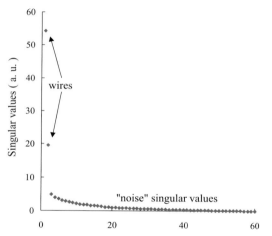

Fig. 10. Singular values of the transfer matrix calculated at 1.5 MHz

so-produced pressure field is measured across the guide at the range of the wires (Fig. 11). For each eigenvector, the wave is focused at the position of one wire. In both cases the residual level is lower than −18 dB and the −6 dB focal width is 1.4 mm. In fact, the width is overestimated because the width of the probe is 0.5 mm; the real focal width is probably around 1.2 mm, which is 9 times thinner than the theoretical free space focal width.

For comparison, the same experiment was performed after removing the guide. In this case the wires are not resolved, and only the first singular value is significant. The pressure pattern is measured for transmission of the first eigenvector; the focal width is 13 mm (Fig. 11). Consequently, the guide allows a focusing at least 10 times thinner than in free space to be achieved.

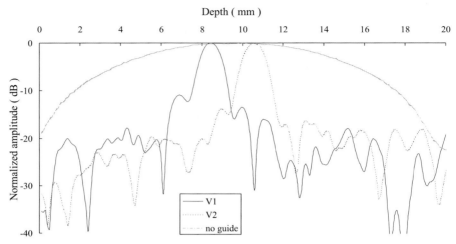

Fig. 11. Pressure pattern measured across the guide at the range of the wires after transmission of the eigenvectors. *Solid and dotted lines* first and second eigenvectors, respectively, obtained with the guide; first *gray line* eigenvector without the guide

The angular directivity of each transducer limits the number of reflections at the guide interfaces that can be recorded. This induces an apodization of the virtual array made of the set of images of the real one. Taking this phenomenon into account, the focal width roughly corresponds to a virtual aperture consisting of 8 pairs of images of the array.

4.2 Detection Near the Interface

In many cases, the detection of a defect near an interface is difficult, especially if the reflectivity coefficient of the interface is close to -1, which is the case for the water–air interface. Indeed, in this situation, the virtual image of the defect with respect to the interface behaves as a source in opposite phase to the defect. The real source and the virtual source interfere in a destructive way so that the reflected signal is very low. The ability of the D.O.R.T. method to detect a wire that is close to a water–air interface was analyzed. The experiment was done in a water waveguide of 35 mm width limited by air at the surface and steel at the bottom. A wire of 0.2-mm diameter is placed inside the guide 400 mm from the array. The wire is moved step by step from the bottom to the surface, and for each position the transfer matrix is measured and decomposed. The two first singular values are display versus distance to the surface in Fig. 12. The first singular value represents the signal level, and the second the noise level. When the wire reaches the bottom, the singular value increases rapidly by a factor of two. The echoes from the scatterer and from its image add constructively. Conversely, when the wire gets to the surface, the singular value decreases rapidly. It remains well separated from the noise singular values until the distance between the wire and the interface reaches $\lambda/5$. This result shows that the distance under which the scatterer is no longer detectable is less than $\lambda/5$ at a range of 400λ.

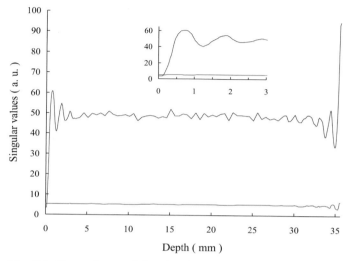

Fig. 12. Dependence of the singular values of the transfer matrix on the distance to the surface

4.3 Detection in a Nonstationary Waveguide

The efficiency of the method in a nonstationary guide with surface waves at the water-air interface is now studied. A vertical plate with horizontal oscillations at 6 Hz produces surface waves with a typical wavelength of 30 mm. The height of the waves is varied using a diaphragm (Fig. 13). The root-mean-square (rms) height of the waves kh_{rms} is varied from 0 to 1.7 mm,

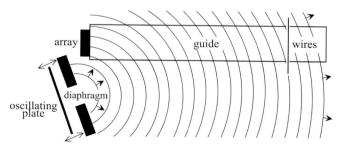

Fig. 13. Experimental setup to produce surface waves

which corresponds to $0 < kh_{\mathrm{rms}} < 10$. The array of transducers is the same as before. Two wires are placed 500 mm from the array and spaced 5 mm.

For each value of kh_{rms}, the transfer matrix is measured and decomposed. While kh_{rms} is lower than 1.5, the two higher singular values are well separated from noise singular values. The corresponding eigenvectors focus at the position of the wires; however, the main lobes are approximately 1.6 times larger and the residual lever 2 times higher than in the absence of waves (Fig. 14).

For high waves ($kh_{\mathrm{rms}} = 10$), the signal singular values are not separated from noise with a single measurement of **K**. However, the average of 10 realizations of the transfer matrix reduces the noise singular values enough to separate the wires. Then it is possible to obtain a selective focusing with a resolution almost 3 times thinner than in free space (Fig. 15).

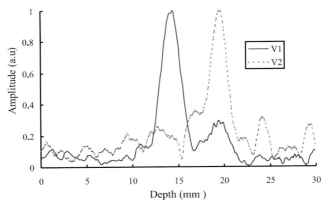

Fig. 14. Pressure field measured for transmission of eigenvectors 1 and 2 calculated with an unaveraged transfer matrix obtained with surface waves of $h_{\mathrm{rms}} = 0.23$ mm

5 Inverse-Scattering Analysis and Target Resonance

The preceding experiments illustrate the efficiency of the D.O.R.T. method to focus selectively on different scatterers through complex media. More generally this method isolates and classifies the scattering centers or secondary sources in the medium and can be used to analyse the scattering from extended objects. In particular, this method applies to scattering by a thin hollow cylinder. Time-reversal techniques with short ultrasonic signals have been applied to such scattering experiment by *Thomas* et al. [9]. They are efficient when the contributions of the various waves can be selected by a time window. It is now shown that even if the waves interfere in time it is possible to separate them using the D.O.R.T. method [20,21].

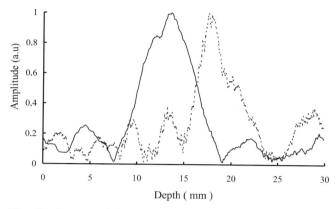

Fig. 15. Pressure field measured for transmission of eigenvectors 1 and 2 calculated from an average of ten realizations of the transfer matrix

5.1 Experiment

The array is linear and made of 96 rectangular transducers similar to those used in Sect. 3. A hollow steel cylinder with a diameter of 20 mm and a thickness of approximately 0.6 mm is placed perpendicular to the array of transducers at a distance of 80 mm symmetrically with respect to the array axis (Fig. 16). For an incident plane wave, a Lamb wave is generated at a given angle of incidence, θ, with respect to the normal to the surface. This angle satisfies the relation $\sin(\theta) = \frac{C_0}{C_\phi}$, where C_0 is the sound velocity in water and C_ϕ is the phase velocity of the Lamb wave.

Consequently, two Lamb waves are generated at points A and B, symmetrical with respect to the incident direction (Fig. 17). While propagating around the cylinder, those two waves radiate backward from the same points, A and B, which behave as secondary sources. The distance d_{AB} between those

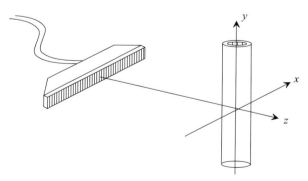

Fig. 16. Experimental setup

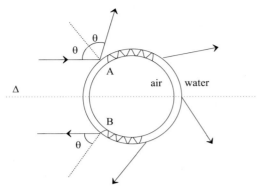

Fig. 17. Generation and radiation of a Lamb wave on a thin hollow cylinder

points is given by

$$d_{\mathrm{AB}} = D \frac{C_0}{C_\phi}, \tag{14}$$

where D is the diameter of the cylinder. Thus, the nature of the wave can be determined by knowing its radiation points. For such a thickness and frequency range, the dispersion curves (Fig. 18) predict that radiation of the three Lamb waves A_0, S_0 and A_1 should be observed.

A short pulse is launched from the center element of the array. The echo of the cylinder is recorded on the 96 elements (Fig. 19). The first wavefront corresponds to the strong specular echo. The signal observed later is the elastic part of the echo. Between 15 ms and 25 ms, two pairs of wavefronts with interference fringes can be distinguished. Those wavefronts correspond to the radiation of two pairs of circumferential waves after one turn around

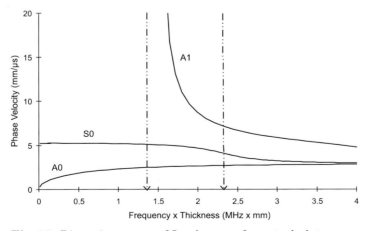

Fig. 18. Dispersion curves of Lamb waves for a steel plate

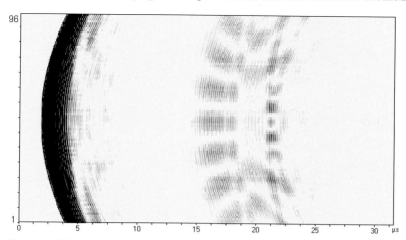

Fig. 19. Echo of the shell received by the 128 transducers after transmission of a short pulse from the center element of the array. The first wavefront is the specular echo, the second the contribution of the S_0 Lamb wave and the third the contribution of the A_0 Lamb wave

the shell. The first one is identified as the S_0 Lamb mode, and the second as the A_0 Lamb mode. Interfering with those well-defined wavefronts is the contribution of the highly dispersive A_1 wave.

5.2 Invariants of the Time-Reversal Process

The two circumferential waves generated at points A and B (Fig. 17) are linked by reciprocity. The first one can be obtained by time reversal of the second one. Both waves are invariant under two successive time-reversal processes; consequently they should be associated to eigenvectors of the time-reversal operator. In fact, due to the symmetry of the problem, they are both associated with the same two eigenvectors: one corresponding to the generation in phase with the two waves, and the other to the generation in opposite phase.

To separate these contributions we now apply the D.O.R.T. method. After the measurement of the 96×96 inter-element impulse responses, the whole process remains numerical. Only the elastic part of the signal is used to calculate the time-reversal operator (between 15 ms and 25 ms). At 3.05 MHz, the diagonalization of the time-reversal operator has six dominant eigenvalues. The modulus of the components of each eigenvector (1 to 6) is represented versus array element in Fig. 20. The interference fringes are easily observed. They are the equivalent at one frequency of the interference pattern observed on the echoes. As in the experiment with two wires (1.4), this means that an eigenvector corresponds to the interference of two coherent point sources.

First and second eigenvectors

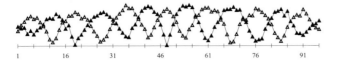

Third and fourth eigenvectors

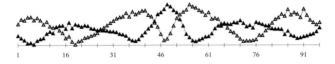

Fifth and sixth eigenvectors

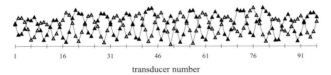

transducer number

Fig. 20. Modulus of the components of the 6 eigenvectors

The numerical back-propagation of each eigenvector allows the distance between the sources to be determined (Fig. 21). Each pair of sources corresponds to one particular Lamb wave. At this frequency, the first and second eigenvectors are associated with the wave S_0, the third and fourth with the wave A_1 and the fifth and sixth with the wave A_0.

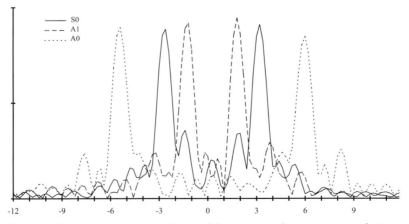

Fig. 21. Directivity patterns obtained by numerical propagation of eigenvectors 1, 3 and 5

The same calculation is done at several frequencies from 2.2 MHz to 4 MHz, so that the dispersion curves for the three waves can be plotted (Fig. 22). These curves are very close to the theoretical curves obtained for a steel plate of thickness of 0.6 mm. In particular, the determination of the cutoff frequency of the wave A_1 allows the thickness of the shell to be found.

One limitation of this method is that the generation points of the circumferential waves need to be spatially resolved. In the case of two waves of close phase velocities, the separation may not be possible. As the phase velocity increases, the two generation points get closer and are no longer resolved by the system. This partly explains the reason why the velocity of the wave A_1 could not be measured closer to the cutoff frequency.

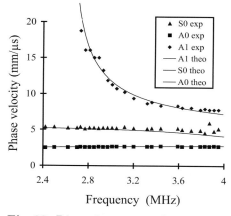

Fig. 22. Dispersion curves: theory and experiment

5.3 Resonance Frequencies of the Shell

The eigenvalue associated with one particular wave depends on the frequency; it is proportional to the level of contribution of the wave to the scattered field. Beside the reponses of the transducers, the generation and radiation coefficients of the wave are responsible for these variations. Moreover if the dynamics and duration of the recorded signals allow several turns of the wave around the shell to be detected, a fast modulation of the corresponding eigenvalue versus frequency is induced, the maxima corresponding to the resonance frequencies of the shell. In the experiment, the wave A_0 is attenuated so fast that only one turn can be observed. But several turns of A_1 and S_0 contribute to the scattered field. To take into account these multiple turns, the time-reversal operator was calculated using 40 ms of signal. Then the eigenvalues of the time-reversal operator were calculated from 2.2 MHz to 3.8 MHz. The first six eigenvalues are represented versus frequency in Fig. 23. The two

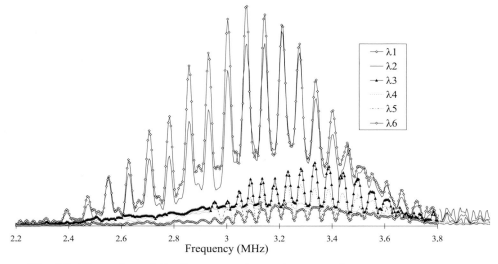

Fig. 23. Eigenvalues of the time-reversal operator versus frequency

curves $\lambda_1(\omega)$ and $\lambda_2(\omega)$ correspond to the wave S_0. Their maxima occur at the resonance frequencies of the shell corresponding to this wave. The width of the peaks is mainly due to the length and dynamic of the recorded signals, which allow only three turns of S_0 wave around the shell to be seen. Similar observations can be made for the wave A_1. This wave is associated with the eigenvalues $\lambda_3(\omega)$ and $\lambda_4(\omega)$ around their corresponding resonance frequencies and with $\lambda_5(\omega)$ and $\lambda_6(\omega)$ near their corresponding anti-resonance frequencies. The resonance peaks are well defined although the contribution of the wave A_1 is weaker than that of S_0.

The D.O.R.T method provides the resonance frequencies due to the waves S_0 and A_1, with the significant advantage that close resonance frequencies can be distinguished.

6 The D.O.R.T. Method in the Time Domain

In the preceding sections, analysis of the transfer function is done frequency by frequency. Only a small part of the information contained in the interelement impulse response functions is used. In fact, the decomposition of the time-reversal operator can be done at any frequency. In order to get temporal signals, it would be natural to calculate the eigenvectors in the whole band of the transducers and to perform an inverse Fourier transform of the eigenvector function of the frequency. However, this operation is nontrivial. The main reason is that the scatterers' reflectivity generally depends on frequency, so that at one frequency the first eigenvector can be associated to one scatterer, while it is associated to another one at another frequency. However, if the strengths of the scatterers are sufficiently different, then the first eigenvector may correspond to the same scatterer in the whole frequency band of the transducers. In this case, it is possible to build temporal signals from the eigenvectors. If the first eigenvector corresponds to one point-like scatterer, then the temporal signal will provide the impulse Green's function connecting the scatterer with the array.

6.1 Construction of the Temporal Green's Functions

The above-mentioned conditions are satisfied in the following example. The array of transducers and the waveguide are the same as in Sect. 1. The range of the scatterers is 400 mm, the distance between the scatterers is 2 mm, and their reflectivities differ by a factor of three in the frequency band of the transducers. The SVD of the transfer matrix is calculated at each frequency of the discrete spectrum from 0.8 to 2.2 MHz. The singular values, distribution versus frequency is shown Fig. 24: two singular values are apart from the 58 noise singular values and well separated from each other.

The impulse response function from the strong scatterer to the array can be reconstructed from the eigenvectors $\sqrt{\lambda_1(\omega)}\mathbf{V}_1(\omega)$. Assuming the reflectivity of the scatterer is independent of the frequency, this response is the temporal Green's function connecting the scatterer to the array convoluted by the acousto-electrical response of the transducer (Fig. 25, top). The same procedure applied to $\sqrt{\lambda_2(\omega)}\mathbf{V}_2(\omega)$ provides the impulse Green's function from the second scatterer to the array (Fig. 25, bottom). This result is of

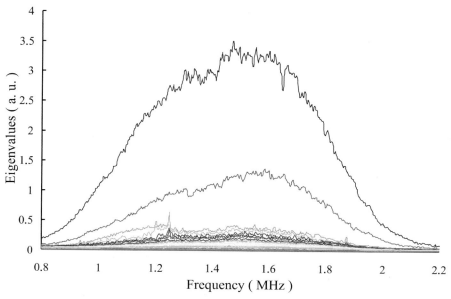

Fig. 24. Singular values of the transfer matrix versus frequency

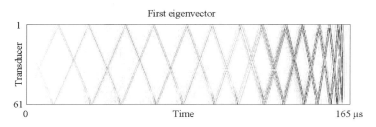

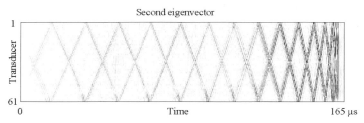

Fig. 25. Signals reconstructed from (**a**) the first eigenvector and (**b**) the second eigenvector. These signals correspond to the impulse response from each wire to the array

particular interest in a complex propagating medium such as a waveguide. Indeed, the low signal-to-noise ratio due to the length of the multiple path and the complexity of the echographic response of scatterer due to the double paths along the guide render the determination of the impulse responses of the scatterers very difficult.

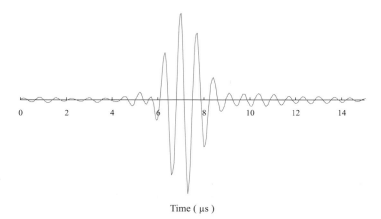

Time (μs)

Fig. 26. Time-domain compression: signal received at the position of the first wire after transmission of the first temporal eigenvector

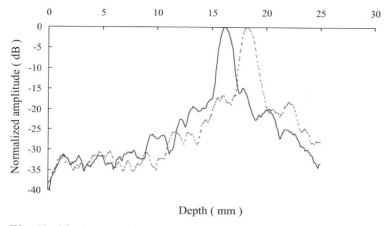

Fig. 27. Maximum of the pressure field measured at the depth of the wires across the guide after transmission of eigenvectors 1 and 2

6.2 Selective Focusing in the Pulse Mode

These signals are then transmitted from the array and the so-produced field is recorded along a line at the initial depth of the wires. One can observe an excellent temporal compression at the position of the wires: the signal received at the wire position is a pulse 3-ms long, while the transmitted signals are 165-ms long (Fig. 26).

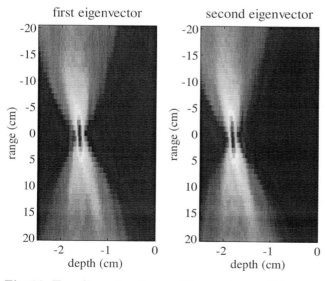

Fig. 28. Two-dimensional map of the maximum of the pressure field measured after transmission of the first and second eigenvectors

The transverse peak pressure pattern of the first and second eigenvectors at the depth of the wire (Fig. 27) can be compared with the one obtained in monochromatic transmission (Fig. 11). The improvement in spatial focusing is undeniable. The secondary lobes decrease to $-30\,\text{dB}$, while in the monochromatic transmission they remained around $-18\,\text{dB}$. Furthermore, the focusing is also excellent in range (Fig. 28).

7 Conclusion

In brief, the D.O.R.T. method is a generalization of the principle of iterative time-reversal mirrors. It allows the different waves contributing to the scattered field to be sorted and provides information on the scattering medium that up to now has not been available. It is a powerful tool for detection and focusing in heterogeneous and multiple target media and more generally for inverse scattering analysis.

The various experimental results shown opened several axes of research that are now under study. The detection of microcalcification in the breast is probably one of the most interesting applications in medical imaging. The D.O.R.T. method can be applied to nondestructive testing in solid waveguides. This is of particular interest for finding defects that are close to the interfaces. The results presented in a nonstationary guide are promising and have motivated further studies of underwater applications such as mine countermeasures. The time-domain D.O.R.T. method could be used in detection to reduce sidelobe effects such as problems of false location.

References

1. M. Fink, C. Prada, F. Wu, Self focusing in inhomogeneous media with time reversal acoustic mirrors, Proc. IEEE Ultrason. Symp. **2**, 681–686 (1989)
2. M. Fink, Time Reversal Mirrors, J. Phys. D **26**, 1333–1350 (1993)
3. C. Prada, J.-L. Thomas, M. Fink, The iterative time reversal process: analysis of the convergence, J. Acoust. Soc. Am. **97**, 62–71 (1995)
4. C. Prada, M. Fink, Eigenmodes of the time reversal operator: a solution to selective focusing in multiple target media, Wave Motion **20**, 151–163 (1994)
5. C. Prada, S. Manneville, D. Spoliansky, M. Fink, Decomposition of the time reversal operator: detection and selective focusing on two scatterers, J. Acoust. Soc. Am. **99**, 2067–2076 (1996)
6. C. Prada, M. Tanter, M. Fink, Flaw detection in solid with the D.O.R.T. method, Proc. IEEE Ultrason. Symp. **2**, 681–686 (1989)
7. G. Bienvenu, L. Kopp, Optimality of high resolution array processing using the eigensystem approach, IEEE Trans. Acoust. Speech Sig. Proc. **31** (1983)
8. R. O. Schmidt, Multiple Emitter Location and Signal Parameter Estimation, IEEE Trans. Ant. and Prop. **AP-34**, 276–281 (1986)
9. B. Baggeroer, W. A. Kuperman, P. N. Mikhalevsky, An overview of matched field methods in ocean acoustics, IEEE J. Ocean. Eng. **18**, 401–424 (1993)

10. Parvulescu, Matched – signal ('MESS') processing by the ocean, J. Acoust. Soc. Am. **98**, 943–960 (1995)
11. C. S. Clay, Optimum time domain signal transmission and source location in a waveguide, J. Acoust. Soc. Am. **81**, 660–664 (1987)
12. S. Li, C. S. Clay, Optimum time domain signal transmission and source location in a waveguide: Experiments in an ideal wedge waveguide, J. Acoust. Soc. Am. **82**, 1409–1417 (1987)
13. D. R. Jackson, D. R. Dowling, Phase conjugation in underwater acoustics, J. Acoust. Soc. Am. **89**, 171–181 (1991)
14. P. Roux, B. Roman, M. Fink, Time-reversal in an ultrasonic waveguide, Appl. Phys. Lett. **70**, 1811–1813 (1997)
15. W. A. Kuperman, W. S. Hodgkiss, H. C. Song, T. Akal, C. Ferla, D. R. Jackson, Phase conjugation in the ocean: Experimental demonstration of an acoustic time-reversal mirror, J. Acoust. Soc. Am. **103**, 25–40 (1998)
16. H. C. Song, W. A. Kuperman, W. S. Hodgkiss, Iterative time reversal in the ocean, J. Acoust. Soc. Am. **105**, 3176–3184 (1999)
17. W. S. Hodgkiss, H. C. Song, W. A. Kuperman, A long-range and variable focus phase-conjugation experiment in shallow water, J. Acoust. Soc. Am. **105**, 1597–1602 (1999)
18. N. Mordant, C. Prada, M. Fink, Highly resolved detection and selective focusing in a waveguide using the D.O.R.T. method, J. Acoust. Soc. Am. **105**, 2634–2642 (1999)
19. J.-L. Thomas, P. Roux, M. Fink, Inverse scattering analysis with an acoustic time-reversal mirror, Phys. Rev. Lett. **72**, 637–640 (1994)
20. C. Prada, J.-L. Thomas, P. Roux, M. Fink, Acoustic time reversal and inverse scattering, Proc. Int. Symp. Inv. Prob. (1994) 309–316
21. C. Prada, M. Fink, Separation of interfering acoustic scattered signals using the invariant of the time-reversal operator. Application to Lamb waves characterization, J. Acoust. Soc. Am. **104**, 801–807 (1998)

Ultrasound Imaging and Its Modeling

Jørgen A. Jensen

Ørsted · DTU, Building 348, Technical University of Denmark
2800 Lyngby, Denmark
jaj@oersted.dtu.dk

Abstract. Modern medical ultrasound scanners are used to image nearly all soft tissue structures in the body. The anatomy can be studied from gray-scale B-mode images, where the reflectivity and scattering strength of the tissues are displayed. The imaging is performed in real time with 20 to 100 images per second. The technique is widely used, since it does not use ionizing radiation and is safe and painless for the patient. This chapter gives a short introduction to modern ultrasound imaging using array transducers. It includes a description of the different imaging methods, the beam-forming strategies used, and the resulting fields and their modeling.

1 Fundamental Ultrasound Imaging

The main units of a modern B-mode imaging system are shown in Fig. 1. A multi-element transducer is used for both transmitting and receiving the pulsed ultrasound field. The central frequency of the transducer can be from 2 to 15 MHz, depending on the use. Often advanced composite materials are used in the transducer, and they can attain a relative bandwidth in excess of 100%. The resolution is, thus, on the order of one to three wavelengths. The mean speed of sound in the tissue investigated varies from 1446 m/s (fat) to 1566 m/s (spleen) [1,2], and an average value of 1540 m/s is used in the scanners. This gives a wavelength of 0.308 mm at 5 MHz and a resolution in the axial direction of 0.1 to 1 mm.

The emission of the beam is controlled electronically as described in Sect. 3 and is for a phased-array system swept over the region of interest in a polar scan. A single focus can be used when transmitting, and the user can select the depth of the focus. The reflected and scattered field is then received by the transducer again and amplified by the time-gain-compensation amplifier. The latter compensates for the loss in amplitude due to the attenuation experienced during propagation of the sound field in the tissue. Typical attenuation values are shown in Table 1. A typical value used, when designing the scanner, is 0.7 dB/(MHz cm), indicating that the attenuation increases exponentially with both depth and frequency. This is the one-way attenuation, and a 5 MHz wave measured at a depth of 10 cm would, thus, be attenuated by 70 dB.

After amplification the signals from all transducer elements are passed to the electronic beamformer, which focuses the received beam. For low-end

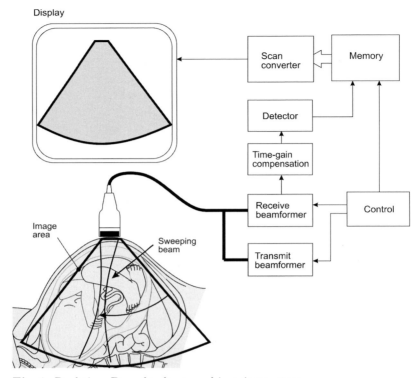

Fig. 1. Real-time B-mode ultrasound imaging system

Table 1. Typical attenuation values for human tissue (assembled from the compilation in [3])

Tissue	Attenuation dB/(MHz cm)
Liver	0.6 – 0.9
Kidney	0.8 – 1.0
Spleen	0.5 – 1.0
Fat	1.0 – 2.0
Blood	0.17 – 0.24
Plasma	0.01
Bone	16.0 – 23.0

scanners this is done through analog delay lines, whereas more modern high-end scanners employ digital signal processing on a sampled version of the signal from all elements. Hereby a continuous focus can be attained giving a very high-resolution image. Often the beamformer can handle 64 to 192 transducer elements, and this is the typical element count in modern scanners. There is a continuous effort to expand the number of channels to improve image resolution and contrast.

The beamformed signal is envelope detected and stored in a memory bank. A scan conversion is then performed to finally show the ultrasound image on a gray-scale display in real time. The images can cover an area of 15 by 15 cm, and a single pulse–echo line then takes $2 \times 0.15/1540 = 195$ ms to acquire. Since an image consists of roughly 100 polar lines, a frame rate of 51 images per second is obtained. Often smaller images are selected to increase the frame rate, especially for blood velocity imaging [4].

A typical ultrasound image for a fetus in the 13th week is shown in Fig. 2.

The head, mouth, legs, and spine are clearly identified in the image. It is also seen that the image has a grainy appearance and that there are no clear

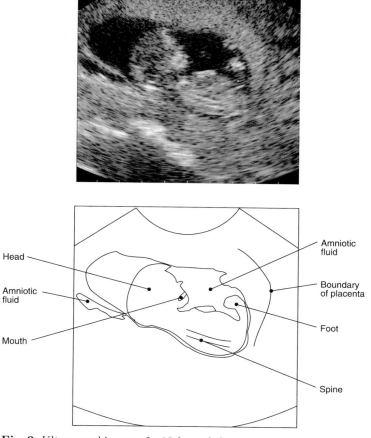

Fig. 2. Ultrasound image of a 13th-week fetus. The markers at the border of the image indicate 1 cm. (Reproduced with permission from Jensen [1996], © Cambridge Univ. Press)

demarcations or reflection between the placenta and the amniotic fluid surrounding the fetus. There are, thus, no distinct reflections from plane boundaries, as they seldom exist in the body. It is the ultrasound field scattered by the constituents of the tissue that is displayed in medical ultrasound, and the medical scanners are optimized to display the scattered signal. The scattered field emanates from small changes in density, compressibility, and absorption from the connective tissue, cells, and fibrous tissue. These structures are much smaller than one wavelength of the ultrasound, and the resulting speckle pattern displayed does not directly reveal physical structure. It is rather the constructive and destructive interference of scattered signals from all the small structures. So it is not possible to visualize and diagnose microstructure, but the strength of the signal is an indication of pathology. A strong signal from liver tissue, making a bright image, is, for example, an indication of a fatty or cirrhotic liver.

As the scattered wave emanates from numerous contributors, it is appropriate to characterize it in statistical terms. The amplitude distribution follows a Gaussian distribution [5] and is, thus, fully characterized by its mean and variance. The mean value is zero, since the scattered signal is generated by differences in the tissue from the mean acoustic properties.

Since the backscattered signal depends on the constructive and destructive interference of waves from numerous small tissue structures, it is not meaningful to talk about the reflection strength of the individual structures. Rather, it is the deviations from the mean density and speed of sound within the tissue and the composition of the tissue that determine the strength of the returned signal. The magnitude of the returned signal is, therefore, described in terms of the power of the scattered signal. Since the small structures re-radiate waves in all directions and the scattering structures might be ordered in some direction, the returned power will, in general, be dependent on the relative position between the ultrasound emitter and receiver. Such a medium is called anisotropic, examples of which are muscle and kidney tissue. By comparison, liver tissue is a fairly isotropic scattering medium, when its major vessels are excluded, and so is blood.

2 Imaging with Arrays

Basically there are three different kinds of images acquired by multi-element array transducers, i.e., linear, convex, and phased, as shown in Figs. 3, 5, and 6. The linear-array transducer is shown in Fig. 3. It selects the region of investigation by firing a set of elements situated over the region. The beam is moved over the imaging region by firing sets of contiguous elements. Focusing when transmitting is achieved by delaying the excitation of the individual elements, so an initially concave beam shape is emitted, as shown in Fig. 4.

The beam can also be focused during reception by delaying and adding responses from the different elements. A continuous focus or several focal

Linear array imaging

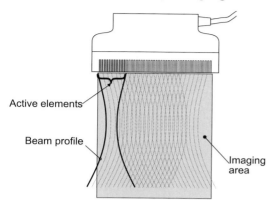

Fig. 3. Linear-array transducer for obtaining a rectangular cross-sectional image

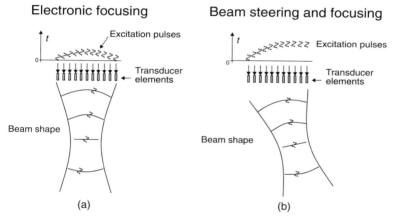

Fig. 4. Electronic focusing and steering of an ultrasound beam. (Reproduced with permission from Jensen [1996], © Cambridge Univ. Press)

zones can be maintained, as explained in Sect. 3. Only one focal zone is possible when transmitting, but a composite image using a set of foci from several transmissions can be made. Often 4 to 8 zones can be individually placed at selected depths in modern scanners. The frame rate is then lowered by the number of transmit foci.

The linear arrays acquire a rectangular image, and the arrays can be quite large to cover a sufficient region of interest (ROI). A larger area can be scanned with a smaller array, if the elements are placed on a convex surface as shown in Fig. 5. A sector scan is then obtained. The method of focusing and beam sweeping during transmittance and reception is the same as for

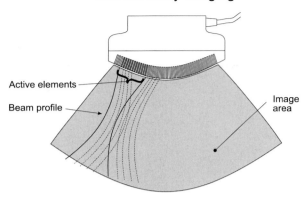

Fig. 5. Convex array transducer for obtaining a polar cross-sectional image

the linear array, and a substantial number of elements (often 128 or 256) is employed.

The convex and linear arrays are often too large to image the heart when probing between the ribs. A small array size can be used and a large field of view attained by using a phased array, as shown in Fig. 6. All array elements are used here both during transmittance and reception. The direction of the beam is steered by electrically delaying the signals to or from the elements, as shown in Fig. 4b. Images can be acquired through a small window, and the

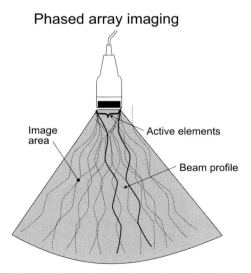

Fig. 6. Phased-array transducer for obtaining a polar cross-sectional image using a transducer with a small footprint

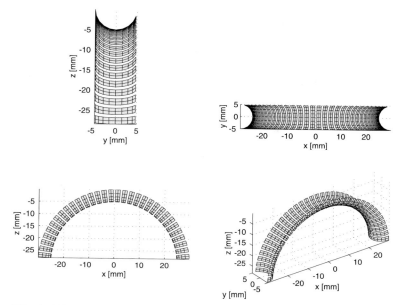

Fig. 7. Elevation-focused convex-array transducer for obtaining a rectangular cross-sectional image, which is focused in the out-of-plane direction. The curvature in the elevation direction is exaggerated for illustration purposes

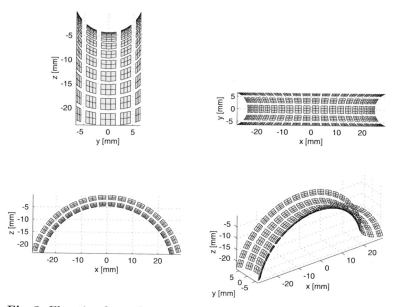

Fig. 8. Elevation-focused convex-array transducer with element division in the elevation direction. The curvature in the elevation direction is exaggerated for illustration purposes

beam rapidly swept over the ROI. The rapid steering of the beam compared to mechanical transducers is of special importance in flow imaging [4]. This has made the phased array the choice for cardiological investigations through the ribs.

More advanced arrays have recently been introduced with the increase in the number of elements and digital beamforming. Especially elevation focusing (out of the imaging plane) is important. A curved surface as shown in Fig. 7 is used for obtaining the elevation focusing essential for an improved image quality. Electronic beamforming can also be used in the elevation direction by dividing the elements in that direction. The elevation focusing when receiving can then be dynamically controlled for, e.g., the array shown in Fig. 8.

3 Focusing

The essence of focusing an ultrasound beam is to align the pressure fields from all parts of the aperture to arrive at the field point at the same time. This can be done through either a physically curved aperture, through a lens in front of the aperture, or by the use of electronic delays for multi-element arrays. All seek to align the arrival of the waves at a given point through delaying or advancing the fields from the individual elements. The delay (positive or negative) is determined using ray acoustics. The path length from the aperture to the point gives the propagation time, and that is adjusted relative to some reference point. The propagation time t_i from the center of the aperture element to the field point is

$$t_i = \frac{1}{c}\sqrt{(x_i - x_f)^2 + (y_i - y_f)^2 + (z_i - z_f)^2}. \tag{1}$$

where (x_f, y_f, z_f) is the position of the focal point, (x_i, y_i, z_i) is the center for the physical element number i, and c is the speed of sound.

A point is selected on the whole aperture as a reference for the imaging process. The propagation time for this is

$$t_c = \frac{1}{c}\sqrt{(x_c - x_f)^2 + (y_c - y_f)^2 + (z_c - z_f)^2}. \tag{2}$$

where (x_c, y_c, z_c) is the reference center point on the aperture. The delay to use on each element of the array is then

$$\Delta t_i = \frac{1}{c}\left(\sqrt{(x_c - x_f)^2 + (y_c - y_f)^2 + (z_c - z_f)^2} \right.$$
$$\left. - \sqrt{(x_i - x_f)^2 + (y_i - y_f)^2 + (z_i - z_f)^2}\right). \tag{3}$$

Notice that there is no limit on the selection of the different points, and the beam can, thus, be steered in any preferred direction.

The arguments here have been given for emission from an array, but they are equally valid during reception of the ultrasound waves due to acoustic reciprocity. At reception it is also possible to change the focus as a function of time and thereby obtain a dynamic tracking focus. This is used by all modern ultrasound scanners. Beamformers based on analog technology make it possible to create several receive foci, and the newer digital scanners change the focusing continuously for every depth while receiving.

The focusing can, thus, be defined through time lines as follows:

From time	Focus at
0	x_1, y_1, z_1
t_1	x_1, y_1, z_1
t_2	x_2, y_2, z_2
⋮	⋮

For each focal zone there is an associated focal point, and the time from this focus is used. The arrival time from the field point to the physical transducer element is used to decide which focus is used. Another possibility is to set the focusing to be dynamic, so that the focus is changed as a function of time and thereby depth. The focusing is then set as a direction defined by two angles and a starting point on the aperture.

Section 4 shows that the side and grating lobes of the array can be reduced by employing apodization of the elements. Again a fixed function can be used when transmitting and a dynamic function when receiving, defined as follows:

From time	Apodize with
0	$a_{1,1}, a_{1,2}, \cdots a_{1,N_e}$
t_1	$a_{1,1}, a_{1,2}, \cdots a_{1,N_e}$
t_2	$a_{2,1}, a_{2,2}, \cdots a_{2,N_e}$
t_3	$a_{3,1}, a_{3,2}, \cdots a_{3,N_e}$
⋮	⋮

Here $a_{i,j}$ is the amplitude-scaling value multiplied onto element j after time instance t_i. Typically a Hamming- or Gaussian-shaped function is used for the apodization. While receiving the width of the function is often increased to compensate for attenuation effects and for keeping the point spread function roughly constant. The F-number defined by

$$F = \frac{D}{L}, \tag{4}$$

where L is the total width of the active aperture and D is the distance to the focus, is often kept constant. More of the aperture is often used for larger depths, and a compensation for the attenuation is thereby partly made. An example of the use of dynamic apodization is given in Sect. 7.

4 Ultrasound Fields

This section derives a simple relation between the oscillation of the transducer surface and the ultrasound field. It is shown that field in the far field can be found by a simple 1-dimensional Fourier transform of the 1-dimensional aperture pattern. This might seem far from the actual imaging situation in the near field using pulsed excitation, but the approach is very convenient for introducing all the major concepts such as main- and sidelobes, grating lobes, etc. It also very clearly reveals information about the relation between aperture properties and field properties.

4.1 Derivation of the Fourier Relation

Consider a simple line source of length L as shown in Fig. 9 with a harmonic particle speed of $U_0 \exp(j\omega t)$. Here U_0 is the vibration amplitude and ω is its angular frequency. The line element of length dx generates an increment in pressure at r' of [6]

$$dp = j\frac{\rho_0 c k}{4\pi r'} U_0 a_p(x) e^{j(\omega t - k r')} dx , \qquad (5)$$

where ρ_0 is density, c is speed of sound, $k = \omega/c$ is the wavenumber, and $a_p(x)$ is an amplitude scaling of the individual parts of the aperture. In the far field ($r \gg L$) the distance from the radiator to the field points is (see Fig. 9):

$$r' = r - x \sin\theta . \qquad (6)$$

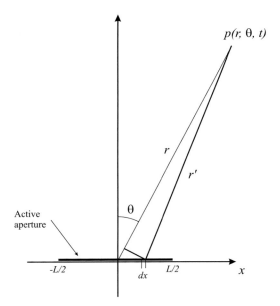

Fig. 9. Geometry for line aperture

The emitted pressure is found by integrating over all the small elements of the aperture

$$p(r, \theta, t) = j\frac{\rho_0 c U_0 k}{4\pi} \int_{-\infty}^{+\infty} a_p(x) \frac{e^{j(\omega t - kr')}}{r'} dx. \tag{7}$$

Notice that $a_p(x) = 0$ if $|x| > L/2$. Here r' can be replaced with r, if the extent of the array is small compared to the distance to the field point ($r \gg L$). Using this approximation and inserting (6) into (7) gives

$$p(r, \theta, t) = j\frac{\rho_0 c U_0 k}{4\pi r} \int_{-\infty}^{+\infty} a_p(x) e^{j(\omega t - kr + kx \sin \theta)} dx$$

$$= j\frac{\rho_0 c U_0 k}{4\pi r} e^{j(\omega t - kr)} \int_{-\infty}^{+\infty} a_p(x) e^{jkx \sin \theta} dx, \tag{8}$$

since ωt and kr are independent of x. Hereby the pressure amplitude of the field for a given frequency can be split into two factors:

$$P_{\text{ax}}(r) = \frac{\rho_0 c U_0 k L}{4\pi r},$$

$$H(\theta) = \frac{1}{L} \int_{-\infty}^{+\infty} a_p(x) e^{jkx \sin \theta} dx,$$

$$P(r, \theta) = P_{\text{ax}}(r) H(\theta). \tag{9}$$

The first factor $P_{\text{ax}}(r)$ characterizes how the field drops off in the axial direction as a factor of distance, and $H(\theta)$ gives the variation of the field as a function of angle. The first term drops off with $1/r$ as for a simple point source; and $H(\theta)$ is found from the aperture function $a_p(x)$. A slight rearrangement gives[1]

$$H(\theta) = \frac{1}{L} \int_{-\infty}^{+\infty} a_p(x) \exp\left(j2\pi x f \frac{\sin \theta}{c}\right) dx$$

$$= \frac{1}{L} \int_{-\infty}^{+\infty} a_p(x) \exp\left(j2\pi x f'\right) dx. \tag{10}$$

This very closely resembles the standard Fourier integral given by

$$G(f) = \int_{-\infty}^{+\infty} g(t) e^{-j2\pi t f} dt,$$

$$g(t) = \int_{-\infty}^{+\infty} G(f) e^{j2\pi t f} df, \tag{11}$$

There is, thus, a Fourier relation between the radial beam pattern and the aperture function, and the normal Fourier relations can be used in order to understand the beam patterns for typical apertures.

---
[1] The term $1/L$ is included to make $H(\theta)$ a unitless number.

4.2 Beam Patterns

The first example is for a simple line source where the aperture function is constant such that

$$a_p(x) = \begin{cases} 1 & |x| \leq L/2 \\ 0 & \text{otherwise} \end{cases}. \tag{12}$$

The angular factor is then

$$H(\theta) = \frac{\sin\left(\pi L f \frac{\sin\theta}{c}\right)}{\pi L f \frac{\sin\theta}{c}} = \frac{\sin\left(\frac{k}{2} L \sin\theta\right)}{\frac{k}{2} L \sin\theta}. \tag{13}$$

A plot of the sinc function is shown in Fig. 10. A single main lobe can be seen with a number of sidelobe peaks. The peaks fall off proportionally to k or f. The angle of the first zero in the function is found at

$$\sin\theta = \frac{c}{Lf} = \frac{\lambda}{L}. \tag{14}$$

The angle is, thus, dependent on the frequency and the size of the array. A large array or a high emitted frequency, therefore, gives a narrow mainlobe.

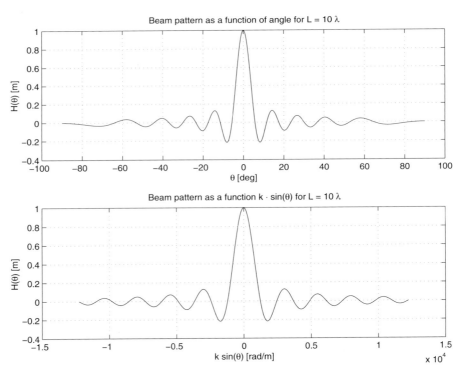

Fig. 10. Angular beam pattern for a line aperture with a uniform aperture function as a function of angle (*top*) and as a function of $k\sin(\theta)$ (*bottom*)

The magnitude of the first sidelobe relative to the mainlobe is given by

$$\frac{H\left[\arcsin\left(\frac{3c}{2Lf}\right)\right]}{H(0)} = L\frac{\sin(3\pi/2)}{3\pi/2}/L = \frac{2}{3\pi}. \tag{15}$$

The relative sidelobe level is, thus, independent of the size of the array and of the frequency, and is solely determined by the aperture function $a_p(x)$ through the Fourier relation. The large discontinuities of $a_p(x)$, thus, give rise to the high sidelobe level, and they can be reduced by selecting an aperture function that is smoother, such as a Hanning window or a Gaussian shape.

Modern ultrasound transducers consist of a number of elements, each radiating ultrasound energy. Neglecting the phasing of the element (see Sect. 3) due to the far-field assumption, the aperture function can be described by

$$a_p(x) = a_{ps}(x) * \sum_{n=-N/2}^{N/2} \delta(x - d_x n), \tag{16}$$

where $a_{ps}(x)$ is the aperture function or apodization for the individual elements, d_x is the spacing (pitch) between the centers of the individual elements, and $N+1$ is the number of elements in the array. Using the Fourier relationship the angular beam pattern can be described by

$$H_p(\theta) = H_{ps}(\theta) H_{per}(\theta), \tag{17}$$

where

$$\sum_{n=-N/2}^{N/2} \delta(x - d_x n) \leftrightarrow H_{per}(\theta) = \sum_{n=-N/2}^{N/2} \exp\left(-j n d_x k \sin\theta\right)$$

$$= \sum_{n=-N/2}^{N/2} \exp\left(-j 2\pi \frac{f \sin\theta}{c} n d_x\right). \tag{18}$$

Summing the geometric series gives

$$H_{per}(\theta) = \frac{\sin\left[(N+1)\frac{k}{2} d_x \sin\theta\right]}{\sin\left(\frac{k}{2} d_x \sin\theta\right)}, \tag{19}$$

which is the Fourier transform of a series of delta functions. This function repeats itself with a period that is a multiple of

$$\pi = \frac{k}{2} d_x \sin\theta,$$

$$\sin\theta = \frac{2\pi}{k d_x} = \frac{\lambda}{d_x}. \tag{20}$$

This repetitive function gives rise to the grating lobes in the field. An example is shown in Fig. 11. The grating lobes are due to the periodic nature of the

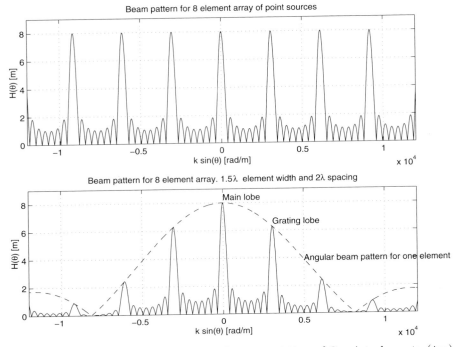

Fig. 11. Grating lobes for array transducer consisting of 8 point elements (*top*) and of 8 elements with a size of 1.5λ (*bottom*). The pitch (distance between the elements) is 2λ

array and correspond to the sampling of a continuous time signal. The grating lobes will be outside a ±90° imaging area if

$$\frac{\lambda}{d_x} = 1,$$
$$d_x = \lambda. \tag{21}$$

Often the beam is steered in a particular direction, and in order to ensure that grating lobes do not appear in the image, the spacing or pitch of the elements is selected to be $d_x = \lambda/2$. This also includes ample margin for the modern transducers that often have a very broad bandwidth.

An array beam can be steered in a particular direction by applying a time delay to the individual elements. The difference in arrival time between elements for a given direction θ_0 is

$$\tau = \frac{d_x \sin \theta_0}{c}. \tag{22}$$

Steering in the direction θ_0 can, therefore, be accomplished by using

$$\sin \theta_0 = \frac{c\tau}{d_x}, \tag{23}$$

where τ is the delay to apply to the signal on the element closest to the center of the array. A delay of 2τ is then applied on the second element and so forth. The beam pattern for the grating lobe is then replaced by

$$H_{\text{per}}(\theta) = \frac{\sin\left[(N+1)\frac{k}{2}d_x\left(\sin\theta - \frac{c\tau}{d_x}\right)\right]}{\sin\left[\frac{k}{2}d_x\left(\sin\theta - \frac{c\tau}{d_x}\right)\right]}. \tag{24}$$

Notice that the delay is independent of frequency, since it is essentially only determined by the speed of sound.

5 Spatial Impulse Responses

The description in the last section is strictly only valid for the far-field, continuous-wave case, whereas the fields employed in medical ultrasound are pulsed and in the near field. Thus, a more accurate and general solution is needed, and this is developed here. The approach is based on the concept of spatial impulse responses developed by *Tupholme* [7] and *Stepanishen* [8,9].

5.1 Fields in Linear Acoustic Systems

It is a well-known fact in electrical engineering that a linear electrical system is fully characterized by its impulse response. Applying a delta function to the input of the circuit and measuring its output characterizes the system. The output $y(t)$ to any kind of input signal $x(t)$ is then given by

$$y(t) = h(t) * x(t) = \int_{-\infty}^{+\infty} h(\theta)x(t-\theta)\mathrm{d}\theta, \tag{25}$$

where $h(t)$ is the impulse response of the linear system and $*$ denotes time convolution. The transfer function of the system is given by the Fourier transform of the impulse response and characterizes the system's amplification of a time-harmonic input signal.

The same approach can be taken to characterize a linear acoustic system. The basic setup is shown in Fig. 12. The acoustic radiator (transducer) on the left is mounted in an infinite, rigid baffle, and its position is denoted by $\mathbf{r}_2$. It radiates into a homogeneous medium with a constant speed of sound c and density ρ_0 throughout the medium. The point denoted by $\mathbf{r}_1$ is where the acoustic pressure from the transducer is measured by a small point hydrophone. A voltage excitation of the transducer with a delta function will give rise to a pressure field that is measured by the hydrophone. The measured response is the acoustic impulse response for this particular system with the given setup. Moving the transducer or the hydrophone to a new position will give a different response. Moving the hydrophone closer to the transducer

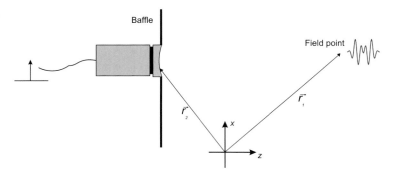

Fig. 12. A linear acoustic system

surface will often increase the signal,[2] and moving it away from the center axis of the transducer will often diminish it. Thus, the impulse response depends on the relative position of both the transmitter and receiver ($r_2 - r_1$); hence it is called a spatial impulse response.

The sound field for a fixed time instance can be obtained by employing Huygens' principle, in which every point on the radiating surface is the origin of an outgoing spherical wave. This is illustrated in Fig. 13. Each of the outgoing spherical waves are given by

$$p_s(r_1, t) = k_p \frac{\delta\left(t - \frac{|r_2 - r_1|}{c}\right)}{|r_2 - r_1|} = k_p \frac{\delta\left(t - \frac{|r|}{c}\right)}{|r|}, \tag{26}$$

where r_1 indicates the point in space, r_2 is the point on the transducer surface, k_p is a constant, and t is the time for the snapshot of the spatial distribution of the pressure. The spatial impulse response is then found by observing the pressure waves at a fixed position in space over time by having all the spherical waves pass the point of observation and summing them. Being on the acoustical axis of the transducer gives a short response whereas an off-axis point yields a longer impulse response as shown in Fig. 13.

5.2 Basic Theory

In this section the exact expression for the spatial impulse response will more formally be derived. The basic setup is shown in Fig. 14. The triangularly shaped aperture is placed in an infinite, rigid baffle on which the velocity normal to the plane is zero, except at the aperture. The field point is denoted

---

[2] This is not always the case. It depends on the focusing of the transducer. Moving closer to the transducer but away from its focus will decrease the signal.

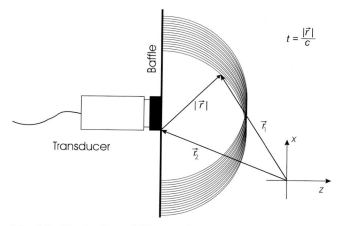

Fig. 13. Illustration of Huygens' principle for a fixed time instance. A spherical wave with a radius of $|\mathbf{r}| = ct$ is radiated from each point on the aperture

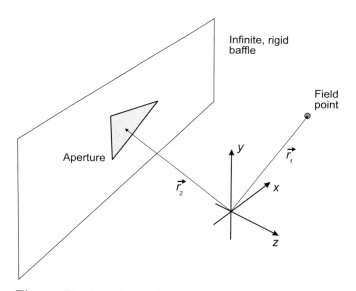

Fig. 14. Position of transducer, field point, and coordinate system

by $\mathbf{r}_1$ and the aperture by $\mathbf{r}_2$. The pressure field generated by the aperture is then found by the Rayleigh integral [10]

$$p(\mathbf{r}_1, t) = \frac{\rho_0}{2\pi} \int_S \frac{\partial v_n(\mathbf{r}_2, t - \frac{|\mathbf{r}_1 - \mathbf{r}_2|}{c})}{\partial t} \frac{1}{|\mathbf{r}_1 - \mathbf{r}_2|} dS, \qquad (27)$$

where v_n is the velocity normal to the transducer surface. The integral is a statement of Huygens' principle that the field is found by integrating the contributions from all the infinitesimally small area elements that make up the aperture. This integral formulation assumes linearity and propagation in a homogeneous medium without attenuation. Further, the radiating aperture is assumed to be flat, so no re-radiation from scattering and reflection takes place. Exchanging the integration and the partial derivative, the integral can be written as

$$p(\boldsymbol{r}_1, t) = \frac{\rho_0}{2\pi} \frac{\partial \int_S \frac{v_n\left(\boldsymbol{r}_2, t - \frac{|\boldsymbol{r}_1 - \boldsymbol{r}_2|}{c}\right)}{|\boldsymbol{r}_1 - \boldsymbol{r}_2|} dS}{\partial t} . \tag{28}$$

It is convenient to introduce the velocity potential ψ that satisfies the equations [11]

$$\boldsymbol{v}(\boldsymbol{r}, t) = -\nabla \psi(\boldsymbol{r}, t),$$
$$p(\boldsymbol{r}, t) = \rho_0 \frac{\partial \psi(\boldsymbol{r}, t)}{\partial t} . \tag{29}$$

Then only a scalar quantity need to be calculated, and all field quantities can be derived from it. The surface integral is then equal to the velocity potential:

$$\psi(\boldsymbol{r}_1, t) = \int_S \frac{v_n\left(\boldsymbol{r}_2, t - \frac{|\boldsymbol{r}_1 - \boldsymbol{r}_2|}{c}\right)}{2\pi |\boldsymbol{r}_1 - \boldsymbol{r}_2|} dS . \tag{30}$$

The excitation pulse can be separated from the transducer geometry by introducing a time convolution with a delta function as follows:

$$\psi(\boldsymbol{r}_1, t) = \int_S \int_T \frac{v_n(\boldsymbol{r}_2, t_2) \, \delta\left(t - t_2 - \frac{|\boldsymbol{r}_1 - \boldsymbol{r}_2|}{c}\right)}{2\pi |\boldsymbol{r}_1 - \boldsymbol{r}_2|} dt_2 dS , \tag{31}$$

where δ is the Dirac delta function.

Assume now that the surface velocity is uniform over the aperture, making it independent of $\boldsymbol{r}_2$, then

$$\psi(\boldsymbol{r}_1, t) = v_n(t) * \int_S \frac{\delta\left(t - \frac{|\boldsymbol{r}_1 - \boldsymbol{r}_2|}{c}\right)}{2\pi |\boldsymbol{r}_1 - \boldsymbol{r}_2|} dS , \tag{32}$$

where $*$ denotes convolution in time. The integral in this equation,

$$h(\boldsymbol{r}_1, t) = \int_S \frac{\delta\left(t - \frac{|\boldsymbol{r}_1 - \boldsymbol{r}_2|}{c}\right)}{2\pi |\boldsymbol{r}_1 - \boldsymbol{r}_2|} dS , \tag{33}$$

is called the spatial impulse response and characterizes the 3-dimensional extent of the field for a particular transducer geometry. Note that this is a function of the relative position between the aperture and the field.

Using the spatial impulse response the pressure is written as

$$p(\boldsymbol{r}_1, t) = \rho_0 \frac{\partial v_\mathrm{n}(t)}{\partial t} * h(\boldsymbol{r}_1, t),\qquad(34)$$

which equals the emitted pulsed pressure for any kind of surface vibration $v_\mathrm{n}(t)$. The continuous wave field can be found from the Fourier transform of (34). The received response for a collection of scatterers can also be found from the spatial impulse response [12,13]. Thus, the calculation of the spatial impulse response makes it possible to find all ultrasound fields of interest.

5.3 Geometric Considerations

The calculation of the spatial impulse response assumes linearity, and any complex-shaped transducer can therefore be divided into smaller apertures, and the response can be found by adding the responses from the subapertures. The integral is, as mentioned before, a statement of Huygens' principle of summing contributions from all areas of the aperture.

An alternative interpretation is found by using the acoustic reciprocity theorem [6]. This states, "If in an unchanging environment the locations of a small source and a small receiver are interchanged, the received signal will remain the same." Thus, the source and receiver can be interchanged. Emitting a spherical wave from the field point and finding the wave's intersection with the aperture also yields the spatial impulse response. The situation is depicted in Fig. 15, where an outgoing spherical wave is emitted from the

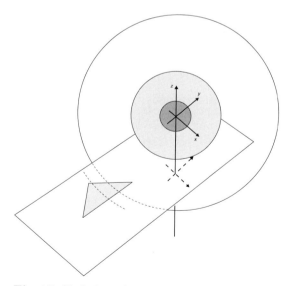

Fig. 15. Emission of a spherical wave from the field point and its intersection of the aperture

origin of the coordinate system. The dashed curves indicate the circles from the projected spherical wave.

The calculation of the impulse response is then facilitated by projecting the field point onto the plane of the aperture. The task is thereby reduced to a 2-dimensional problem, and the field point is given as a (x, y) coordinate set and a height z above the plane. The 3-dimensional spherical waves are then reduced to circles in the x–y plane with the origin at the position of the projected field point, as shown in Fig. 16.

The spatial impulse response is thus determined by the relative length of the part of the arc that intersects the aperture. It is the crossing of the projected spherical waves with the edges of the aperture that determines the spatial impulse responses. This fact is used for deriving equations for the spatial impulse responses in the next section.

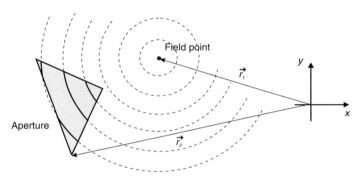

Fig. 16. Intersection of spherical waves from the field point by the aperture when the field point is projected onto the plane of the aperture

5.4 Calculation of Spatial Impulse Responses

The spatial impulse response is found from the Rayleigh integral derived earlier,

$$h(\boldsymbol{r}_1, t) = \int_S \frac{\delta\left(t - \frac{|\boldsymbol{r}_1 - \boldsymbol{r}_2|}{c}\right)}{2\pi \, |\boldsymbol{r}_1 - \boldsymbol{r}_2|} \mathrm{d}S. \tag{35}$$

The task is to project the field point onto the plane coinciding with the aperture and then to find the intersection of the projected spherical wave (the circle) with the active aperture, as shown in Fig. 16.

Rewriting the integral into polar coordinates gives

$$h(\boldsymbol{r}_1, t) = \int_{\Theta_1}^{\Theta_2} \int_{d_1}^{d_2} \frac{\delta\left(t - \frac{R}{c}\right)}{2\pi R} r \, \mathrm{d}r \, \mathrm{d}\Theta, \tag{36}$$

where r is the radius of the projected circle and R is the distance from the field point to the aperture given by $R^2 = r^2 + z_p^2$. Here z_p is the field point height above the x–y plane of the aperture. The projected distances d_1 and d_2 are determined by the aperture and are the distances closest to and furthest away from the aperture; Θ_1 and Θ_2 are the corresponding angles for a given time (see Fig. 17).

Introducing the substitution $2R\mathrm{d}R = 2r\mathrm{d}r$ gives

$$h(\boldsymbol{r}_1, t) = \frac{1}{2\pi} \int_{\Theta_1}^{\Theta_2} \int_{R_1}^{R_2} \delta\left(t - \frac{R}{c}\right) \mathrm{d}R \, \mathrm{d}\Theta. \tag{37}$$

The variables R_1 and R_2 denote the edges closest to and furthest away from the field point. Finally, using the substitution $t' = R/c$ gives

$$h(\boldsymbol{r}_1, t) = \frac{c}{2\pi} \int_{\Theta_1}^{\Theta_2} \int_{t_1}^{t_2} \delta(t - t') \mathrm{d}t' \, \mathrm{d}\Theta. \tag{38}$$

For a given time instance, the contribution along the arc is constant and the integral gives

$$h(\boldsymbol{r}_1, t) = \frac{\Theta_2 - \Theta_1}{2\pi} c, \tag{39}$$

when the circle arc is assumed to be intersected only once by the aperture. The angles Θ_1 and Θ_2 are determined by the intersection of the aperture and the projected spherical wave, and the spatial impulse response is, thus, solely determined by these intersections, when no apodization of the aperture is used. The response can therefore be evaluated by keeping track of the intersections as a function of time.

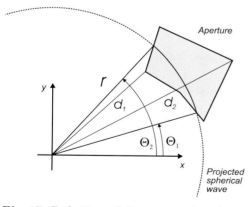

Fig. 17. Definition of distances and angles in the aperture plan for evaluating the Rayleigh integral

5.5 Examples of Spatial Impulse Responses

The first example shows the spatial impulse responses from a 3×5 mm rectangular element for different spatial positions 5 mm from the front face of the transducer. The responses are found from the center of the rectangle ($y = 0$) and out in steps of 2 mm in the x-direction to 6 mm away from the center of the rectangle. A schematic of the situation is shown in Fig. 18 for the on-axis response. The impulse response is zero before the first spherical wave reaches the aperture. Then the response stays constant at a value of c. The first edge of the aperture is met, and the response drops off. The decrease with time is increased when the next edge of the aperture is reached, and the response becomes zero when the projected spherical waves are all outside the area of the aperture.

A plot of the results for the different lateral field positions is shown in Fig. 19. It can be seen how the spatial impulse response changes as a function of relative position from the aperture.

The second example shows the response from a circular, flat transducer. Two different cases are shown in Fig. 20. The top graph shows the traditional spatial impulse response when no apodization is used, so that the aperture vibrates as a piston. The field is calculated 10 mm from the front face of the transducer, starting at the center axis of the aperture. Twenty-one responses for lateral distances of 0 to 20 mm off-axis are shown. The same calculation is repeated in the bottom graph, where a Gaussian apodization has been imposed on the aperture. The vibration amplitude is a factor of $1/\exp(4)$ less at the edges of the aperture than at the center. It is seen how the apodization reduces some of the sharp discontinuities in the spatial impulse response, which can reduce the sidelobes of the field.

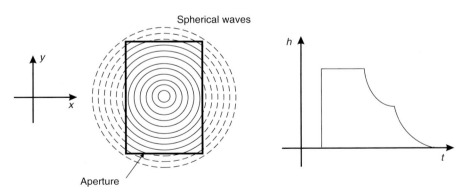

Fig. 18. Schematic of field from rectangular element (reproduced with permission from Jensen [1996], © Cambridge Univ. Press)

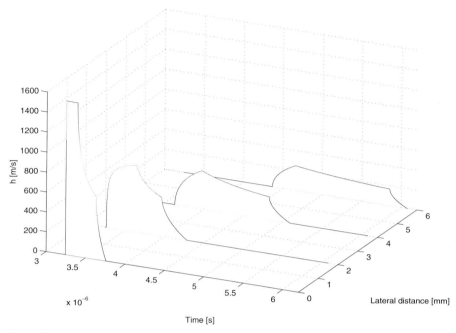

Fig. 19. Spatial impulse response from a rectangular aperture of 4×5 mm for different lateral positions

5.6 Pulse–Echo Fields

The scattered field and signal received by the transducer can also be described using the spatial impulse response. The received signal from the transducer is [13]

$$p_\mathrm{r}(\boldsymbol{r},t) = v_\mathrm{pe}(t) \underset{t}{\star} h_\mathrm{pe}(\boldsymbol{r},t) \underset{r}{\star} f_\mathrm{m}(\boldsymbol{r}), \tag{40}$$

where $\star_r$ denotes spatial convolution and $\star_t$ denotes temporal convolution. v_pe is the pulse–echo impulse, which includes the transducer excitation and the electromechanical impulse response during emission and reception of the pulse. f_m accounts for the inhomogeneities in the tissue due to density and speed of sound perturbations, which give rise to the scattered signal. h_pe is the pulse–echo spatial impulse response that relates the transducer geometry to the spatial extent of the scattered field. Explicitly written out, these terms are

$$\begin{aligned} v_\mathrm{pe}(t) &= \frac{\rho}{2c^2} E_\mathrm{m}(t) \underset{t}{\star} \frac{\partial^3 v(t)}{\partial t^3}, \\ f_\mathrm{m}(\boldsymbol{r}_1) &= \frac{\Delta \rho(\boldsymbol{r})}{\rho} - \frac{2\Delta c(\boldsymbol{r})}{c}, \\ h_\mathrm{pe}(\boldsymbol{r},t) &= h_t(\boldsymbol{r},t) * h_r(\boldsymbol{r},t). \end{aligned} \tag{41}$$

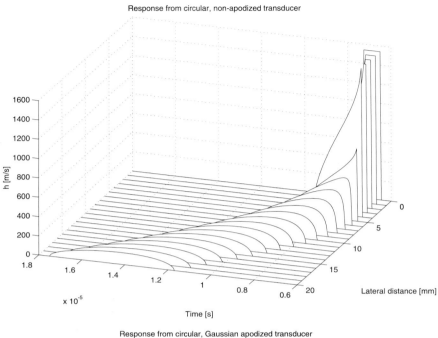

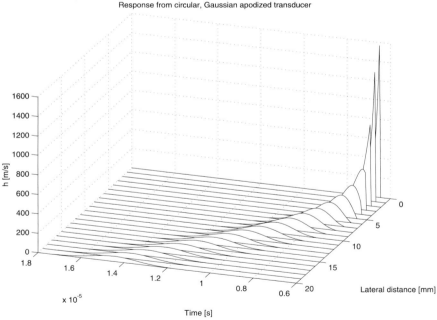

Fig. 20. Spatial impulse response from a circular aperture. *Top*: without apodization of the aperture; *bottom*: with a Gaussian apodization function. The radius of the aperture is 5 mm, and the field is calculated 10 mm from the transducer surface

Here $\Delta\rho$ are the perturbations in density and Δc the perturbations in speed of sound; $h_t(\boldsymbol{r},t)$ and $h_r(\boldsymbol{r},t)$ are the spatial impulse responses for the transmitting and receiving apertures, respectively; $E_{\rm m}(t)$ is the electromechanical impulse response of the transducer during reception. So the received response can be calculated by finding the spatial impulse response for the transmitting and receiving transducer and then convolving with the impulse response of the transducer. A single RF line in an image can be calculated by summing the response from a collection of scatterers in which the scattering strength is determined by the density and speed of sound perturbations in the tissue. Homogeneous tissue can thus be made from a collection of randomly placed scatterers that have a scattering strength with a Gaussian distribution, where the variance of the distribution is determined by the backscattering cross-section of the particular tissue.

6 Fields from Array Transducers

Most modern scanners use arrays to generate and receive the ultrasound fields. These fields are quite simple to calculate, when the spatial impulse response for a single element is known. This is the approach used in the Field II program [14], and this section will extend the spatial impulse response to multi-element transducers and will elaborate on some of the features derived for the fields in Sect. 4.

Since ultrasound propagation is assumed to be linear, the individual spatial impulse responses can simply be added. If $h_{\rm e}(\boldsymbol{r}_{\rm p},t)$ denotes the spatial impulse response for the element at position $\boldsymbol{r}_{\rm i}$ and the field point $\boldsymbol{r}_{\rm p}$, then the spatial impulse response for the array is

$$h_{\rm a}(\boldsymbol{r}_{\rm p},t) = \sum_{i=0}^{N-1} h_{\rm e}(\boldsymbol{r}_{\rm i},\boldsymbol{r}_{\rm p},t), \tag{42}$$

assuming all N elements are identical.

Let us assume that the elements are very small and the field point is far away from the array, so $h_{\rm e}$ is a Dirac function. Then

$$h_{\rm a}(\boldsymbol{r}_{\rm p},t) = \frac{k}{R_{\rm p}} \sum_{i=0}^{N-1} \delta\left(t - \frac{|\boldsymbol{r}_{\rm i} - \boldsymbol{r}_{\rm p}|}{c}\right), \tag{43}$$

where $R_{\rm p} = |\boldsymbol{r}_{\rm a} - \boldsymbol{r}_{\rm p}|$, k is a constant of proportionality, and $\boldsymbol{r}_{\rm a}$ is the position of the array. Thus, $h_{\rm a}$ is a train of Dirac pulses. If the spacing between the elements is d_x, then

$$h_{\rm a}(\boldsymbol{r}_{\rm p},t) = \frac{k}{R_{\rm p}} \sum_{i=0}^{N-1} \delta\left(t - \frac{|\boldsymbol{r}_{\rm a} + id_x\boldsymbol{r}_{\rm e} - \boldsymbol{r}_{\rm p}|}{c}\right), \tag{44}$$

where $\boldsymbol{r}_{\rm e}$ is a unit vector pointing in the direction along the elements. The geometry is shown in Fig. 21.

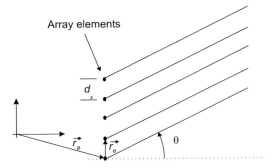

Fig. 21. Geometry of linear array

The difference in arrival time between elements far from the transducer is

$$\Delta t = \frac{d_x \sin \Theta}{c}. \qquad (45)$$

The spatial impulse response is, thus, a series of Dirac pulses separated by Δt:

$$h_a(\boldsymbol{r}_p, t) \approx \frac{k}{R_p} \sum_{i=0}^{N-1} \delta\left(t - \frac{R_p}{c} - i\Delta t\right). \qquad (46)$$

The time between the Dirac pulses and the shape of the excitation determines whether signals from individual elements add or cancel out. If the separation of arrival times corresponds to exactly one or more periods of a sine wave, then they are in phase and add constructively. Thus, peaks in the response are found for

$$n\frac{1}{f} = \frac{d_x \sin \Theta}{c}. \qquad (47)$$

The main lobe is found for $\Theta = 0$, and the next maximum in the response is found for

$$\Theta = \arcsin\left(\frac{c}{f d_x}\right) = \arcsin\left(\frac{\lambda}{d_x}\right). \qquad (48)$$

For a 3-MHz array with an element spacing of 1 mm, this amounts to $\Theta = 31°$, which will be within the image plane. The received response is, thus, affected by scatterers positioned 31° off the image axis, and they will appear in the lines acquired as grating lobes. The first grating lobe can be moved outside the image plane if the elements are separated by less than a wavelength. Usually, half a wavelength separation is desirable, as this gives some margin for a broadband pulse and beam steering.

The beam pattern as a function of angle for a particular frequency can be found by Fourier transforming h_a:

$$H_\mathrm{a}(f) = \frac{k}{R_\mathrm{p}} \sum_{i=0}^{N-1} \exp\left[-j2\pi f \left(\frac{R_\mathrm{p}}{c} + i\frac{d_x \sin\Theta}{c}\right)\right]$$

$$= \exp\left(-j2\pi f \frac{R_\mathrm{p}}{c}\right) \frac{k}{R_\mathrm{p}} \sum_{i=0}^{N-1} \exp\left(-j2\pi f \frac{d_x \sin\Theta}{c}\right)^i$$

$$= \frac{\sin\left(\pi f \frac{d_x \sin\Theta}{c} N\right)}{\sin\left(\pi f \frac{d_x \sin\Theta}{c}\right)} \exp\left(-j\pi f(N-1)\frac{d_x \sin\Theta}{c}\right) \frac{k}{R_\mathrm{p}}$$

$$\times \exp\left(-j2\pi f \frac{R_\mathrm{p}}{c}\right). \tag{49}$$

The terms $\exp\left(-j2\pi f \frac{R_\mathrm{p}}{c}\right)$ and $\exp\left[-j\pi f(N-1)\frac{d_x \sin\Theta}{c}\right]$ are constant phase shifts and do not influence the amplitude of the beam profile. Thus, the amplitude of the beam profile is

$$|H_\mathrm{a}(f)| = \left|\frac{k}{R_\mathrm{p}} \frac{\sin(N\pi \frac{d_x}{\lambda}\sin\Theta)}{\sin(\pi \frac{d_x}{\lambda}\sin\Theta)}\right|, \tag{50}$$

which is consistent with the previously derived result.

Several factors change the beam profile for real, pulsed arrays compared with the analysis given here. First, the elements are not points, but rather are rectangular elements with an off-axis spatial impulse response markedly different from a Dirac pulse. Therefore, the spatial impulse responses of the individual elements will overlap, and exact cancellation or addition will not take place. Second, the excitation pulse is broadband, which again influences the sidelobes. The influence of these factors is shown in a set of simulations in the next section.

7 Examples of Ultrasound Fields

The field examples are generated using computer phantoms and the Field II simulation program, which is based on the spatial impulse approach [14,15].

The first synthetic phantom consists of a number of point targets placed a distance of 5 mm apart starting at 15 mm from the transducer surface. A linear sweep image of the points is then made, and the resulting image is compressed to show a 40-dB dynamic range. This phantom is suited for showing the spatial variation of the point spread function for a particular transducer, focusing, and apodization scheme.

Twelve examples using this phantom are shown in Fig. 22. The top graphs show imaging without apodization and the bottom graphs show images, when a Hanning window is used for apodization when transmitting and receiving.

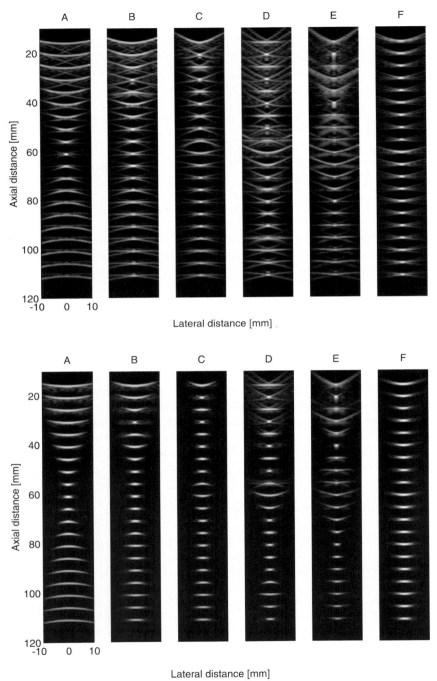

Fig. 22. Point-target phantom imaged for different setups of transmission and reception focusing and apodization. See text for an explanation of the setup

A 128-element transducer with a nominal frequency of 3 MHz was used. The element height was 5 mm, the width was a wavelength, and the kerf was 0.1 mm. The excitation of the transducer consisted of 2 periods of a 3-MHz sinusoid with a Hanning weighting, and the impulse response of both the emit and receive aperture was also a 2-cycle, Hanning-weighted pulse. In the graphs A – C, 64 of the transducer elements were used for imaging, and the scanning was done by translating the 64 active elements over the aperture and focusing in the proper points. In graphs D and E 128 elements were used, and the imaging was done solely by moving the focal points.

Graph A uses only a single focal point at 60 mm for both emission and reception. Graph B also uses reception focusing every 20 mm starting from 30 mm. Graph C further adds emission focusing at 10, 20, 40 and 80 mm. Graph D applies the same focal zones as graph C, but uses 128 elements in the active aperture.

The focusing scheme used for graphs E and F applies a new reception profile every 2 mm. For analog beamformers this is a small zone size. For digital beamformers it is a large zone size. A digital beamformer can be programmed for each sample, and thus "continuous" beamtracking can be obtained. In imaging systems focusing is used to obtain high-detail and high-contrast resolution preferably constant for all depths. This is not possible, so compromises must be made. As an example graph F shows the result for multiple transmission zones and reception zones, like graph E, but now a restriction is put on the active aperture. The size of the aperture is controlled to have a constant F-number (depth of focus in tissue divided by width of aperture), 4 for transmission and 2 for reception, by dynamic apodization. This gives a more homogeneous point spread function throughout the full depth, especially for the apodized version. However, it can be seen that the composite transmission can be improved in order to avoid the increased width of the point spread function at, e.g., 40 and 60 mm.

The next phantom consists of a collection of point targets, five cyst regions, and five highly scattering regions. This can be used for characterizing the contrast-lesion detection capabilities of an imaging system. The scatterers in the phantom are generated by finding their random position within a $60 \times 40 \times 15$ mm cube, and then ascribing a Gaussian-distributed amplitude to the scatterers. If the scatterer resides within a cyst region, the amplitude is set to zero. Within the high-scatter region, the amplitude is multiplied by 10. The point targets have a fixed amplitude of 100, whereas the standard deviation of the Gaussian distributions is 1. A linear scan of the phantom was performed with a 192-element transducer, using 64 active elements with a Hanning apodization in transmission and reception. The element height was 5 mm, the width was a wavelength and the kerf was 0.05 mm. The pulses where the same as those used for the point phantom discribed above. A single transmission focus was placed at 60 mm, and a reception focusing was done at 20 mm intervals from 30 mm from the transducer surface. The resulting image

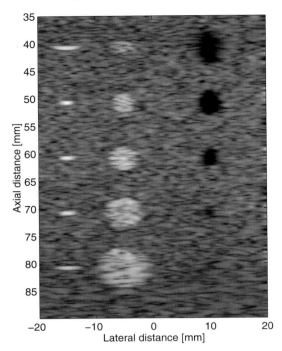

Fig. 23. Computer phantom with point targets, cyst regions, and strongly reflecting regions

for 100 000 scatterers is shown in Fig. 23. A homogeneous speckle pattern is seen along with all the features of the phantom.

8 Summary

Modern ultrasound scanners have attained a very high image quality through the use of digital beamforming. The delays on the individual transducer elements and their relative weight or apodization are changed continuously as a function of depth. This yields near-perfect focused images for all depths and has increased the contrast in the displayed image, thus benefiting the diagnostic value of ultrasonic imaging. The development of the focusing strategies is nearly exclusively based on linear acoustics, and the high success of the approach attests to the validity of linear acoustics. It is thus appropriate to characterize medical ultrasound systems using linear acoustics. This chapter has developed a complete linear description of all the fields encountered in medical ultrasound imaging. The various imaging methods were described, and then the concept of spatial impulse responses was developed. This could be used for describing both emitted and pulse–echo fields for both pulse-emission and continuous-wave systems using linear systems theory. Examples of the influence of digital beamforming and apodization were also given.

References

1. S. A. Goss, R. L. Johnston, F. Dunn, Comprehensive compilation of empirical ultrasonic properties of mammalian tissues, J. Acoust. Soc. Am. **64**, 423–457 (1978)
2. S. A. Goss, R. L. Johnston, F. Dunn, Compilation of empirical ultrasonic properties of mammalian tissues II. J. Acoust. Soc. Am. **68**, 93–108 (1980)
3. M. J. Haney, W. D. O'Brien. Temperature dependency of ultrasonic propagation properties in biological materials In *Tissue Characterization with Ultrasound*, ed. by J. F. Greenleaf (CRC, Boca Raton 1986)
4. J. A. Jensen, *Estimation of Blood Velocities Using Ultrasound: A Signal Processing Approach* (Cambridge University Press, New York 1996)
5. R. F. Wagner, S. W. Smith, J. M. Sandrick, H. Lopez, Statistics of speckle in ultrasound B-scans, IEEE Trans. Son. Ultrason. **30**, 156–163 (1983)
6. L. E. Kinsler, A. R. Frey, A. B. Coppens, J. V. Sanders, *Fundamentals of Acoustics* (Wiley, New York 1982)
7. G. E. Tupholme, Generation of acoustic pulses by baffled plane pistons. Mathematika **16**, 209–224 (1969)
8. P. R. Stepanishen, The time-dependent force and radiation impedance on a piston in a rigid infinite planar baffle, J. Acoust. Soc. Am. **49**, 841–849 (1971)
9. P. R. Stepanishen, Transient radiation from pistons in an infinite planar baffle, J. Acoust. Soc. Am. **49**, 1629–1638 (1971)
10. A. D. Pierce, *Acoustics, An Introduction to Physical Principles and Applications* (Acoust. Soc. Am., New York 1989)
11. P. M. Morse, K. U. Ingard, *Theoretical Acoustics* (McGraw-Hill, New York 1968)
12. P. R. Stepanishen, Pulsed transmit/receive response of ultrasonic piezoelectric transducers, J. Acoust. Soc. Am. **69**, 1815–1827 (1981)
13. J. A. Jensen, A model for the propagation and scattering of ultrasound in tissue, J. Acoust. Soc. Am. **89**, 182–191 (1991)
14. J. A. Jensen, Field: A program for simulating ultrasound systems, Med. Biol. Eng. Comp. **4** Suppl. 1, Part 1, 351–353 (1996)
15. J. A. Jensen, N. B. Svendsen, Calculation of pressure fields from arbitrarily shaped, apodized, and excited ultrasound transducers, IEEE Trans. Ultrason. Ferroelec. Freq. **39**, 262–267 (1992)

Nondestructive Acoustic Imaging Techniques

Volker Schmitz

Fraunhofer Institute for Nondestructive Testing, Universität Saarbrücken,
66123 Saarbrücken, Germany
schmitz@izfp.fhg.de

Abstract. Acoustic imaging techniques are used in the field of nondestructive testing of technical components to measure defects such as lack of side wall fusion or cracks in welded joints. Data acquisition is performed by a remote-controlled manipulator and a PC for the mass storage of the high-frequency time-of-flight data at each probe position. The quality of the acoustic images and the interpretation relies on the proper understanding of the transmitted wave fronts and the arrangement of the probes in pulse–echo mode or in pitch-and-catch arrangement. The use of the Synthetic Aperture Focusing Technique allows the depth-dependent resolution to be replaced by a depth-independent resolution and the signal-to-noise ratio to be improved. Examples with surface-connected cracks are shown to demonstrate the improved features. The localization accuracy could be improved by entering 2-dimensional or 3-dimensional reconstructed data into the environment of a 3-dimensional CAD drawing. The propagation of ultrasonic waves through austenitic welds is disturbed by the anisotropic and inhomogeneous structure of the material. The effect is more or less severe depending upon the longitudinal or shear wave modes. To optimize the performance of an inspection software tool, a 3-dimensional CAD-Ray program has been implemented, where the shape of the inhomogeneous part of a weld can be simulated together with the grain structure based on the elastic constants. Ray-tracing results are depicted for embedded and for surface-connected defects.

1 Introduction

Safety standards have the task of specifying those safety-related requirements which shall be met with regard to the state of science and technology to avoid damage arising from the construction or operation of technical components. Nondestructive examinations of the internal and external surfaces and of the body of the component contribute to safe operation through the entire lifetime of components such as pipes, nozzles, pressure vessels, pump casings, turbines and other technical structures.

Examinations methods with regard to cracks in the surface or near-surface regions include magnetic or liquid penetrant examination procedures or ultrasonic methods with special techniques such as acoustic surface waves, mode-conversion effects, angular mirror effect or the use of dual search units with longitudinal waves. With respect to volumetric examination, radiographic or ultrasonic techniques are used for thicker walls. Mostly the ultrasonic procedures use a single probe technique with a straight or angle beam technique, tandem or pitch-and-catch techniques, special probes such as focusing probes in the immersion technique or phased-array probes to steer or focus a beam electronically. The first approach in data acquisition is to set up a time window, rectify the signal and store the highest amplitude of all the received pulses together with the time-of-flight (TOF) value. The next step is to store all maxima together with all TOF values. Other methods follow the amplitude behavior of one pulse while the ultrasonic probe moves along the surface of the component. This gives the registration length, which in some way is correlated with the actual flaw length or flaw depth.

In most cases, the evaluation of amplitudes to assess the fitness for a purpose by the fracture mechanics calculation will give insufficient results. To provide the fracture mechanics calculation with more quantitative results about flaw sizes, acoustic imaging techniques or "reconstruction techniques", such as Ultrasonic Holography or Synthetic Aperture Focusing Technique (SAFT), are used. The paper discusses the work that has been performed during the last 30 years to improve the state of the art of acoustic imaging techniques for NDT (nondestructive testing) purposes.

2 The Nondestructive Testing Task

Ultrasonic methods have to evaluate position, size and shape of possible material defects and to characterize the defect as being planar or volumetric. The frequency range which shall be selected has to be optimized for the different materials. In the case of concrete material, used to construct bridges or tunnels, the frequency range is below 100 kHz; pipelines or pressure vessels made of ferritic or austenitic steel are tested in a frequency range between 1 MHz and 10 MHz, and ceramic material at frequencies up to 100 MHz. The arrangement of the sensors is bistatic or monostatic. In NDT monostatic means that one probe transmits and receives the ultrasonic pulses in a pulse–echo arrangement. A bistatic arrangement allows two probes to be placed on one side of the specimen, either in forward or in backward scattering positions or in transmission on the opposite side of the specimen.

A typical data acquisition and evaluation system consists of a manipulator which is adapted to the component: central mast manipulator, pipe scanner, x-y-scanner or an immersion tank. The ultrasonic transmitter/receiver is connected to the probe which is moved by the manipulator; the ultrasonic data are stored as a function of probe position after analog-to-digital conversion on a mass memory board. A typical PC-based data acquisition system – where the ultrasonic cards are a part of a personal computer – is shown in Fig. 1.

In most ultrasonic testing applications, one relies upon the amplitude information. Basically two methods are used, the DGS evaluation according to distance–gain–size curves and the contour estimation as B or C scans (side view or top view). The DGS method is recommended by many codes and standards, especially for manual inspection, and is an important tool when

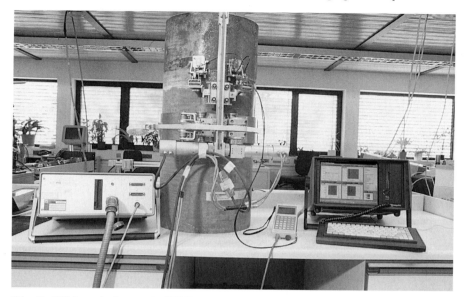

Fig. 1. PC-based ultrasonic NDT equipment with remote-controlled pipe scanner

the defects are smaller than the ultrasonic wavelength and a tomographic method cannot be used. In the case of the contour estimation for horizontally lying defects, a perpendicular insonification with longitudinal waves is used; for defects oriented almost perpendicular to the surface, an oblique insonification with shear waves under 45°, 60° or 70° is used. For both methods, the size of the image depends not only on ultrasonic beam parameters such as the insonfication and divergency angle, but on defect parameters such as shape, orientation and roughness too.

Therefore we will discuss only those methods which use, in addition to the amplitude, the TOF information and which have full access to the complete nonrectified high-frequency signal at each probe position. The quality of acoustic images relies on the quality of the data which are obtained by forward or back scattering or by geometrical reflection. Therefore, it is of interest to understand the interaction between acoustic pulses and the defect. A representative example is a surface connected crack (Fig. 2). The numerical method to model the elastodynamic (including mode-conversion effects) interaction is the elastodynamic finite integration technique (EFIT), which was developed and optimized by *Langenberg* [1].

The so-called creeping wave is a subsurface wave (SSLW), with the propagation speed of the longitudinal wave, which is forward scattered at the lower crack tip, forms a cylindrical wavefront and is back-reflected towards the probe. Additional head waves and shear waves will contribute to a complex high-frequency A scan signal and, after applying a reconstruction algorithm,

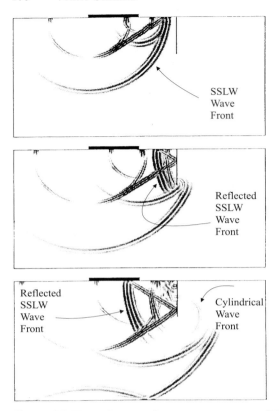

Fig. 2. EFIT prediction of a "creeping wave" (SSLW) interaction with a surface-connected crack

will cause artefacts (defects at positions where no real defects exist) which have to be explained as due to the EFIT simulation.

3 The Inverse Problem

In the following we will discuss the SAFT Formalism, which uses the high-frequency data collected at a 1- or 2-dimensional scanned area across the surface of the component being tested. The original ultrasonic applications of SAFT followed quite directly from the radar literature by *Prine* [2], and *Burckhardt* et al. [3]. The first application of SAFT was to improve the lateral resolution of airborne radar mapping systems. For airborne data-collection systems, the antenna size is necessarily limited to a fraction of the size of the airplane. Therefore, the airplane is used to scan the antenna over a large area. This allows a large effective aperture to be synthesized and consequently the resolution to be improved.

The first digital implementation of SAFT in ultrasound was carried out at the University of Michigan [4] and was continued by SwRI [5] and Battelle Northwest [6]. Broadband pulses, direct data recording and coherent summation are used to demonstrate the feasibility of digital SAFT and – with the improved lateral resolution with respect to other NDT methods – to obtain a better evaluation of size, location and orientation of a defect in components such as pressure vessels and oil- or water-carrying pipelines. Line-SAFT has been implemented to a PC by *Schmitz* [7] to obtain 2-dimensional side view reconstructions. Since 1990 research institutions have used the SAFT algorithm more and more in different versions, e.g., 2- and 3-dimensional, in the time and frequency domains, in pulse–echo, pitch-and-catch and tandem arrangements.

SAFT testing devices have been continuously improved for sizing defects during the inservice inspection of nuclear power plants. Basic studies of the influence of different scan or probe parameters on the quality of reconstruction have been performed by IZFP [8], using the underlying theory from the University of Kassel [9]. In the following, we give a heuristic "derivation" of the ultrasonic NDT inversion scheme SAFT as it was originally proposed; placement of (linearized) inverse scattering into a mathematical theory can be found in *Langenberg* [10] and *Mayer* [11].

Consider a point scatterer located at the coordinate origin, which is represented by the equivalent source q_c,

$$q_c(\boldsymbol{R}',\omega) \approx \delta(\boldsymbol{R}')p_i(\boldsymbol{R}',\omega) \tag{1}$$

with the incident field coming from a point-source transducer at point R_0 on a measurement plane S_M, i.e.,

$$p_i(\boldsymbol{R}',\omega) = F(\omega)\frac{e^{jk|\boldsymbol{R}_0-\boldsymbol{R}'|}}{4\pi|\boldsymbol{R}_0-\boldsymbol{R}'|} \tag{2}$$

where $F(\omega)$ stands for the frequency spectrum of the transducer and k is the wavenumber of the embedding material. Therefore we have

$$q_c(\boldsymbol{R}',\omega) \approx \delta(\boldsymbol{R}')\frac{e^{jk\boldsymbol{R}_0}}{4\pi\boldsymbol{R}_0}, \tag{3}$$

which results in a scattered field at some observation point $\boldsymbol{R}$:

$$p_s(\boldsymbol{R},\omega) \approx F(\omega)\frac{e^{jkR_0}e^{jkR}}{4\pi R_0 4\pi R}. \tag{4}$$

Very often, ultrasonic imaging is performed in pulse–echo mode: the transmitting and receiving transducers coincide. Hence, for $\boldsymbol{R} = \boldsymbol{R}_0$ we obtain

$$p_s(\boldsymbol{R}_0,\omega) \approx F(\omega)\frac{e^{2jkR_0}}{R_0^2}, \tag{5}$$

and, consequently, via Fourier inversion,

$$p_{\text{s}}(\boldsymbol{R}_0, t) \approx \frac{F\left(t - \frac{2R_0}{c}\right)}{R_0^2}. \tag{6}$$

The transmitted pulse shape appears as a replica in the "measured" signal, which is, in turn, shifted by twice the travel time from the transmitter/receiver to the scatterer.

All point scatterers located on a sphere centered at $\boldsymbol{R}_0$ with radius R_0 yield the same "measured" signal (the so-called A scan). To reconstruct the location of the point scatterer, SAFT proposes the following: "invert" the A scan onto every point on the sphere around $\boldsymbol{R}_0$ with radius R_0 and proceed similarly for all points $\boldsymbol{R}_0$ on the measurement surface. This superposition results in a "focussing" at the intersection point of all spheres where the scatterer is supposed to reside. This version of SAFT is called the "A-scan-driven approach". Obviously, 3-dimensional data – two scan coordinates and one time coordinate – are required to produce 3-dimensional focussing. Another view of SAFT – the pixel-driven approach – uses arguments as follows:

Let S_M be, for example, a planar measurement surface with distance d from the point scatterer, i.e., $R_0 = \sqrt{x_0^2 + y_0^2 + d^2}$, where x_0, y_0 and d are the cartesian coordinates of $\boldsymbol{R}_0$. Assume further that: $F(t) = \delta(t)$, then the non zero x_0, y_0, t data are given by the hyperboloid $c^2 t^2 / 4 - x_0^2 - y_0^2 = d^2$, which is called the diffraction surface. An inversion equation is then proposed as follows:

$$o(\boldsymbol{R}) = \iint_{S_\text{M}} p_{\text{s}}\left(x_0, y_0, d, t = \frac{2\,|\boldsymbol{R} - \boldsymbol{R}_0|}{c}\right) \mathrm{d}x_0 \mathrm{d}y_0. \tag{7}$$

For every point $\boldsymbol{R}$ in the "reconstruction space", the integration extends over hyperboloids belonging to the point under consideration, $\boldsymbol{R}$; but only one hyperboloid, the data hyperboloid, results in a significant amplitude contribution and hence it "reconstructs" the point scatter $o(\boldsymbol{R}) \approx \delta(\boldsymbol{R})$. Generally, the inversion scheme SAFT applies the above ideas to real defect geometries and nonplanar measurement surfaces; this algorithm has been very successful particularly in NDT.

Major points which contribute to the success of the SAFT imaging scheme are the image resolution, the signal-to-noise ratio (SNR) improvement, and the localization accuracy.

4 Special Features of SAFT

As *Cutrona* [12] has demonstrated, the SAFT resolution is independent of the ultrasonic frequency and is instead related to the physical size of the probe which will be moved during the scanning operation.

4.1 Lateral Resolution

In the far-field approximation, the radiation pattern of an array is the radiation pattern of a single element multiplied by an array factor. The array factor has significantly narrower beamwidths than the radiation patterns of the elements of the array. The half-power beam width β in radians of the array factor of such an antenna can be obtained from the length of a physical antenna L_{eff} and the wavelength λ. The factor of 2 arises when the same element radiates and receives signals:

$$\beta_{\text{eff}} = 0.5\lambda/L_{\text{eff}}. \tag{8}$$

Let D represent the horizontal aperture of the physical transducer carried by a manipulator. The width of the horizontal beam at range R gives the maximum value for the length of synthetic aperture that can be used at that range. Since the beamwidth of such an antenna is given by the ratio of the wavelength λ to its horizontal aperture D, the length of this synthetic antenna aperture is given by

$$L_{\text{eff}} = R\lambda/D. \tag{9}$$

The linear resolution in azimuth δ is the product of the effective beamwidth with (8) and the range R:

$$\delta = \beta_{\text{eff}} R. \tag{10}$$

If (8) and (9) are combined with (10), one obtains

$$\delta = \frac{\lambda}{2L_{\text{eff}}} R = \frac{\lambda R}{2} \frac{D}{R\lambda} = \frac{D}{2}. \tag{11}$$

This indicates that the resolution is independent of the wavelength and becomes finer if the physical antenna is a small probe. In B-scan imaging, the resolution would become worse with increasing depth z according to

$$\delta = 1.33 z \lambda/D. \tag{12}$$

An ideal representation of a small contact technique probe would be a focal probe in an immersion technique where the distance of the probe to the surface corresponds to the focal length of the probe.

Figure 3 gives an overview of the resolution capabilities. In conventional ultrasonic testing, known as a B scan, we obtain a depth-dependent resolution; the depth-independent SAFT resolution for components with plane surfaces differs from that for cylindrical components. For cylindrical components, especially if the defects lie close to the rotation axis, the resolution is better than the value derived by *Cutrona* [12].

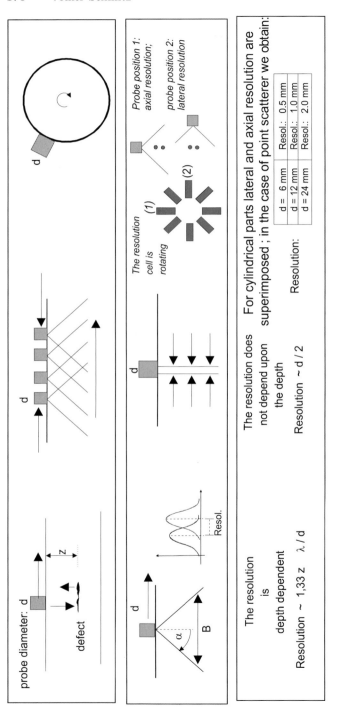

Fig. 3. Resolution capabilities of B scan and SAFT in plane and cylindrical components

4.2 Signal-to-Noise Ratio Improvement by SAFT

It is well known that the signal processing technique of the SAFT reconstruction process can be described as a TOF-corrected superposition of high-frequency signals with an SNR improvement proportional to the square-root of independent probe positions. Under favorable circumstances, an SNR improvement of 20 dB can be achieved, and signals which are buried in noise can be recovered and used for the detection of otherwise barely visible defects. This is demonstrated in Fig. 4 for an austenitic pipe of 50-mm thickness with a circumferential weld. A vertically oriented crack has been placed beneath the center line of the weld between 20 mm and 30 mm depth. With an ultrasonic frequency of 2.25 MHz and an insonification angle of 60°, the upper and lower tip of the crack could be imaged due to the back-scattering behavior.

Figure 4 shows the SAFT side view starting from the upper surface down to the back wall. For reasons of clarity, the seam weld has been sketched. The two arrows indicate the position of the two crack tips. Using the color-coded look-up table, the SNRs of the upper and lower tip were measured to be 16.2 dB and 21.4 dB, respectively. If we compare these values with the SNR in the high-frequency A scan, the SAFT procedure leads to an SNR improvement of 7.4 dB and 11.2 dB, respectively.

4.3 Localization Accuracy

Besides lateral resolution and SNR improvement, the acoustic imaging relies on the reliability of the exact positioning of the defect relative to the surfaces of the component or the weld. In remote-controlled ultrasonic inspections with contact-technique probes, the nominal insonification angle of the angle beam probe is used. To avoid incorrect positioning of a defect in a B-scan image, the real ultrasonic insonification angle is checked using a suitable test block. But real-life industrial components do not always have smooth surfaces, and the probe may be locally changed in its orientation due to a locally varying surface profile. The decision as to whether a surface-connected or embedded crack has been detected depends upon the achieved localization accuracy. SAFT does not use the nominal insonification angle and is – due to the TOF-corrected positioning of the amplitudes in all pixels which have a constant TOF distance to the probe – only dependent on the exact knowledge of the sound-field propagation velocity.

Let us take a usual situation where we have to find the depth extension of a vertical surface-connected crack and where the probe can only be placed at one side of the component. The example in Fig. 5 shows a 30-mm crack in a 58-mm-thick steel plate. A 4-MHz contact technique probe has been used to insonify with a nominal angle of 70°. The image spots in the SAFT image have constant size in the direction perpendicular to 70°. The image spots of the B scan show an increasing width proportional to the distance from the probe, which reflects the spreading of the sound field. Comparing both

176 Volker Schmitz

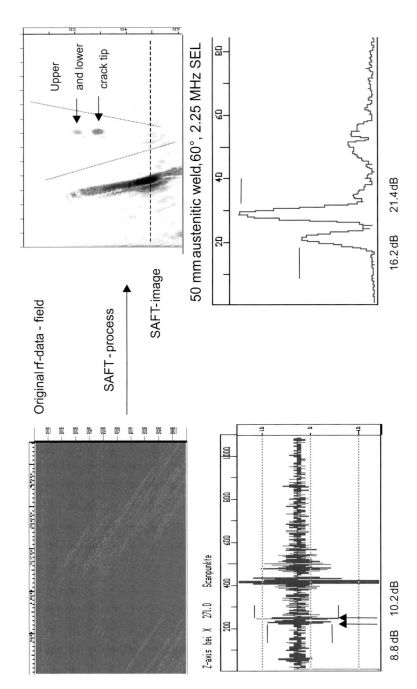

Fig. 4. Signal-to-noise ratio improvement with SAFT

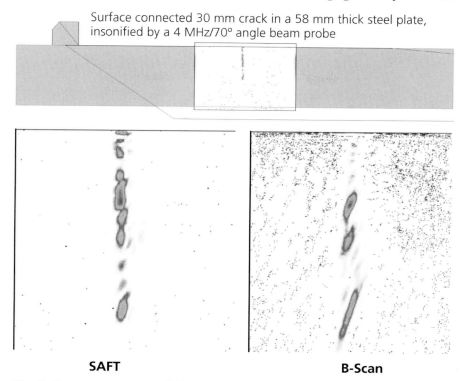

Fig. 5. B-scan image versus SAFT reconstruction of a surface-connected crack

images, confidence in the evaluation of the crack as a surface-connected crack has been improved by the SAFT algorithm.

4.4 Pulse–Echo/Pitch-and-Catch Reconstruction

In most ultrasonic inspection techniques, the pulse–echo procedure is applied; the ultrasonic pulse is transmitted and received by the same probe. In geophysics, the common-offset arrangement is used: the receiver is separated from the transmitter and moved simultaneously. The Time-of-Flight Diffraction (TOFD) Technique uses two angled-beam probes with longitudinal waves. The transmitting probe insonifies the region of interest, which may be the welded zone; the forward-scattered waves are received by the receiving probe which points towards that zone (Fig. 6). The characteristic feature of this arrangement is that forward-scattered signals are used for defect detection and sizing. In contrast, the pulse–echo arrangement uses the reflected signal for detection and additional signal processing for sizing. We will now discuss the performance capability of the two methods shown in Fig. 7.

In searching mode, the two TOFD probes are moved simultaneously and parallel to a weld. In analyzing mode, they are moved towards the weld as

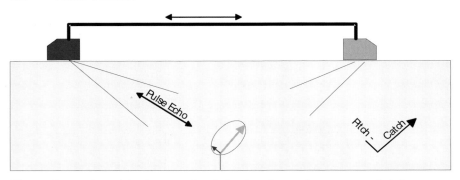

Fig. 6. Forward- and back-scattering behavior of a surface-connected defect

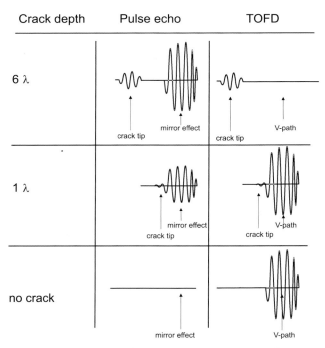

Fig. 7. Detectability of surface-connected cracks for pulse–echo and TOFD probe arrangements as a function of crack depth extension

in an pulse–echo arrangement. In both cases – pulse–echo and TOFD – the high-frequency ultrasonic data are mapped versus the scan path. This allows a SAFT reconstruction algorithm to be written for the TOFD probe arrangement and the result to be compared with that of the pulse–echo technique (see Fig. 8).

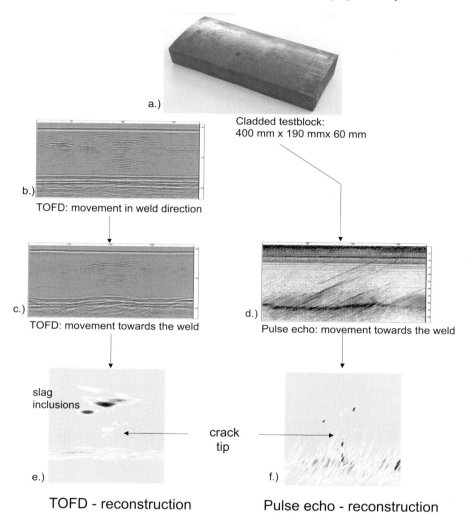

Fig. 8a–f SAFT reconstructions for pitch-and-catch and pulse–echo probe arrangements

In pulse–echo, a large surface-connected crack – 6λ – will cause a large signal due to the mirror effect; a smaller tip echo signal will be separated due to the different TOF. In TOFD, the crack tip will cause a signal, but the surface-connected defect will not allow a signal to be received on the V path. With decreasing flaw size, the crack-tip signal moves towards the mirror-effect signal. Both amplitudes become smaller. In the TOFD procedure, the crack-tip signal moves into the V path signal, which becomes larger, with the effect that it may not be distinguishable in the TOFD-data presentation.

If there are no surface-connected cracks, pulse–echo shows no signal due to the geometrical forward propagation of the ultrasonic pulse; TOFD will show a large signal due to the forward-propagated signal on the V path. Overall, TOFD will clearly indicate large extended surface-connected cracks. The quality of the crack-tip images obtained by pulse–echo or TOFD data will depend upon how much energy of the crack tip is forward or back scattered.

We can compare the TOFD original data in Fig. 8c with those of the pulse–echo technique in Fig. 8d. The final analysis with the TOFD procedure is performed in Fig. 8c, whereas the pulse–echo data have to be reconstructed by SAFT to obtain the image in Fig. 8f. In the case of a complex defect configuration, the TOF curves in Fig. 8c are difficult to assign to defect parameters such as crack-tip or slag inclusions. This will be facilitated if we add an additional step – a SAFT reconstruction for the pitch-and-catch arrangement – to the TOFD data. A comparison between the two SAFT reconstructions shows strong signals for the slag inclusions in Fig. 8e but a weak crack-tip image compared to Fig. 8f. Despite the fact that for large surface-connected cracks TOFD and pulse–echo should detect the crack tip very well, the energy which is forward scattered is small compared to that back scattered.

4.5 Acoustic Imaging in a 3-dimensional CAD Environment

One of the classical tasks of ultrasonic NDT, the detection, sizing and characterization of material damages such as cracks in welds, lack of fusion, foreign material or impact damages, can be improved by scanning the surface of the component under inspection in two dimensions and applying to the data set a 3-dimensional (3D) imaging algorithm. This imaging algorithm has been implemented in a 3D version in the frequency domain on a Apollo work station at the University of Kassel and in the time domain on a Pentium personal computer at the Fraunhofer Institute, Saarbrücken. In normal industrial applications, it has been shown that the pseudo-3D SAFT reconstructions are sufficient to evaluate the depth extension of defects. For more highly sophisticated defect sizing and characterization purposes, in an laboratory exercise on a simple test block with three rows of flat-bottom holes (Fig. 9), we will show that, in isotropic material, 3D SAFT can be extremely precise concerning localization accuracy and sizing. The localization accuracy is checked by entering the reconstructed data into the 3D-CAD drawing of the component.

The holes are drilled from the back side up to 50 mm below the surface. They have diameters of 1 mm, 3 mm and 10 mm and are separated by 1 mm, 2 mm, ..., 10 mm. To relate these distances to the lateral resolution capability, a 4-MHz longitudinal wave probe has been selected for an inspection perpendicular to the surface. Horizontal distances, depths and sizes of the imaged holes match perfectly with the true situation in the test block.

Using the capability to enter reconstructed data into a CAD drawing of the component allows, with respect to quantitative nondestructive imaging,

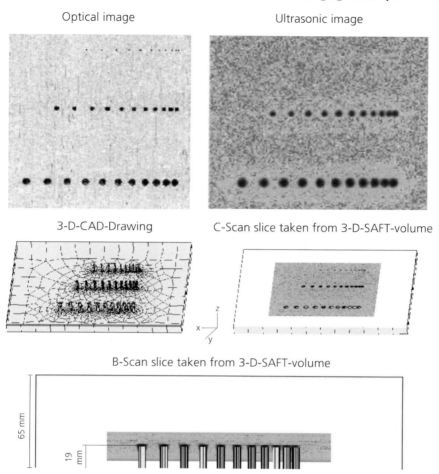

Fig. 9. Implementation of a 3D-reconstructed SAFT data set into a 3D-CAD drawing

the main features of a defect to be extracted by presenting top and side views of the image. In the experiment, we acquire data by the pulse–echo technique on a piece of a pipe with a circumferential weld. The inner radius of the pipe is 450 mm and the thickness 60 mm. As depicted in Fig. 10, the sensor is moved in pulse–echo mode towards the weld, the high-frequency data are stored and the data from each scan line are processed into a SAFT image. After an incremental side step in direction, the new scan line data are stored and reconstructed and all acoustic images stored in a 3D data file.

The evaluation of the acoustic images in the x-, y- and z-directions allows it to be stated that the defect is half-elliptical, centered in the middle of the weld and vertically oriented. This defect is sketched in the lower part of Fig. 10.

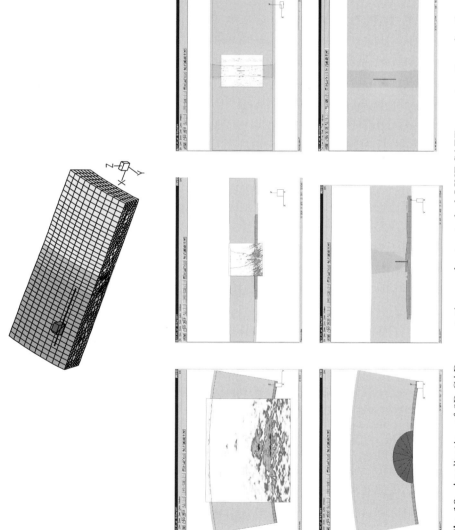

Fig. 10. Application of 3D-CAD presentation modes to stacked LINE-SAFT reconstructions for the ultrasonic B-, D-, and C-scan presentation directions and the schematic evaluation of the half-penny-shaped crack

5 Imaging in Transversally Isotropic Material – Ray Tracing

The propagation of ultrasonic waves through austenitic welds is disturbed by the anisotropic and inhomogeneous structure of the material. The effect is more or less severe depending upon the kind of wave which is used for inspection. To optimize the performance of an inspection, a software tool is needed to support the selection of suitable probes, insonification direction or scan paths. Therefore, a ray-tracing algorithm, "3-dimensional-CAD-Ray", has been implemented to follow in three dimensions longitudinal and shear-wave propagation from the base material through the cladding and the reflection at the border of a component or at implemented defects.

Resulting from the complexity of real-world, stainless-steel welds and the ability to solve basic equations, a model is used which describes a unidirectional weld structure [13]. Relative to the Cartesian coordinate system, the grain orientation and the group velocity can be chosen arbitrarily (Fig. 11). The direction of the phase velocity differs from the direction of the group velocity. The material boundaries are described by a set of the planar triangular or rectangular surface elements that are generated by a parametric CAD-software tool for a simple geometry such as a plate, a pipe or a turbine shaft or by AUTOCAD or AUTO-SKETCH for more complex shapes. The isotropic regions are characterized by two sound velocities for shear and longitudinal waves. The anisotropic properties of the material are defined by five elastic constants and the material density.

Inhomogeneous material properties are described by empirical functions of the grain orientation vector:

$$G_x = x^\eta, \quad G_y = 0, \quad G_z = D + E_z, \tag{13}$$

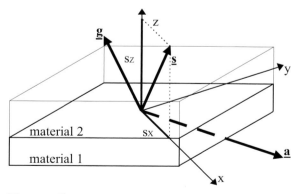

Fig. 11. Grain orientation a, direction of phase velocity s, and direction of group velocity (energy-transport direction) g relative to the Cartesian coordinate system

where D, E and η are constants. The procedure to calculate the local ray propagation direction has to be adhered to as follows:

1. In the current ray position, find the group-velocity direction using the phase-velocity direction and the grain orientation.
2. Step one time interval forward in the group-velocity direction and check that no region change has occurred (a region may be a weld or the basic material).
3. If the region changes, reduce the time interval to reach the interface between regions and continue with step 4.
4. In the new position, define an artificial interface to refract the ray. This interface is selected to be perpendicular to the ray propagation direction (group velocity).
5. Look for the new grain orientation and calculate the new phase-velocity direction.
6. Return to step 1 and repeat until the fixed ray's length or the fixed number of time intervals is reached.

5.1 Sound Propagation Through a V Weld with Defects

The benefit of ray tracing depends strongly upon the ability to visualize not only the center ray of the transmitted beam, but also the rays inside the cone of the transmitted sound field simultaneously. For this purpose, it is required to have appropriate tools [14,15] for the design of the component, the design of the weld configuration, the selection of the grain orientation pattern and the scan path with the different search unit. In Fig. 12, the information about the shape of the weld and the grain structure was obtained from a micrograph.

The grain orientation distribution was approached by a formula where in the middle of the weld the grains are oriented vertically and turn slowly into a direction perpendicular to the seam weld.

The simulation in Fig. 13 shows – according to experience from the field – that the vertically polarized shear wave is strongly influenced by the grain structure. In an homogeneous material, the rays would propagate towards the defect, but due to the grain orientation distribution and the elastic properties of the weld material, the rays are turned upwards. It is of interest that in the case of longitudinal waves the rays are bent downwards but less dramatically compared to the case of the SV wave. The horizontally polarized shear waves (SH) behave similar to the longitudinal waves. If the orientation of the lack of side wall fusion, elastic constants or the grain field would be changed systematically, a parametric study could be performed to analyze these effects with respect to detectability.

A similar simulation was performed for a surface-connected crack placed into the right seam at the root of the weld (Fig. 14). The longitudinal wave and the horizontally polarized shear wave detect the crack due to the mirror effect. The efficiency of the mirror effect depends strongly on the shape of

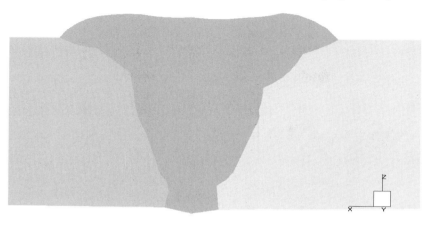

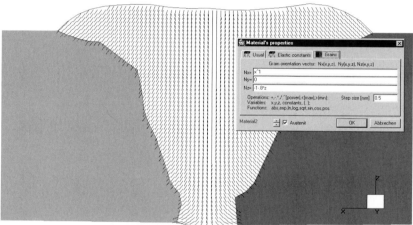

Fig. 12. V weld with defects. Pipe diameter: 425 mm; wall thickness: 90 mm. Shape and grain structure taken from a micrograph

the root of the weld and could only be used in practical application if the root itself has been grinded.

5.2 A 10-Layer Approximated Austenitic Weld

The validity of a model has to be verified by experiment. A test specimen with unidirectional weld structure, where the grain orientation is parallel to the surface, is used [15]. Comparison was made with a transmit/receive search unit arrangement. Longitudinal waves are transmitted perpendicular to the surface into the material, and the receiver scans the opposite surface

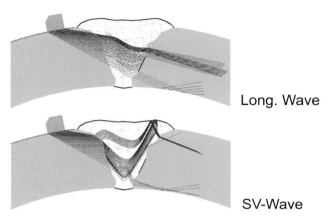

Fig. 13. Possible interactions with a lack of side wall fusion of 20 mm depth extension

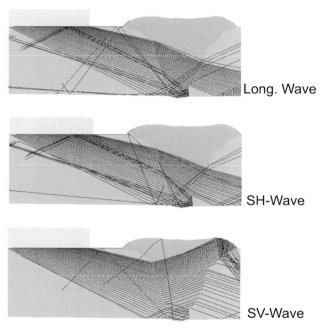

Fig. 14. Response of different ultrasonic waves with a surface-connected crack

to receive possible distortions from the sound field. The search unit location relative to the weld center line is varied. The rays which are reflected at the back wall form a circular shape nearly the same size as that obtained in the experiment. After partial penetration into the transversely isotropic material,

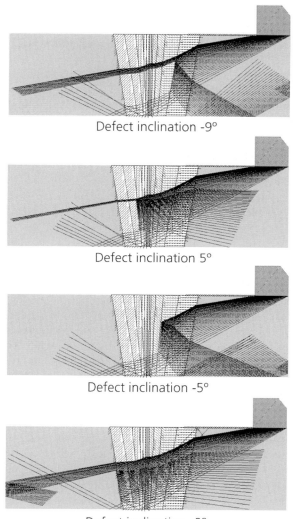

Fig. 15. Simulation of different defect positions between welded slices in a 10-layer austenitic stainless-steel structure; 70° insonification with longitudinal waves at a fixed probe position

the shape becomes more and more stretched into an elliptical shape. A more realistic austenitic stainless-steel weld structure has been chosen in Fig. 15.

The weld has been put together piecewise using 10 layers. Each layer has a uniform grain structure of different inclination. The center line of the weld contains a thin layer with a vertically oriented grain structure. Towards both sides, the grain orientation gets more and more horizontal. A lack-of-fusion defect of 20-mm depth extension has been placed between different layers to check how the grain orientation influences detectability. The longitudinal-wave transmitting probe is held at a fixed position; the insonfication angle was 70°.

Acoustic imaging in transversally isotropic material needs reconstruction software which takes into account the different behaviors of the sound propagation for vertically or horizontally polarized shear waves or for longitudinal waves. In the above-mentioned software package, the reconstruction for the SAFT algorithm in an homogeneous material should be replaced by a new algorithm which takes into account the five elastic constants and the grain orientation distribution. Today such an algorithm has not yet been developed for applications in the field of nondestructive testing.

6 Summary

Important features of the Synthetic Aperture Focusing Technique in its 2-dimensional or 3-dimensional versions have been described and some results presented. The results are as good as the assumptions involved are valid. Some of those assumptions are as follows: an homogeneous isotropic frequency-independent material, a point-source transducer, infinitely broadband ultrasonic signals and an infinitely large aperture. The scatterer should be weak scattering. Many of these assumptions are not fulfilled in nondestructive-testing acoustic imaging. Therefore, the present algorithm has to be re-evaluated. One of the important steps would be to include mode-conversion effects using a vectorial imaging technique.

An increased volume of research has been performed over the past 20 years to develop and implement SAFT-UT (SAFT for Ultrasonic Testing). Piezo-electric, electromagnetic and air- or water-coupled probes can be used. In the USA SAFT technology has been formally integrated into ASME Code Section V, Article 4, Appendix E, which addresses computerized UT imaging systems. Pulse–echo and Pitch-and-Catch configurations and Tandem technique data can be processed by SAFT. The frequency range covers application 75 kHz for concrete up to 100 MHz for ceramic material. In an example, SAFT improved the SNR for crack-tip detection by up to 16 dB. The nominal insonification angle does not need to be known to position the contour of the component, weld seams or defects correctly.

The propagation of ultrasound through austenitic stainless-steel weld material or through austenitic cladding is affected by the anisotropic and inhomogeneous structure of these materials. A software package has been developed to understand the phenomena of wave propagation, to extract parameters for an optimized inspection and to support the interpretation of ultrasonic inspection data. The accuracy of the model has been checked on a simple model with horizontally oriented grains. More work has to be performed for more realistic weld structures, and the rays have to be weighted by their amplitudes.

References

1. R. Marklein, *Numerische Verfahren zur Modellierung von akustischen, elektromagnetischen, elastischen und piezoelektrischen Wellenausbreitungsproblemen im Zeitbereich basierend auf der Finiten Integrationstechnik* (Shaker, Aachen 1997)
2. D. W. Prine, Synthetic Aperture Ultrasonic Imaging, In *Proc. Engineering Applications of Holography Symp.* (Society of Photo-optical Instrumentation Engineers, Los Angeles, 1972) p. 287
3. C. B. Burckhardt, P. A. Grandchamp, H. Hoffmann, Methods for Increasing the Lateral Resolution of B-Scan, In *Acoustical Imaging*, Vol. 5, ed. by P. S. Green (Plenum, New York 1974)
4. J. R. Frederick, J. A. Seydel, R. C. Fairchild. Improved Ultrasonic Non-Destructive Testing of Pressure Vessels, NUREG-0007-2 (University of Michigan, Ann Arbor 1976)
5. J. L. Jackson, Program for Field Validation of the Synthetic Aperture Focusing Technique for Ultrasonic Testing (SAFT-UT) – Analysis Before Test, NUREG/CR-0288 (Southwest Research Institute, San Antonio 1978)
6. S. R. Doctor, G. J. Schuster, L. D. Reid, T. E. Hall, Real-Time 3-D SAFT-UT System Evaluation and Validation, NUREG/CR-6344 PNNL-10571 (Pacific Northwest National Laboratory, Richland 1996)
7. V. Schmitz, Line Synthetic Aperture Focusing Technique, IZFP Report 001 (IZFP, Saarbrücken 1981)
8. V. Schmitz, W. Müller, K. J. Langenberg, HOLOSAFT II, Final Report 920206 (FHGIZFP, Saarbrücken 1992)
9. K. J. Langenberg, M. Brandfaß, R. Hannemann, C. Hofmann, T. Kaczorowski, J. Kostka, R. Markelein, K. Mayer, A. Pitsch, Inverse scattering with acoustic, electromagnetic and elastic waves as applied in nondestructive evaluation, In *Wavefield Inversion*, ed. by A. Wirgin (Springer, Vienna 1999)
10. K. J. Langenberg: Applied inverse problems, In *Basic Methods of Tomography and Inverse Problems*, ed. by P. C. Sabatier (Adam Hilger, Bristol 1987)
11. K. Mayer, R. Marklein, K. J. Langenberg, T. Kreutter, Three-dimensional imaging system based on fourier transform synthetic aperture focussing technique, Ultrasonics **28**, 241 (1990)
12. L. J. Cutrona, Comparison of sonar system performance achievable using synthetic aperture focusing techniques with the performance available by more conventional means, J. Acoust. Soc. Am. **58**, 336–348 (1975)
13. J. A. Ogilvy, Computerised ultrasonic ray tracing in austenitic steels, Nondestruct. Test. Int. **18**, 67–77 (1985)
14. V. Schmitz, F. Walte, S. V. Chakhlov, 3D Ray Tracing in Austenite Materials, Nondestruct. Test. Eng. Int. **32**, 201–213 (1999)
15. V. Schmitz, F. Walte, S. V. Chakhlov, Modeling of sound fields through austenitic welds, In *Proc. First Int. Conf. NDE in Relation to Structural Integrity for Nuclear and Pressurized Components*, Amsterdam (1998)

Seismic Anisotropy Tomography

Jean-Paul Montagner

Institut Universitaire de France, Département de Sismologie CNRS UMR 7580,
Institut de Physique du Globe, Paris, France
jpm@ipgq.jussieu.fr

Abstract. The main breakthrough in seismology during the last ten years is related to the emergence and development of more and more sophisticated 3-dimensional imaging techniques, usually named seismic tomography, from the local scale up to global scale of the Earth. The progress has been made possible by the rapid developments in seismic instrumentation, in electronics, and by the extensive use of massive computational facilities. However, in contrast to usual experiments in physics, geophysicists cannot control all the conditions and must use natural sources. Consequently, most global tomographic models suffer severe limitations due to imperfect data coverage and theoretical approximations. It is usually assumed that the propagating elastic medium is isotropic, which is shown to be a poor approximation. We show in this paper how to take account of the anisotropy of the Earth's materials. The consequence is that, by including other geological constraints, we are able to map not only the 3-dimensional temperature heterogeneities but also the flow field within the convecting mantle. The complete tomographic technique, which includes the resolution of a forward problem and of an inverse problem, is described. It is important to emphasize the fact that in order to check the reliability of a tomographic model it is necessary to calculate the errors and the resolution associated with the model by considering the structure of the data space (errors and correlations) and the parameter space (*a posteriori* errors, covariance function, resolution). However, despite the increasing quality of seismograms provided by modern digital networks (GEOSCOPE, IRIS, etc.), the lateral resolution at the global scale is limited to about 1000 km and the installation of ocean-bottom observatories constitutes a new challenge for the next century. The next step is to apply to data recent theoretical developments in order to use all the information provided by seismic waveforms. Then, we will receive new insight into anisotropic and anelastic parameters within the Earth, and also within other solid materials.

1 Introduction

For several decades, the main thrust of seismologists was to retrieve reference spherically symmetric Earth models. However, due to the improvement of instruments and measurement techniques, some lateral heteogeneities were evidenced [1]. More than 10 years ago, the first global 3-dimensional tomographic models of both kinds of elastic waves, P and S waves, were published [2,3,4,5,6]. Since that time, many new tomographic models have been published and a large family of techniques has been made available. This important progress was made possible by the extensive use of computers which can handle very large data sets and by the availability of good-quality digital seismograms recorded by seismic networks such as the International Deployment of accelerometers [7], the Global Digital Seismograph Network [8] and more recently GEOSCOPE [9] and IRIS [10]. However, most tomographic techniques only make use of the phase information in seismograms and very few of the amplitude, even when using seismic waveforms [2]. It can be shown from a theoretical and practical point of view (see, for example, [11]) that it is easier to explain the phase of seismic signals than their amplitude.

Therefore, global tomographic models have been improved by an increase in the number of data and by more general parameterizations, now including anisotropy (radial anisotropy in *Nataf* et al. [6]; general slight anisotropy in *Montagner* [12] and *Montagner* and *Tanimoto* [13,14]; and to a lesser extent anelasticity [15,16,17]). However, there is still a major step to perform: a complete account of the amplitude anomalies in the most general case. There have been some attempts to do so on a global scale [18,19] and on a regional scale, following the works of *Snieder* [20,21]. But, from a practical point of view, inverse problems of very large size are induced and the parameter space must be limited to that part which is isotropic.

In this contribution, it is shown how to take account of seismic anisotropy. Contrary to the common belief, anisotropy is not a second-order effect and neglecting it can bias seismic velocities V_P and V_S. The practical implementation of the inverse problem is presented as well as how to take account of errors in data and how to calculate the resolution and a posteriori covariance functions of parameters. Finally, some geodynamic applications are shown.

2 The Anatomy of Seismograms

Seismology is an observational field which is based on the exploitation of recordings of the displacement (or acceleration) of the Earth induced by earthquakes. It is a very old field: the first Chinese "seismoscope" dates back one century before Christ. Much progress with regard to seismic instrumentation was made during the last century, although the principle of the seismometer did not change dramatically.

2.1 Progress in Instrumentation

The seismometer is based on the relative movement of a mass coupled to the Earth's motion through a spring or a pendulum (Fig. 1). Such a system is characterized by a natural oscillation frequency. However, due to the existence of particularly large permanent seismic noise between 1 s and 10 s, there were, for a long time, two different fields in seismology: short-period seismology devoted to periods smaller than 1 s and long-period seismology for periods longer than 10 s.

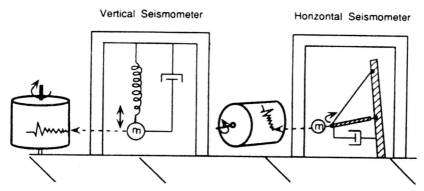

Fig. 1. The basic principle of a vertical seismometer with its mass, its spring and damping system

With the development of electronics and force-feedback systems, this gap was filled in the 1980s. It is now possible to record the 3 components of seismic displacement and the dynamic range and frequency range of seismometers (from 0.1 s up to 1000 s) have been increased. A negative feedback loop enables application of a force proportional to the inertial mass displacement in order to cancel its relative motion. This approach extends the bandwidth and the linearity of the seismometer. Since the development of STS1 by *Wielandt* and *Streckeisen* [22], a new generation of seismometers applying this simple principle has appeared. They share common characteristics: they are now light (less than 1 kg), have low power consumption and are very sensitive, robust and reliable. They can detect seismic displacement much smaller than the interatomic distance (10^{-10} m) in the seismic frequency band. An example of such a seismometer is presented in Fig. 2 [23]. They are designed to operate in very hostile environments such as Mars and the seafloor, and they are planned to be used for planetary exploration [24] and for future ocean-bottom stations [25].

The quality of seismograms is no longer related to the intrinsic qualities of seismometers but is primarily largely dependent on the noise level of the station. A correct station must display a noise level between the low-noise model and the high-noise model of *Peterson* [26] (Fig. 3). The seismic noise is minimum in the period range 20−100 s and at very short periods below 1 s. These 2 ranges are separated by the microseismic peak which is due to the complex interaction between fluid Earth (ocean and atmosphere) and solid Earth. The existence of this peak explains why there were traditionally two fields in seismology. It is only since the development of broadband seismometers with a high dynamic range (> 128 dB) that both fields have been merged into one field, broadband seismology.

Broadband 3-component highly dynamic seismometers have been installed at more than 200 stations around the world. Their deployment, the definition

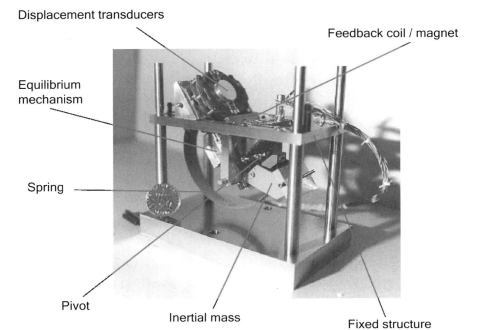

Fig. 2. A modern broadband seismometer with its feedback system [23]

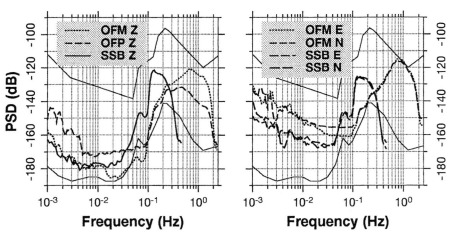

Fig. 3. Broadband seismic noise between 0.1 s and 1000 s, recorded in SSB (Continental GEOSCOPE station) and on the seafloor during the Sismobs-OFM experiment in May 1992 [27]. The 2 continuous lines correspond to the low-noise and high-noise models [26]

of a standard format (SEED) and the free availability of data are coordinated through the FDSN (Federation of Digital Seismograph Networks, Fig. 4). It is worthwhile noticing that the whole community of seismologists chose to share their data. However, despite these international efforts, there are still many areas on the Earth's surface devoid of broadband seismic stations. These regions are primarily located in southern hemisphere and more particularly in oceanic areas where no island is present. Therefore, an international effort is ongoing, coordinated through the ION. (International Ocean Network), to promote the installation of geophysical ocean-bottom observatories in order to fill the enormous gaps in station coverage [28].

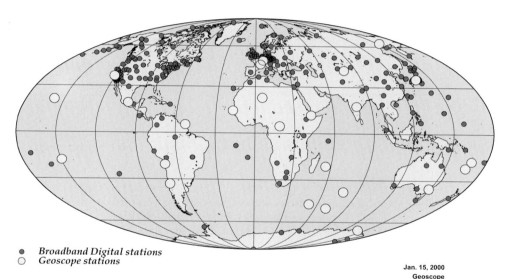

Fig. 4. Distribution of the broadband standardized stations of the FDSN network

2.2 Body Waves, Surface Waves and Normal Modes

A seismic record is characterized by its natural complexity, looking like a "chaotic" series of oscillations, well above seismic noise. The basic job of a seismologist is to unravel the succession of impulsive arrivals and to distinguish the body waves from the following complex dispersed wavetrains, the surface waves. Figure 5 shows an example of 3-component broadband seismograms for an earthquake located in Chile, of magnitude 7.3, recorded at the GEOSCOPE station of Canberra, Australia. The horizontal-component seismograms (north–south and east–west) have been rotated into longitudinal

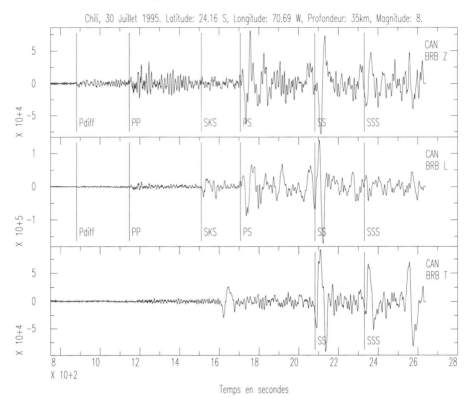

Fig. 5. Example of 3-component broadband seismogram recorded at the GEOSCOPE station CAN (Canberra, Australia) for a Chilean earthquake (−24.16°S, −70.69°W, depth 20 km) of magnitude 7.3, which occurred on July 30, 1995. The time scale is in hundreds of seconds

and transverse components in order to separate different kinds of body waves and surface waves. The seismic noise can be quantitatively assessed by considering the level of unexplained signal before the first arrival of the P waves. Most of the largest impulses can be explained by reflected or refracted waves at the major seismic discontinuities, i.e. the surface of the Earth (PP, SS, SSS, PS), the core–mantle boundary (Pdiff, SKS, etc.) at 2900 km depth and the ICB (inner core boundary). The nomenclature of all these body waves is explained in Fig. 6 and reflects the propagation history of waves in the 3 main layers of the Earth, i.e., the mantle, the outer core and the inner core.

Surface waves arrive after body waves and are the most energetic waves at large distances and at long periods. Figure 7 shows seismograms for the same earthquake as that in Fig. 5, but for a longer and low-pass-filtered time series ($T > 100$ s). Two kinds of surface waves are observed: Love waves on the transverse component corresponding to SH-guided waves, and Rayleigh

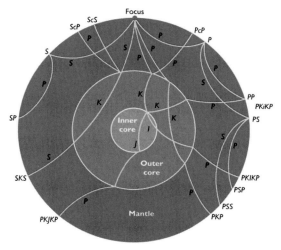

Fig. 6. Seismic rays through the Earth: explanation of the nomenclature

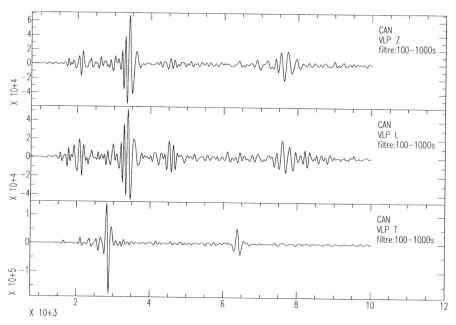

Fig. 7. Low-pass-filtered ($T > 100$ s) seismogram recorded at the GEOSCOPE station CAN for the same earthquake as that in Fig. 5

waves on the vertical and longitudinal components resulting from the complex coupling of P- and SV waves. Group velocities are approximately 4.3 km/s for Love waves and 4 km/s for Rayleigh waves.

When considering even longer time series (more than 1 day) following very large earthquakes (magnitude larger than 7), it is possible to display the free oscillations of the Earth. Figure 8a shows an example of 2 d of the vertical seismic record at SCZ (Santa Cruz station, California) of the large Kuril earthquake which occurred in October 1994. The spectrum calculated for the first 3 h of record in 3 components (Fig. 8b) does not present any specific characteristics. It is primarily reflecting the convolution of the source time function with the transfer function of the Earth between the epicenter and the station. However, if the spectrum is calculated on a 36 h time series (Fig. 8c), very narrow peaks appear. They can be explained by the constructive interference between stationary waves traveling in opposite directions. These well-defined frequencies are named eigenfrequencies and are characteristic of the structure of the Earth in the same way as sounds are characteristic of a guitar string. They constitute the basis of Earth spectroscopy. The modes on the transverse component (toroidal modes related to Love waves and SH waves) are different from the ones visible on the vertical and longitudinal component (spheroidal modes related to Rayleigh waves resulting from the coupling between P and SV waves). We will briefly review how the eigenfrequencies and the corresponding eigenfunctions can be calculated in a spherically symmetric reference Earth model.

2.3 Normal-Mode Theory

The basic equation which governs the displacement $\boldsymbol{u}(\boldsymbol{r},t)$ is the elastodynamics equation:

$$\rho_0 \frac{d^2 u_i}{dt^2} = \sum_j \sigma_{ij,j} + F_{\mathrm{S}i} + F_{\mathrm{E}i}, \tag{1}$$

where $F_{\mathrm{S}i}$ and $F_{\mathrm{E}i}$ represent respectively the whole ensemble of applied inertial and external forces (see [29] for a complete description of all terms). Generally, by neglecting the advection term, this equation can be written in a simple way:

$$(\rho_0 \partial_{tt} - H_0)\boldsymbol{u}(\boldsymbol{r},t) = \boldsymbol{F}(\boldsymbol{r}_\mathrm{S},t), \tag{2}$$

where H is an integro-differential operator and $\boldsymbol{F}$ expresses all forces applied to the source point $\boldsymbol{r}_\mathrm{S}$ at time t. We assume that $\boldsymbol{F}$ is equal to 0 for $t < 0$. In the elastic case, there is a linear relationship between σ_{ij} and the strain tensor ϵ_{kl}; $\sigma_{ij} = \sum_{kl} \Gamma_{ijkl}\epsilon_{kl}$ (+ terms related to the initial stress). Γ_{ijkl} is a fourth-order tensor. By using the different symmetry conditions $\Gamma_{ijkl} = \Gamma_{jikl} = \Gamma_{ijlk} = \Gamma_{klij}$, the tensor $\boldsymbol{\Gamma}$ is shown to have 21 independent elastic moduli.

When we are looking for the free oscillations of the Earth, we assume that $\boldsymbol{F} = 0$. The solution $\boldsymbol{u}(\boldsymbol{r},t)$ of (2) can be calculated for a spherically symmetric nonrotating reference Earth model, M_0, associated with the operator

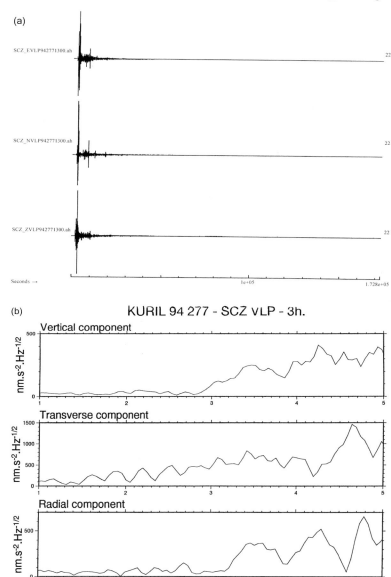

Fig. 8. Spectra of the seismograms corresponding to 2 different time series for the Kuril Earthquake, October 1994, recorded at the Santa Cruz station in California. (**a**) 2 d time series (**b**) Spectrum for the first 3 h of the record

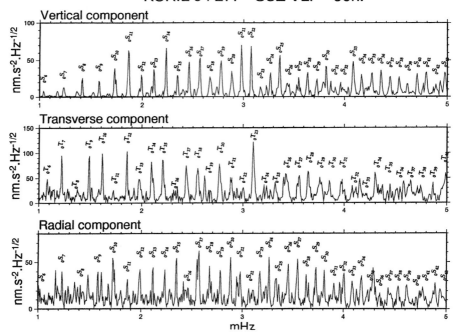

Fig. 8. (c) spectrum for the first 36 h

H_0, according to

$$\rho_0 \partial_{tt} \mathbf{u}(\mathbf{r},t) = H_0 \mathbf{u}(\mathbf{r},t) \,. \tag{3}$$

The eigenvalues of the operator H_0 are equal to $-\rho_0 {}_n\omega_\ell^2$ where ${}_n\omega_\ell$ is the eigenfrequency characterized by 2 quantum numbers n and ℓ, respectively termed radial and angular orders. The corresponding eigenfunctions ${}_n\mathbf{u}_\ell^m(\mathbf{r},t)$ are dependent on 3 quantum numbers n, ℓ, m, where m is the azimuthal order, with the following property: $-\ell \leq m \leq \ell$. Therefore, for a given eigenfrequency ${}_n\omega_\ell$ calculated in a spherically symmetric Earth model, $2\ell + 1$ eigenfunctions can be defined. The eigenfrequency ${}_n\omega_\ell$ is said to be degenerated, with a degree of degeneracy of $2\ell + 1$. There is a complete formal similarity with the calculation of the energy levels of the atom of hydrogen in quantum mechanics. The eigenfunctions ${}_n\mathbf{u}_\ell^m(\mathbf{r},t)$ of the operator H_0 are orthogonal and normalized. The displacement ${}_n\mathbf{u}_\ell^m(\mathbf{r},t)$ associated with the mode n, ℓ, m can be written

$$_n\mathbf{u}_\ell^m(\mathbf{r},t) = {}_n\mathbf{D}_\ell Y_\ell^m(\theta,\phi) e^{\mathrm{i}\,{}_n\omega_\ell t} \,, \tag{4}$$

where

$$_n\mathbf{D}_\ell = {}_nU_\ell(r)\mathbf{r} + {}_nV_\ell(r)\nabla_1 + {}_nW_\ell(r)(-\mathbf{r}\times\nabla_1) \,, \tag{5}$$

where $_nU_\ell$, $_nV_\ell$, and $_nW_\ell$ are the radial eigenfunctions of spheroidal and toroidal modes. $Y_\ell^m(\theta,\phi)$ are spherical harmonics normalized according to *Edmonds* [30] and ∇_1 is the surface gradient operator on the unit sphere.

The important point is that the basis of the functions defined by (4) is complete. Thus, any displacement at the surface of the Earth can be expressed as a linear combination of these eigenfuctions:

$$\boldsymbol{u}(\boldsymbol{r},t) = \sum_{n,\ell,m} {}_na_\ell^m {}_n\boldsymbol{u}_\ell^m(\boldsymbol{r},t).$$

Therefore, these eigenfunctions can be used to calculate the displacement at any point $\boldsymbol{r}$, at time t, due to a point-force system $\boldsymbol{F}$ at point $\boldsymbol{r}_S$ and a step time function, which is a good starting model for earthquakes. The solution of (2) is given by

$$\boldsymbol{u}(\boldsymbol{r},t) = \sum_{n,\ell,m} -{}_n\boldsymbol{u}_\ell^m(\boldsymbol{r}) \frac{\cos {}_n\omega_\ell t}{{}_n\omega_\ell^2} e^{\frac{-{}_n\omega_\ell t}{2Q}} ({}_n\boldsymbol{u}_\ell^m \cdot \boldsymbol{F})_S \qquad (6)$$

The source term $({}_n\boldsymbol{u}_\ell^m\, \boldsymbol{F})_S$ can be replaced using Green's theorem:

$$({}_n\boldsymbol{u}_\ell^m \cdot \boldsymbol{F})_S = (\mathbf{M}:\boldsymbol{\epsilon})_S,$$

where $\mathbf{M}$ and ϵ are respectively the seismic moment tensor and the deformation tensor. Both tensors are symmetric. Since (6) is linear in $\mathbf{M}$, it can be easily generalized to more complex spatial and temporal source functions, and can be rewritten as

$$\boldsymbol{u}(\boldsymbol{r},t) = \boldsymbol{G}(\boldsymbol{r},\boldsymbol{r}_S,t,t_S)\mathbf{M}(\boldsymbol{r}_S,t_S)$$

where $\boldsymbol{G}(\boldsymbol{r},\boldsymbol{r}_S,t,t_S)$ is the Green operator of the medium. The normal-mode theory is routinely used for calculating synthetic seismograms at long periods ($T \geq 40\,\mathrm{s}$), and the agreement between synthetic and observed seismograms is quite good, as is shown in Fig. 9.

However, it can be seen that some frequency-dependent time shift is still present between the observed and synthetic seismogram. The goal of seismology is to explain this anomaly: its presence means that some of the hypotheses considered for calculating the synthetic seismograms are insufficient. The simplest way to explain the observed time delay is by removing the assumption that the Earth is spherically symmetric, or in other words that there are lateral heterogeneities between the source and the receiver. The next issue consists of characterizing these lateral heterogeneities. Since there is agreement between synthetic and observed seismograms at long periods, we can reasonably infer that the amplitude of heterogeneities is small. Behind the surface wave train, a long coda is usually observed, interpreted as scattered waves. However, when filtering out the periods shorter than $40\,\mathrm{s}$, this coda vanishes, which means that the scattering effect is only present in the

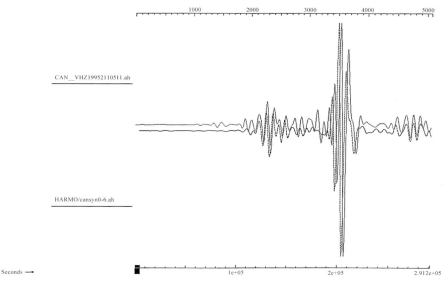

Fig. 9. Observed and synthetic seismograms calculated for the Chilean earthquake of July 1995 at the GEOSCOPE station CAN

shallowest regions of the Earth (primarily the crust) and that it is probably negligible at largest depths. Therefore, we make the hypothesis that the scale of lateral heterogeneities is large compared with the wavelength. A second hypothesis which has to be discussed is the isotropic nature of the Earth's materials. Actually, it is a poor hypothesis, because seismic anisotropy can undoubtedly be observed at different scales.

3   An Anisotropic Earth

Different geophysical fields are involved in the investigation of the manifestations of anisotropy of Earth materials: mineral physics and geology for the study of the microscopic scale, and seismology for scales larger than typically 1 km.

3.1   Seismic Anisotropy at All Scales

The different observations related to anisotropy at different scales are reviewed in *Montagner* [31].

3.1.1   Microscopic Scale

The different minerals present in the upper mantle are strongly anisotropic (*Peselnick* [32]). The difference of velocity between the fast axis and the slow axis is larger than 20% for olivine, the main constituent of the upper mantle. Other important constituents such as orthopyroxene or clinopyroxene are anisotropic as well (> 10%) (see *Anderson* [33] and *Babuska* and *Cara* [34], for reviews). However, some other constituents, such as garnet, display a cubic crystallographic structure which presents a low anisotropy. Consequently, the petrological models which are assemblages of different minerals are less anisotropic than pure olivine. The amount of anisotropy is largely dependent on the percentage of these different minerals and on the mechanisms which align the crystallographic axes in preferred orientations. For example, the anisotropy of the pyrolitic model, mainly composed of olivine and orthopyroxene [35], will depend on the relative orientation of the crystallographic axes of different constituents [36]. However, through the mechanisms of lattice-preferred orientation, its anisotropy can be larger than 10% [37]. For competing petrological models such as piclogite [38,39], where the percentage of olivine (garnet) is smaller (larger), the amount of anisotropy is smaller (about 5%). At microscopic scales, we can conclude that earth materials in the upper mantle are strongly anisotropic, but that the anisotropy tends to decrease as depth is increasing.

At slightly larger scales, the scale of rock samples, several studies of anisotropy have been undertaken. Dunite, which is almost pure olivine, displays a high anisotropy [40]. Moreover, this anisotropy is coherent in whole massifs of ophiolites over several tens of kilometers [41,42]. Some attempts have been undertaken to numerically model seismic anisotropy within convecting cells [43]. At larger wavelengths, anisotropy is also present and can be investigated from seismic observations.

3.1.2 Macroscopic Scale

Different and independent seismic data sets make it evident that the effect of anisotropy is not negligible in the explanation of the propagation of seismic waves inside the Earth. Early evidence was the discrepancy between Rayleigh and Love wave dispersion [44,45] and the azimuthal dependence of P_n velocities [46]. Azimuthal variations are now well documented for different areas in the world for body waves and surface waves.

For body waves, this kind of information primarily results from the investigation of splitting in teleseismic shear waves such as SKS [47,48,49,50], ScS [51,52] and S [53,54]. Since these pioneering papers, many studies have confirmed the existence of splitting of S waves. These waves are shown to provide excellent lateral resolution, if observations are restricted to the deep upper mantle (i.e., below the crust). Among these different observations, the splitting information derived from SKS is the less ambiguous and has been used extensively in teleseismic anisotropy investigations [48,49,55,56,57]. The drawback of this technique is that it is almost impossible to locate at depth the anisotropic area.

Surface waves are also well suited for investigating upper-mantle anisotropy. Two kinds of observable anisotropy can be considered: The first of these is the radial anisotropy which results from the discrepancy between Love and Rayleigh waves, also named the "polarization" anisotropy. In order to remove this discrepancy, it is necessary to consider a transversely isotropic model with a vertical symmetry axis. This kind of anisotropy is characterized by five elastic parameters plus density [44]. However, *Levshin* and *Ratnikova* [58] showed that lateral heterogeneity can lead to a Rayleigh–Love discrepancy and that we must be cautious in the interpretation of this discrepancy in terms of the anisotropic model. On a global scale, *Nataf* et al. [5,6] have derived, by the simultaneous inversion of Rayleigh- and Love-wave dispersion, the geographical distributions of radial S-wave anisotropy at different depths assuming transverse isotropy with a vertical symmetry axis. The second kind of observable anisotropy is the azimuthal anisotropy which is directly derived from the azimuthal variation of phase velocity. It was observed for the first time on surface waves in Nazca plate by *Forsyth* [59]. Several global and regional models have been derived for both kinds of anisotropy [60,61]. *Tanimoto* and *Anderson* [62] obtained a global distribution of the Rayleigh wave azimuthal anisotropy at long periods. On a regional scale, several tomographic investigations report the existence of azimuthal anisotropy in the Indian Ocean [12,63], in the Pacific Ocean [64,65], in the Atlantic Ocean [66], in Africa [67] and in Central Asia [68,69]. *Lévêque* and *Cara* [70] and *Cara* and *Lévêque* [71] used higher mode data to display anisotropy under the Pacific Ocean and North America down to at least 300 km.

However, "polarization" (or radial) and azimuthal anisotropy are two different manifestations of the same phenomenon: the anisotropy of the upper mantle. *Montagner* and *Nataf* [72] designed a technique which makes it possible to simultaneously explain these two forms of seismically observable anisotropy. The principles of this technique will be described in Sect. 3.2 for the most general case of anisotropy (on condition that it is weak). The method can be slightly simplified by introducing only one symmetry axis (seven independent anisotropic parameters) and it was termed "vectorial tomography" [104]. It was applied to the investigation of the Indian Ocean [124] and Africa [67]. These different investigations showed that, paradoxically, *a parameterization with anisotropy requires less parameters than a parameterization with only isotropic terms*. In contrast to body waves, surface waves enable anisotropy at depth to be located, but, so far, its lateral resolution (several thousands of kilometers) is very poor.

Finally, anisotropy has not only an effect on the phase of seismograms, but also on amplitude. Due to the coupling between surface-wave modes, one of the less ambiguous effects is the fact that Love waves can be present on the vertical and the radial components [125]. They can also be found in the observations of polarization anomalies too large to be explained by deviations due to isotropic heterogeneities [77,126].

Therefore, seismic anisotropy cannot be considered to be a second-order effect. It is present at different scales and at different depths. We note that it tends to decrease as wavelength increases, from 20% at the microscopic scale down to 1–2% at very large wavelengths. Several conditions must be fulfilled in order to observe anisotropy at long periods and large wavelengths. The material must be microscopically anisotropic, and there must be some efficient mechanism of preferred orientation which aligns the fast axes (or slow axis) of minerals in the flow field. There must be, in addition, an efficient strain field, with long-wavelength coherency, for spatial wavelengths Λ_S such that $\Lambda_S \gg \lambda$ (where λ is the wavelength of the wave). This kind of condition is usually encountered in many areas all around the world. Body-wave anisotropy and surface-wave anisotropy can be related in the simple case of anisotropy characterized by a horizontal symmetry axis [78].

Therefore, from a seismological point of view, there is no longer doubt that the upper mantle is anisotropic. We will now explain how to simultaneously explain the different observations of surface-wave anisotropy, radial anisotropy and azimuthal anisotropy.

3.2 First-Order Perturbation Theory in the Planar Case

We will only consider the propagation of surface waves in the planar case, but it can be easily extended to the spherical Earth [79]. In the simple planar case (fundamental modes, no coupling between branches of Rayleigh and Love waves), the frequency shift, for a constant wavenumber k can be written as follows:

$$\delta\omega|_k = \frac{1}{2\omega} \frac{\langle \boldsymbol{u}_k | \mathbf{E}^T \boldsymbol{\Gamma} \mathbf{E} | \boldsymbol{u}_k \rangle}{\langle \boldsymbol{u}_k | \rho_0 \boldsymbol{u}_k \rangle} \tag{7}$$

where $\mathbf{E}$ and $\boldsymbol{\Gamma}$ are respectively the deformation and elastic tensors, $|\boldsymbol{u}_k\rangle$ the eigenfunctions as defined in (4), where k corresponds to the multiplet of the 3 quantum numbers (n, ℓ, m). Let us follow the same approach as *Montagner* and *Nataf* [72] for calculating $\delta\omega|_k$, where we consider the propagation of the fundamental modes of Love and Rayleigh waves in an arbitrarily stratified half-space in which a right-handed Cartesian coordinate system (x, y, z) is defined (Fig. 10)

The half-space is assumed to be homogeneous and may be described by its density $\rho(z)$ and its fourth-order elastic tensor $\Gamma(z)$ with 21 independent elastic coefficients. All these parameters are assumed to be independent of the x and y coordinates (z is the vertical component). This condition will be released in the next section. The unperturbed medium is assumed to be isotropic, with an elastic tensor $\Gamma_0(z)$. In this medium, the two cases of Love and Rayleigh wave dispersion can be successively considered.

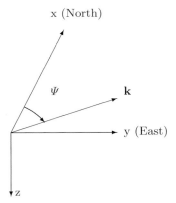

Fig. 10. Definition of the Cartesian coordinate system (x, y, z) used in the calculations. Ψ is the azimuth with respect to north of the wave vector

The unperturbed Love wave displacement is of the form

$$\boldsymbol{u}(\boldsymbol{r}, t) = \begin{pmatrix} -W(z) \sin \Psi \\ W(z) \cos \Psi \\ 0 \end{pmatrix} \exp\{i[k(x \cos \Psi + y \sin \Psi) - \omega t]\}, \tag{8}$$

where $W(z)$ is the scalar depth eigenfunction for Love waves, k is the horizontal wavenumber, and Ψ is the azimuth of the wavenumber k measured clockwise from north.

The unperturbed Rayleigh wave displacement is of the form

$$\boldsymbol{u}(\boldsymbol{r}, t) = \begin{pmatrix} V(z) \cos \Psi \\ V(z) \sin \Psi \\ iU(z) \end{pmatrix} \exp\{i[k(x \cos \Psi + y \sin \Psi) - \omega t]\}, \tag{9}$$

where $V(z)$ and $U(z)$ are the scalar depth eigenfunctions for Rayleigh waves. The associated strain tensor $\epsilon(\boldsymbol{r}, t)$ is defined by:

$$\epsilon_{ij}(\boldsymbol{r}, t) = 1/2(u_{i,j} + u_{j,i}), \tag{10}$$

where j denotes the differentiation with respect to the jth coordinate. The medium is perturbed from $\Gamma_0(z)$ to $\Gamma_0(z) + \gamma(z)$, where $\gamma(z)$ is small compared to $\Gamma_0(z)$, but quite general in the sense that there is no assumption for the kind of anisotropy. This means that in this approximation we can still consider quasi-Love modes and quasi-Rayleigh modes [104]. From Rayleigh's principle, the first-order perturbation $\delta V(\boldsymbol{k})$ in phase-velocity dispersion is [124,125]

$$\delta V(\boldsymbol{k}) = \frac{V}{2\omega^2} \frac{\int_0^\infty \gamma_{ijkl} \epsilon_{ij} \epsilon_{kl}^* dz}{\int_0^\infty \rho_0 u_k u_k^* dz}, \tag{11}$$

where u_i and ϵ_{ij} are respectively the displacement and the strain for the unperturbed half-space and the asterisk denotes complex conjugation.

Because of the symmetry of the tensors $\gamma(z)$ and ϵ, we use the simplified index notation c_{ij} and ϵ_i for the elements γ_{ijkl} and ϵ_{ij}, but the number n_{ij} of coefficients γ_{ijkl} for each c_{ij} must be taken into account. The simplified index notation for the elastic tensor γ_{ijkl} is defined in a coordinate system (x_1, x_2, x_3) by

$$\gamma_{ijkl} \longrightarrow c_{pq} \begin{cases} \text{if } i = j \Rightarrow p = i \\ \text{if } k = l \Rightarrow q = k \\ \text{if } i \neq j \Rightarrow p = 9 - i - j \\ \text{if } k \neq l \Rightarrow q = 9 - k - l \end{cases} . \tag{12}$$

This kind of transformation enables the fourth-order tensor $\boldsymbol{\gamma}(3 \times 3 \times 3 \times 3)$ to be related to to a matrix $\mathbf{c}(6 \times 6)$. The same simplified index notation can be applied to the components of the strain tensor ϵ_{ij}, transforming the second-order tensor $\boldsymbol{\epsilon}(3 \times 3)$ into a vector with 6 components. However, one must be careful, because several γ_{ijkl} correspond to a given c_{pq}, and γ_{ijkl} must be replaced by $n_{pq}c_{pq}$, where n_{pq} is the number of γ_{ijkl} giving the same c_{pq}. Therefore, (11) expressing Rayleigh's principle can be rewritten as

$$\delta V(\boldsymbol{k}) = \frac{V}{2\omega^2} \frac{\int_0^\infty \sum_{ij} n_{ij} c_{ij} \epsilon_i \epsilon_j^*}{\int_0^\infty \rho_0 u_k u_k^* \mathrm{d}z} \mathrm{d}z . \tag{13}$$

We only detail the calculations for Love waves.

3.2.1 Love Waves

By using previous expressions for $\boldsymbol{u}(\boldsymbol{r}, t)$ (9) and $\epsilon_{ij}(\boldsymbol{r}, t)$ (10), the various expressions of strain are:

$$\begin{cases} \epsilon_1 = -\mathrm{i} \cos \Psi \sin \Psi \, kW \\ \epsilon_2 = \mathrm{i} \cos \Psi \sin \Psi \, kW \\ \epsilon_3 = 0 \\ \epsilon_4 = (1/2) \cos \Psi \, W' \\ \epsilon_5 = (-1/2) \sin \Psi \, W' \\ \epsilon_6 = (1/2)(\cos^2 \Psi - \sin^2 \Psi) \, kW \end{cases} , \tag{14}$$

where $W' = \frac{\mathrm{d}W}{\mathrm{d}r}$. In Table 1, the different terms for $n_{ij} c_{ij} \epsilon_i \epsilon_j^*$ are given. We note that when $c_{ij} \epsilon_i \epsilon_j^*$ is a purely imaginary complex, its contribution to $\delta V(k, \Psi)$ is null.

When all the contributions are summed, the different terms $\cos^k \Psi \sin^l \Psi$ are such that $k + l$ is even, which is not surprising in light of the reciprocity principle. Therefore, each term can be developed as a Fourier series in Ψ with

Table 1. Calculation of the various $c_{ij}\epsilon_i\epsilon_j$ for Love waves, with the simplified index notation, $\alpha = \cos\Psi$; $\beta = \sin\Psi$

n	ij	$c_{ij}\epsilon_i\epsilon_j$
1	11	$c_{11}\alpha^2\beta^2 k^2 W^2$
1	22	$c_{22}\alpha^2\beta^2 k^2 W^2$
1	33	0
2	12	$-c_{12}\alpha^2\beta^2 k^2 W^2$
2	13	0
2	23	0
2	24	0
4	14	$c_{14}(-i\alpha^2\beta)\frac{kWW'}{2}$
4	15	$c_{15}(i\alpha^2\beta)\frac{kWW'}{2}$
4	16	$c_{16}(-\alpha\beta)(\alpha^2-\beta^2)\frac{k^2W^2}{2}$
4	24	$c_{24}(-i\alpha^2\beta)\frac{kWW'}{2}$
4	25	$c_{25}(-i\alpha\beta^2)\frac{kWW'}{2}$
4	26	$c_{26}(\alpha\beta)(\alpha^2-\beta^2)\frac{k^2W^2}{2}$
4	34	0
4	35	0
4	36	0
4	44	$c_{44}\alpha^2\frac{W'^2}{4}$
8	45	$c_{45}(-\alpha\beta)\frac{W'^2}{4}$
8	46	$c_{46}(-i\alpha)(\alpha^2-\beta^2)\frac{kWW'}{2}$
4	55	$c_{55}\beta^2\frac{W'^2}{4}$
8	56	$c_{56}(i\beta)(\alpha^2-\beta^2)\frac{kWW'}{2}$
4	66	$c_{66}(\alpha^2-\beta^2)\frac{k^2W^2}{4}$

only even terms. Finally it is found that:

$$\delta V_L(k,\Psi) = \frac{V}{2\omega^2 L_0}\int_0^\infty dz \left[k^2W^2\left(\frac{1}{8}(c_{11}+c_{22}-2c_{12}+4c_{66})\right) \right.$$
$$+ W'^2\left(\frac{1}{2}(c_{44}+c_{55})\right)$$
$$+ \cos 2\Psi\, W'^2\left(\frac{1}{3}(c_{44}-c_{55})\right) - \sin 2\Psi\, W'^2 c_{45}$$
$$- \cos 4\Psi\, k^2W^2\left(\frac{1}{8}(c_{11}+c_{22}-2c_{12}-4c_{66})\right)$$
$$\left. + \sin 4\Psi\, k^2W^2\left(\frac{1}{2}(c_{26}-c_{16})\right) \right]. \tag{15}$$

In the case of a transversely isotropic medium with a vertical symmetry axis (also named a radial anisotropic medium), we have: $c_{11} = c_{22} = \delta A$, $c_{33} = \delta C$, $c_{12} = \delta(A-2N)$, $c_{13} = c_{23} = \delta F$, $c_{44} = c_{55} = \delta L$, $c_{66} = \delta N$ and $c_{14} = c_{24} = c_{15} = c_{25} = c_{16} = c_{26} = 0$. The local azimuthal terms vanish and (15) reduces to

$$\delta V_\mathrm{L}(k,\Psi) = \frac{1}{2V_\mathrm{L}L_0} \int_0^\infty \left(W^2 \delta N + \frac{W'^2}{k^2}\delta L\right) \mathrm{d}z \tag{16}$$

Therefore, the same expressions as in *Takeuchi* and *Saito* [126] are found in the case of radial anisotropy. The 0-Ψ term of (15) corresponds to the averaging over azimuth Ψ, which provides the equivalent transversely isotropic model with a vertical symmetry axis by setting

$$\delta N = \frac{1}{8}(c_{11} + c_{22}) - \frac{1}{4}c_{12} + \frac{1}{2}c_{66},$$

$$\delta L = \frac{1}{2}(c_{44} + c_{55}).$$

If we call C_{ij} the elastic coefficients of the total elastic tensor, we can set

$$N = \rho V_\mathrm{SH}^2 = \frac{1}{8}(C_{11} + C_{22}) - \frac{1}{4}C_{12} + \frac{1}{2}C_{66},$$

$$L = \rho V_\mathrm{SV}^2 = \frac{1}{2}(C_{44} + C_{55}).$$

According to (15), the first-order perturbation in Love-wave phase velocity $\delta V_\mathrm{L}(k, \Psi)$ can then be expressed as

$$\delta V_\mathrm{L}(k, \Psi) = \frac{1}{2V_{0_\mathrm{L}}(k)} [L_1(k) + L_2(k)\cos 2\Psi + L_3(k)\sin 2\Psi \\ + L_4(k)\cos 4\Psi + L_5(k)\sin 4\Psi], \tag{17}$$

where

$$L_0(k) = \int_0^\infty \rho W^2 \mathrm{d}z,$$
$$L_1(k) = \frac{1}{L_0}\int_0^\infty (W^2 \delta N + \frac{W'^2}{k^2}\delta L)\mathrm{d}z,$$
$$L_2(k) = \frac{1}{L_0}\int_0^\infty -G_\mathrm{c}(\frac{W'^2}{k^2})\mathrm{d}z,$$
$$L_3(k) = \frac{1}{L_0}\int_0^\infty -G_\mathrm{s}(\frac{W'^2}{k^2})\mathrm{d}z,$$
$$L_4(k) = \frac{1}{L_0}\int_0^\infty -E_\mathrm{c}W^2\mathrm{d}z,$$
$$L_5(k) = \frac{1}{L_0}\int_0^\infty -E_\mathrm{s}W^2\mathrm{d}z.$$

3.2.2 Rayleigh Waves

The same procedure holds for the local Rayleigh-wave phase-velocity perturbation δV_R, starting from the displacement given previously [72]:

$$\delta V_\mathrm{R}(k, \Psi) = \frac{1}{2V_{0_\mathrm{R}}(k)} [R_1(k) + R_2(k)\cos 2\Psi + R_3(k)\sin 2\Psi \\ + R_4(k)\cos 4\Psi + R_5(k)\sin 4\Psi], \tag{18}$$

where

$$R_0(k) = \int_0^\infty \rho(U^2 + V^2)\mathrm{d}z\,,$$
$$R_1(k) = \frac{1}{R_0}\int_0^\infty [V^2\delta A + \frac{U'^2}{k^2}\delta C + \frac{2U'V}{k}\delta F + (\frac{V'}{k} - U)^2\delta L]\mathrm{d}z\,,$$
$$R_2(k) = \frac{1}{R_0}\int_0^\infty [V^2 B_\mathrm{c} + \frac{2U'V}{k}H_\mathrm{c} + (\frac{V'}{k} - U)^2 G_\mathrm{c}]\mathrm{d}z\,,$$
$$R_3(k) = \frac{1}{R_0}\int_0^\infty [V^2 B_\mathrm{s} + \frac{2U'V}{k}H_\mathrm{s} + (\frac{V'}{k} - U)^2 G_\mathrm{s}]\mathrm{d}z\,,$$
$$R_4(k) = \frac{1}{R_0}\int_0^\infty E_\mathrm{c} V^2 \mathrm{d}z\,,$$
$$R_5(k) = \frac{1}{R_0}\int_0^\infty E_\mathrm{s} V^2 \mathrm{d}z\,.$$

The 13 depth-dependent parameters $A, C, F, L, N, B_\mathrm{c}, B_\mathrm{s}, H_\mathrm{c}, H_\mathrm{s}, G_\mathrm{c}, G_\mathrm{s}, E_\mathrm{c}$ and E_s are linear combinations of the elastic coefficients C_{ij} and are explicitly given as follows:

- Constant term (0-Ψ-azimuthal term, independent of azimuth):

$$A = \rho V_\mathrm{PH}^2 = \frac{3}{8}(C_{11} + C_{22}) + \frac{1}{4}C_{12} + \frac{1}{2}C_{66}\,,$$
$$C = \rho V_\mathrm{PV}^2 = C_{33}\,,$$
$$F = \frac{1}{2}(C_{13} + C_{23})\,,$$
$$L = \rho V_\mathrm{SV}^2 = \frac{1}{2}(C_{44} + C_{55})\,,$$
$$N = \rho V_\mathrm{SH}^2 = \frac{1}{8}(C_{11} + C_{22}) - \frac{1}{4}C_{12} + \frac{1}{2}C_{66}\,.$$

- 2-Ψ-azimuthal term:

$\cos 2\Psi$	$\sin 2\Psi$
$B_\mathrm{c} = \frac{1}{2}(C_{11} - C_{22})$	$B_\mathrm{s} = C_{16} + C_{26}$
$G_\mathrm{c} = \frac{1}{2}(C_{55} - C_{44})$	$G_\mathrm{s} = C_{54}$
$H_\mathrm{c} = \frac{1}{2}(C_{13} - C_{23})$	$H_\mathrm{s} = C_{36}$

- 4-Ψ-azimuthal term:

$\cos 4\Psi$	$\sin 4\Psi$
$E_\mathrm{c} = \frac{1}{8}(C_{11} + C_{22}) - \frac{1}{4}C_{12} - \frac{1}{2}C_{66}$	$E_\mathrm{s} = \frac{1}{2}(C_{16} - C_{26})\,.$

The indices 1 and 2 refer to horizontal coordinates (1: north; 2: east), and index 3 refers to the vertical coordinate. ρ is the density, V_PH and V_PV are respectively horizontal and vertical P-wave velocities, and V_SH and V_SV are respectively horizontal and vertical S-wave velocities. We must bear in mind that the anisotropic parameters A, C, L and N can be retrieved from measurements of the P- and S-wave velocities propagating perpendicular or parallel to the axis of symmetry. Some of the previous combinations have already been derived in the expressions that describe the azimuthal dependence of body waves (for example [73]) in a weakly anisotropic medium:

$$\rho V_\mathrm{P}^2 = A + B_\mathrm{c} \cos 2\Psi + B_\mathrm{s} \sin 2\Psi + E_\mathrm{c} \cos 4\Psi + E_\mathrm{s} \sin 4\Psi,$$
$$\rho V_\mathrm{SP}^2 = N - E_\mathrm{c} \cos 4\Psi - E_\mathrm{s} \sin 4\Psi,$$
$$\rho V_\mathrm{SR}^2 = L + G_\mathrm{c} \cos 2\Psi + G_\mathrm{s} \sin 2\Psi.$$

Therefore, (17) and (18) define the forward problem in the framework of a first-order perturbation theory. We will see in the next section how to solve the inverse problem. That means that, ideally, surface waves have the ability to provide 13 elastic parameters, which emphasizes the enormous potential of surface waves in terms of geodynamical and petrological studies. However, from a practical point of view, data do not have the resolving power to invert for so many parameters. *Montagner* and *Anderson* [37] proposed using constraints from petrology in order to reduce the parameter space. Actually, they found that some of the parameters display large correlations independent of the petrological model used. Two extreme models were used to derive these correlations, the pyrolite model [35] and the piclogite model [38,39,80]. In the inversion process, the smallest correlations between parameters of both models are kept. This approach was already followed by *Montagner* and *Anderson* [81] to derive an average reference Earth model. These simple linear combinations of the elastic tensor components were first displayed by *Montagner* and *Nataf* [72], and they enable the two seismically observable effects of anisotropy on surface waves, polarization anisotropy [82] and azimuthal anisotropy [59], to describe in a simple way.

In conclusion, the 0-Ψ term corresponds to the average over all azimuths and involves 5 independent parameters, A, C, F, L and N, which express the equivalent transverse isotropic medium with a vertical symmetry axis (more simply named radial anisotropy). The other azimuthal terms (2-Ψ and 4-Ψ) depend on 4 groups of 2 parameters, B, G, H and E, respectively describing the azimuthal variation of A, L, F and N.

The other important point in these expressions is that they provide the partial derivatives for the radial and azimuthal anisotropy of surface waves. These partial derivatives of the different azimuthal terms with respect to the elastic parameters can be easily calculated by using a radial anisotropic reference Earth model, such as PREM [83]. The corresponding kernels and their variation at depth are detailed and discussed for the fundamental mode in *Montagner* and *Nataf* [72]. The partial derivatives of the eigenperiod $_0T_\ell$ with respect to parameter p, $\frac{p}{T}\frac{\partial T}{\partial p}$ can easily be converted into phase velocity partial derivatives by using

$$\frac{p}{V}\left(\frac{\partial V}{\partial p}\right)_T = -\frac{V}{U}\frac{p}{T}\left(\frac{\partial T}{\partial p}\right)_k.$$

For example, the parameters G_c and $G_{\hat{s}}$ have the same kernel as parameter L (related to V_SV) as shown by comparing R_1, R_2 and R_3. The calculation of kernels shows that Love waves are almost insensitive to V_{SV}, and Rayleigh waves are almost insensitive to V_{SH}. Rayleigh waves are the most sensitive to SV waves. However, as pointed out by *Anderson* and *Dziewonski* [84], the influence of P waves (through parameters A and C) can be very large in an anisotropic medium. The influence of density is also very large for Love and Rayleigh waves, but as shown by *Takeuchi* and *Saito* [126], it is largely decreased when seismic velocities are inverted for, instead of elastic moduli and density. We will now show how to implement such a theory from a practical but general point of view, and how to design a tomographic technique in order to invert for the 13 different anisotropic parameters.

4 Tomography of Anisotropy

Tomography is a generic term used by seismologists for naming a technique able to image the 3-dimensional structure of an object. The object is usually illuminated by a large number of rays, and its structure is recovered by an inversion procedure. A good description of this object means that we can find its correct location in space and its physical properties (amplitude and spectral content). A tomographic technique necessitates solution of a forward problem and an inverse problem at the same time. By using the results of the previous section, we will successively consider how to set the forward problem and how it is used to retrieve a set of parameters by inversion.

4.1 Forward Problem

We have first to define the data space $\boldsymbol{d}$ and the parameter space $\boldsymbol{p}$. It is assumed that a functional $\boldsymbol{g}$ relating $\boldsymbol{d}$ and $\boldsymbol{p}$ can be found such that:

$\boldsymbol{d} = \boldsymbol{g}(\boldsymbol{p})$, where $\boldsymbol{d}$ is the set of data (which samples the data space), and $\boldsymbol{p}$ the set of parameters.

4.1.1 Data Space: d

The basic data set is made of seismograms $\boldsymbol{u}(t)$. We can try to directly match the waveform in the time domain or we can work in the Fourier domain, by separating phase and amplitude on each component:

$$u_i(t) = \int_{-\infty}^{\infty} A_i(\omega) e^{i(\omega t - \phi_i)} d\omega \, .$$

The approach consisting of fitting the seismic waveform is quite general but, from a practical point of view, it does not necessarily correspond to the simplest choice. In an heterogeneous medium, the calculation of amplitude effects makes it necessary to calculate the coupling between different multiplets, which is very time consuming. When working in the Fourier domain, we can consider different time windows and separately match the phase of different seismic trains, body waves and surface waves. An example of a data seismogram and synthetic seismograms obtained by normal-mode summation with the different higher modes was shown in Fig. 9. The fundamental wavetrain is well separated from other modes at a large epicentral distance. The part of the seismogram corresponding to higher modes is more complex and shows a mixing of these modes in the time domain. Therefore, from a practical point of view, the fitting of the fundamental mode wavetrain will not cause any problem and has been widely used in global mantle tomography.

The use of a higher-mode wavetrain and the separation of overtones is much more difficult. The first attempts were performed by *Nolet* [85], *Cara* [86], *Okal* and *Bong-Jo* [87] and *Dost* [87] by applying a spatial filtering method. Different techniques based on waveform inversion of fundamental and higher-mode surface waves were also designed in the following years [88,89,90]. Unfortunately, all these techniques can only be applied to areas where dense arrays of seismic stations are present, i.e., in North America and Europe. By using a set of seismograms recorded at one station but corresponding to several earthquakes located in a small source area, *Stutzmann* and *Montagner* [91] showed how to separate the different higher modes. A similar approach was recently followed by *Van Heijst* and *Woodhouse* [92]. We only detail in this paper the technique which was designed for fitting the fundamental mode wavetrain, and the reader is referred to *Stutzmann* and *Montagner* [91] and *Van Heijst* and *Woodhouse* [92] to find the higher-mode dispersion properties.

We take advantage of the fact that, according to the Fermat's principle, the phase-velocity perturbation is only dependent to the second order on path perturbations, whereas amplitude perturbations are dependent, to the first order, on these perturbations, which implies that the eigenfunctions must be recalculated at each iteration. Therefore, the phase is a more robust observable than the amplitude. The amplitude $A(\omega)$ depends in a complex manner upon seismic moment tensor, attenuation, scattering, focusing effects, station calibration and near-receiver structure, whereas the phase $\phi(\omega)$ is readily related to lateral heterogeneities of seismic velocity and anisotropic parameters. The data set that we will consider is composed of propagation times (or phase-velocity measurements for surface waves) along paths: $\boldsymbol{d} = \left\{ \frac{\Delta}{V(T)} \right\}$.

A second important ingredient in the inverse-problem formulation is the "structure" of the data space. It is expressed through its covariance function (continuous case) or covariance matrix (discrete case) of data $\mathbf{C_d}$. When data d_i are independent, $\mathbf{C_d}$ is diagonal and its elements are the square of the errors on data σ_{d_i}.

On the other hand, the phase of a seismogram at time t is decomposed as follows: $\phi = \boldsymbol{k r} + \phi'_0$, where $\boldsymbol{k}$ is the wavevector, and ϕ'_0 is the initial phase, which is the sum of several terms: $\phi_0 = \phi_0 + \phi_S + \phi_I$. ϕ_S is the initial source phase, ϕ_0 is related to the number of polar phase shifts and ϕ_I is the instrumental phase; ϕ can be measured on seismograms by Fourier transform. We usually assume that ϕ_S is correctly given by the centroid moment tensor solution. For a path between epicenter S and receiver R with an epicentral distance $Delta$, the phase ϕ is given by

$$\phi = \frac{\omega \Delta}{V_{\text{obs}}} + \phi_0 + \phi_S + \phi_I \tag{19}$$

or

$$\phi - \phi_0' = \frac{\omega \Delta}{V_{\text{obs}}(T)} = \int_E^R \frac{\omega ds}{V(T, \theta, \phi)}, \tag{20}$$

where the integral is understood to be between the epicenter E and the receiver R. Following the results of the previous section, different approximations are implicitly made when using this expression of the phase:

- Large angular order $\ell \gg 1$, but not too large (scattering problems). From a practical point of view, that means that measurements are performed in the period range $50\,\text{s} < T < 300\,\text{s}$.
- Geometrical optics approximation: if λ is the wavelength of the surface wave at period T, and Λ_S the spatial wavelength of heterogeneity: $\Lambda_S \gg \lambda = VT \Rightarrow \Lambda_S \gtrsim 2000\,\text{km}$.
- Slight anisotropy and heterogeneity: $\frac{\delta V}{V} \ll 1$. According to *Smith* and *Dahlen* [124] for the plane case (18,19), the local phase velocity can be decomposed as a Fourier series of the azimuth Ψ:

$$\frac{\delta V(T, \theta, \phi)}{V(T, \theta, \phi)} = A_0 + A_1 \cos 2\Psi + A_2 \sin 2\Psi + A_3 \cos 4\Psi + A_4 \sin 4\Psi. \tag{21}$$

Each azimuthal term $A_i(T, \theta, \phi)$ can be related to the set of parameters p_i (density + 13 elastic parameters):

$$\frac{\Delta}{V_{\text{obs}}(T)} - \frac{\Delta}{V_0(T)} = -\sum_{j=0}^{2} \sum_{i=1}^{13} \int_E^R \frac{ds}{V_0} \int_0^a \left[\left(\frac{p_i}{V} \frac{\partial V}{\partial p_i} \right)_j \frac{\delta p_i(\boldsymbol{r})}{p_i} \cos(2j\Psi) \right.$$
$$\left. + \left(\frac{p_i}{V} \frac{\partial V}{\partial p_i} \right)_j \frac{\delta p_i(\boldsymbol{r})}{p_i} \sin(2j\Psi) \right] \frac{dz}{\Delta h}. \tag{22}$$

However, many terms in (22) are equal to zero, since not all of the parameters are present in each azimuthal term. Following the approach of *Snieder* [93], the approximations that have been made mean that the perturbed medium is at the same time smooth and weak.

4.1.2 Parameter Space: $p(r)$

It is quite important to thoroughly consider the structure of the parameter space. First of all, it is necessary to define which parameters are required to explain the data set, and how many physical parameters can be effectively inverted for, in the framework of the theory that is considered. For example, if the Earth is assumed to be elastic, laterally heterogeneous but isotropic, only three independent physical parameters, V_P, V_S and density ρ (or the elastic moduli λ, $\mu + \rho$), can be inverted for, from surface waves. In a transversely isotropic medium with a vertical symmetry axis [44,126], the number of independent physical parameters is six (5 elastic moduli + density). In the most general case of weak anisotropy, 14 physical parameters (13 combinations of elastic moduli + density) can be inverted for from surface waves. Therefore, the number of "physical" parameters p_i is dependent on the underlying theory which is used to explain the data set.

Once the number of "physical" independent parameters is defined, we must define how many "spatial" (or geographical) parameters are required to describe the 3-dimensional distributions $p_i(r, \theta, \phi)$. This is a difficult problem because the number of "spatial" parameters which can be reliably retrieved from the data set is not necessarily sufficient to provide a correct description of $p_i(r, \theta, \phi)$. The correct description of $p_i(r, \theta, \phi)$ is dependent on its spectral content; for example, if $p_i(r, \theta, \phi)$ is characterized by very large wavelengths, only a small number of spatial parameters is necessary, but if $p_i(r, \theta, \phi)$ presents very small-scale features, the number of spatial parameters will be very large. In any case, it is necessary to assess the range of possible variations for $p_i(r, \theta, \phi)$ in order to provide some bounds on the parameter space. This is done through a covariance function of parameters in the continuous case (or a covariance matrix for the discrete case). These a priori constraints can be provided by other fields in geoscience, geology, mineralogy, numerical modeling, etc.

Consequently, a tomographic technique must not be restricted to the inversion of the parameters $\boldsymbol{p} = \{p_i(r, \theta, \phi)\}$ that are searched for, but must include the calculation of the final covariance function (or matrix) of parameters C_p. That means that the retrieval of parameters is contingent to the resolution and the errors of the final parameters and is largely dependent on the resolving power of the data [94,95,96].

Finally, the functional g, which expresses the theory relating the data space to the parameter space, is also subject to uncertainty. In order to be completely consistent, it is necessary to define the domain of validity of the theory and to assess the error σ_T associated with the theory. *Tarantola* and *Valette* [97] showed that the error σ_T is simply added to the error on data σ_d.

4.2 Inverse Problem

Equation (22), expressing the first-order perturbation theory of the forward problem in the linear case, can be simply written

$$\boldsymbol{d} = \mathbf{G}\boldsymbol{p},$$

where $\mathbf{G}$ is a matrix (or a linear operator) composed of Fréchet derivatives of $\boldsymbol{d}$ with respect to $\boldsymbol{p}$, which has the dimensions $n_\mathrm{d} \times n_\mathrm{p}$ (number of data × number of parameters). This matrix usually is not square and many different techniques in the past have been used to invert $\mathbf{G}$. In any case, the inverse problem involves finding an inverse for the functional $\boldsymbol{g}$, which we will denote $\tilde{g}^{-1}$, without considering the way it is obtained, such that:

$$\boldsymbol{p} = \tilde{g}^{-1}(\boldsymbol{d}).$$

To solve the inverse problem, different algorithms can be used. A quite general algorithm has been derived by *Tarantola* and *Valette* [97]:

$$\begin{aligned}\boldsymbol{p} - \boldsymbol{p}_0 &= (\mathbf{G}^\mathrm{T}\mathbf{C}_\mathrm{d}^{-1}\mathbf{G} + \mathbf{C}_{p_0}^{-1})^{-1}\mathbf{G}^\mathrm{T}\mathbf{C}_\mathrm{d}^{-1}[\boldsymbol{d} - \boldsymbol{g}(\boldsymbol{p}) + \mathbf{G}(\boldsymbol{p} - \boldsymbol{p}_0)] \\ &= \mathbf{C}_{p_0}\mathbf{G}^\mathrm{T}(\mathbf{C}_\mathrm{d} + \mathbf{G}\mathbf{C}_{p_0}\mathbf{G}^\mathrm{T})^{-1}[\boldsymbol{d} - \boldsymbol{g}(\boldsymbol{p}) + \mathbf{G}(\boldsymbol{p} - \boldsymbol{p}_0)],\end{aligned} \quad (23)$$

where C_d is the covariance matrix of data, C_{p_0} the covariance function of parameters $\boldsymbol{p}$, and G is the Frechet derivative of the operator $\boldsymbol{g}$ at point $\boldsymbol{p}(\boldsymbol{r})$. This algorithm can be made more explicit by writing it in its integral form:

$$\boldsymbol{p}(\boldsymbol{r}) = \boldsymbol{p}_0(\boldsymbol{r}) + \sum_i \sum_j \int_V \mathrm{d}\boldsymbol{r}' \mathbf{C}_{p_0}(\boldsymbol{r},\boldsymbol{r}') G_i(\boldsymbol{r}')(S^{-1})_{ij} F_j, \quad (24)$$

with $S_{ij} = C_{d_{ij}} + \int_V \mathrm{d}\boldsymbol{r}_1 \mathrm{d}\boldsymbol{r}_2\, G_i(\boldsymbol{r}_1) C_{p_0}(\boldsymbol{r}_1, \boldsymbol{r}_2) G_j(\boldsymbol{r}_2)$ and $F_j = d_j - g_j(\boldsymbol{p} + \int_V \mathrm{d}\boldsymbol{r}"\, G_j(\boldsymbol{r}")\,[\boldsymbol{p}(\boldsymbol{r}") - \boldsymbol{p}_0(\boldsymbol{r}")]$.

This algorithm can be iterated and enables the solution of slightly nonlinear problems, which is the case for inversion at depth. In the case of a large data set, *Montagner* and *Tanimoto* [14] showed how to handle the inverse problem by making a series expansion of the inverse of matrix $\mathbf{S}$. One of the advantages of this technique is that it can be applied indifferently to regional studies or global studies. In the case of imperfect spatial coverage of the area under investigation, it does not display ringing phenomena commonly observed when a spherical harmonics expansion is used [98].

Different strategies can be followed to invert for the 3-dimensional models $\boldsymbol{p}(\boldsymbol{r})$, because the size of the inverse problem is usually enormous in practical applications. For the example of mantle tomography, a *minimum* parameter space will be composed of 13 (+density) physical parameters multiplied by 30 layers (if the mantle is divided into 30 independent layers). If geographical distributions of parameters are searched for up to degree 36 (lateral resolution around 1000 km), that implies a number of about 600 000 independent parameters. Such a problem is still very hard to handle from a computational point of view. A simple approach for solving this problem consists of dividing the inversion procedure into 2 steps (see *Montagner* [79] for a description of this approach).

The choice of the parameterization is also very important, and different possibilities can be considered:

- Discrete basis of functions: For a global study, the natural basis is composed of the spherical harmonics for the horizontal variations. For the radial variations, polynomial expansions can be used (see, for example, [99] for Tshebyshev polynomials). Another possibility is to divide the Earth into cells of various size according to the resolution expected from the path coverage. The cell decomposition is as valid for global investigations as for regional studies.
- Continuous function $\mathbf{p}(\mathbf{r})$. In this case, the function is directly inverted for. Since the number of parameters is infinite, it is necessary to define a covariance function of parameters $\mathbf{C}_{p_0}(\mathbf{r},\mathbf{r}')$. For the horizontal variations, we can use a Von Mises distribution [100,101] for initial parameters $\mathbf{p_0}(\mathbf{r})$:

$$\mathbf{C}_{p_0}(\mathbf{r},\mathbf{r}') = \sigma_p(\mathbf{r})\sigma_p(\mathbf{r}') \exp \frac{\cos \Delta_{rr'} - 1}{L_{\text{cor}}^2},$$

where L_{cor} is the correlation length which will define the smoothness of the final model. This kind of distribution is well suited for studies on a sphere and is asymptotically equivalent to a Gaussian distribution when $L_{\text{cor}} \ll a$ (a: radius of the Earth). When different azimuthal terms distributions are searched for, it is possible to define cross-correlated covariance functions of parameters $\mathbf{C}_{p_i,p_j}(\mathbf{r},\mathbf{r}')$, but it can be assumed that the different terms of the Fourier expansion in azimuth correspond to orthogonal functions, so the cross-correlated terms outside the diagonal can be taken to be equal to zero.

For the inversion at depth, since the number of physical parameters is very large, it is difficult to assume that physical parameters are uncorrelated. Then, the different terms of the covariance function C_p between parameters p_1 and p_2 at radii r_i and r_j can be defined as follows:

$$\mathbf{C}_{p_1,p_2}(r_i,r_j) = \sigma_{p_1}\sigma_{p_2}\zeta_{p_1,p_2} \exp \left[(r_i - r_j)^2/2L_{r_i}L_{r_j}\right], \qquad (25)$$

where ζ_{p_1,p_2} is the correlation between physical parameters p_1 and p_2 inferred, for instance, from different petrological models [37] and L_{r_i}, L_{r_j} are the radial correlation lengths which enable the inverse model to be smoothed.

The a posteriori covariance function is given by

$$\mathbf{C}_p = \mathbf{C}_{p_0} - \mathbf{C}_{p_0}\mathbf{G}^{\mathrm{T}}(\mathbf{C}_d + \mathbf{G}\mathbf{C}_{p_0}\mathbf{G}^{\mathrm{T}})^{-1}\mathbf{G}\mathbf{C}_{p_0} = (\mathbf{G}^{\mathrm{T}}\mathbf{C}_d^{-1}\mathbf{G} + \mathbf{C}_{p_0}^{-1})^{-1}. \quad (26)$$

The resolution R of parameters can be calculated as well. It corresponds to the impulsive response of the system:

$$p = \tilde{g}^{-1}d = \tilde{g}^{-1}g\,p' = \mathbf{R}\,p'.$$

If the inverse problem is perfectly solved, **R** is the identity function or matrix. However, the following expression of resolution is only valid in the linear case [124]:

$$\mathbf{R} = \mathbf{C}_{p_0}\mathbf{G}^{\mathrm{T}}(\mathbf{C}_d + \mathbf{G}\mathbf{C}_{p_0}\mathbf{G}^{\mathrm{T}})^{-1}\mathbf{G} = (\mathbf{G}^{\mathrm{T}}\mathbf{C}_d\mathbf{G} + \mathbf{C}_{p_0})^{-1}\mathbf{G}^{\mathrm{T}}\mathbf{C}_d^{-1}\mathbf{G}. \quad (27)$$

It is interesting to note, that the local resolution of parameters is imposed by both the correlation length and the path coverage, in contrast to the *Backus–Gilbert* [94,95] approach, which primarily depends on the path coverage. The effect of a damping factor in the algorithm to smooth the solution is equivalent to the introduction of a simple covariance function on parameters weighted by the errors on data [102]. When the correlation length is chosen to be very small, the algorithms of *Backus–Gilbert* [95,96] and *Tarantola* and *Valette* [97] are equivalent.

By considering the a posteriori covariance function and the resolution, it is possible to assess the reliability of the hypotheses made about the independence of parameters. For example *Tanimoto* and *Anderson* [62] and *Montagner* and *Jobert* [103] showed that there is a trade-off between azimuthal terms and constant terms in the case of poor azimuthal coverage. For inversion at depth, *Nataf* et al. [6] display the trade-off between physical parameters V_{PH}, V_{SV}, ξ, ϕ and η when only Rayleigh- and Love-wave 0-Ψ terms are used in the inversion process.

4.3 Practical Implementation

The complete anisotropic tomographic procedure has been implemented for different regional and global studies. From petrological and mineralogical considerations, *Montagner* and *Nataf* [104] and *Montagner* and *Anderson* [37,81] showed that the predominant terms of phase-velocity azimuthal expansion are the 0-Ψ and 2-Ψ terms for Rayleigh waves and 0-Ψ and 4-Ψ terms for Love waves. *Montagner* and *Nataf* [104] showed that the best-resolved parameters are $L = \rho V_{\mathrm{SV}}^2$, $N = \rho V_{\mathrm{SH}}^2$ and G_c, G_s and E_c, E_s, which respectively express the azimuthal variations of V_{SV} and V_{SH}.

4.4 Geophysical Applications

The number of applications of seismic tomography is very large. Seismic tomography is the most efficient approach to visualize, at the same time, seismic velocities and anisotropy heterogeneities, which can in turn, be related to temperature, flow directions and petrological heterogeneities. Therefore, the different fields interested in seismic tomography are geodynamics, gravimetry, geochemistry and tectonics. We will briefly review what kind of information can be provided by tomographic models.

4.4.1 Geodynamics

The most popular application of large-scale tomographic models is the understanding of mantle convection. Seismic velocity anomalies can be converted, under some assumptions, into temperature anomalies, density anomalies but also into chemical or mineralogical heterogeneities. Since the work *Hager* et al. [106], numerous studies have been devoted to the correlation between 3-dimensional seismic velocity structure, the dynamic topography, the geoid and gravimetric anomalies. The main advantage of anisotropy measurements is to provide the principal directions of the strain rate tensor related to flow directions [123]. Therefore, the simultaneous use of seismic velocity and anisotropy heterogeneities enables temperature and petrological heterogeneities, and their directions of flow, to be spatially located. *Montagner* and *Nataf* [104] presented a method for inverting a local symmetry axis. *Montagner* and *Jobert* [103] and *Hadiouche* et al. [67] have been able to plot the 3-dimensional distribution of this symmetry axis in the Indian Ocean and in Africa, respectively. *Montagner* [107] presented what seismic global tomographic models can tell us about mantle convection and what robust features can be determined from the different available models.

Most tomographic models agree that down to about 300 km the deep structure is closely related to plate tectonics and continental distribution. Simultaneously to the SV-wave velocity, three anisotropic parameters are well resolved: the ξ parameter expresses the relative variation of $V_{\rm SV}$ versus $V_{\rm SH}$, providing the tendency for the flow to be radial or horizontal, and the G and Ψ_G parameters express the azimuthal variation of $V_{\rm SV}$. Figure 11 presents two vertical cross-sections for $V_{\rm SV}$ (Fig. 11a) and ξ (Fig. 11b) from the AUM model [14], which illustrates the most robust features of the upper mantle models published so far. In the upper depth range (down to 200 km), plate boundaries are slow: ridges and back-arc areas are slow, shields are fast and seismic velocity in oceanic areas increases with the age of the seafloor. Comparison between Figs. 11a and 11b shows that both maps are poorly correlated. That suggests that they are conveying independent, but complementary information.

Figure 12 displays horizontal cross-sections at 200 km depth for $V_{\rm SV}$, ξ and G. The amplitude of SV-wave azimuthal anisotropy (G parameter) shows an average value of about 2% below oceanic areas (Fig. 12c). There is good correlation between seismic azimuthal anisotropy and plate velocity directions given by *Minster* and *Jordan* [108]. For interpretation in terms of mantle convection, it is important to consider these three maps simultaneously. For example, the existence of a maximum in radial anisotropy in the depth range 200–300 km below shields suggests that the roots of shields are located in this depth range, whereas some fast velocity anomalies are still present below some continents [31,109]. As depth increases, the amplitude of heterogeneities rapidly decreases, some trends tend to vanish, and some distinctive features appear: fast ridges are still slow but slow ridges are hardly

(a) *VS Velocity: Radial cross-section 30 deg.*

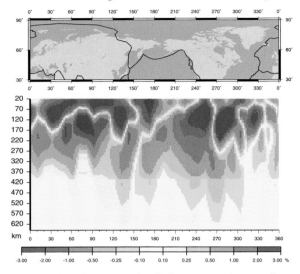

(b) *Xi Anisotropy: Radial cross-section 30 deg.*

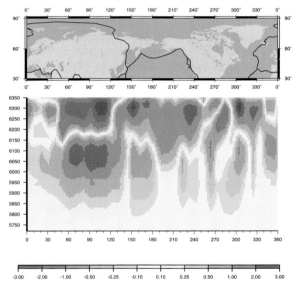

Fig. 11. Two examples of tomographic images obtained by simultaneous inversion of isotropic parameters and anisotropic parameters (AUM model of [14]). Cross-sections are taken at 30°N. (**a**) SV-wave velocity. Since most velocity anomalies reflect temperature anomalies, the color scale reflects hot and cold regions at this depth. The isolines are separated by 1%. (**b**) ξ. In terms of interpretation, warm colors reflect the radial or subradial flow (upwelling or downwelling) and cold colors are related to horizontal or subhorizontal flow

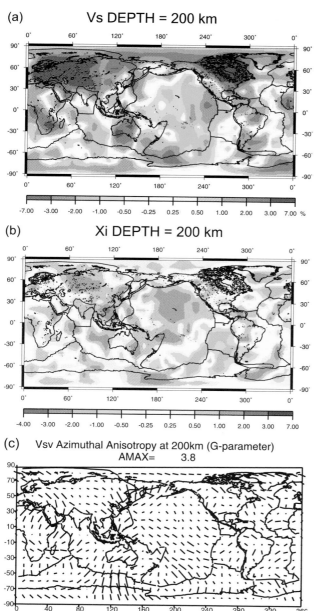

Fig. 12. Two examples of tomographic images obtained by simultaneous inversion of isotropic parameters and anisotropic parameters (AUM model of [14]). (**a**) SV-wave velocity at 200 km depth. (**b**) $\xi = N/L - 1$ radial anisotropy at 200 km depth. (**c**) SV-wave azimuthal anisotropy expressed by parameter G at 200 km. There is good correlation between the directions of maximum velocity and tectonic plate velocity directions

visible, and back-arc regions are no longer systematically slow. Large portions of fast ridges are offset with respect to their surface signatures. Below 300 km, a high velocity body below the western Pacific can be related to subducting slabs. In order to enable a quantitative comparison with other geophysical observables, tomographic models are usually expanded in spherical harmonics according to

$$f(r,\theta,\phi) = \sum_{\ell=0}^{\ell_{\max}} \sum_{m=-\ell}^{m=\ell} a_\ell^m(r) Y_\ell^m(\theta,\phi),$$

where r, θ, ϕ are the spherical coordinates at r and $Y_\ell^m(\theta,\phi)$ is the spherical harmonic of angular order ℓ and azimuthal order m. Another important parameter is the power spectrum $P_\ell(r)$, which provides the amplitude of anomalies at different degrees ℓ at different depths r.

In the first 300–400 km, the power spectrum regularly decreases with decreasing wavelength. This decrease can be described by a ℓ^{-1} law [105]. At greater depth, in the transition zone, degree-2 [100] and to a less extent, degree-6 distributions become predominant. It is also found that degree-4 radial anisotropy [14,16] is the most important degree for this parameter. A simple flow pattern with two upgoing and two downgoing large-scale flows can be invoked to simply explain the predominance of these different degrees [111]. Therefore, below the apparent complexity of plate tectonics, mantle convection is surprisingly simply organized in the transition zone. Between 400 and 1000 km, these large-scale flows are not independent of the circulation in the first 400 km but are related to the most tectonically active zones (fast ridges and slabs). This simple flow pattern, usually called a degree-2 pattern, is also present in the lower mantle but offset with respect to the one in the transition zone.

It is also suggested that the tomographic degree 6 is not independent of the deep degree 2 but might be a consequence of this simple flow pattern [111,112]. Since the hotspot distribution displays a large degree 6 [113], the good correlation between hotspots and seismic degree 6 favors an origin of most of hotspots in the transition zone. However, 2 superplumes in the Central Pacific and Central Africa have their origin in the lower mantle. As shown by *Vinnik* et al. [114] in the Pacific Ocean, these superplumes might feed other plumes, by a branching effect in the transition zone. The same branching effect was evidenced below the lithosphere in Africa [67], where the plume below the Horn of Africa is feeding other small plumes in western Africa. However, a complete understanding of the structure of plumes is still missing. The application of new techniques based on the scattering effects of plumes on surface waves [115] or body waves [116] should provide some answers to this difficult issue.

There seems to be a global decoupling in the mantle, between upper structure and lower structure around 800–1000 km [107], or around 1700 km [127], but there is some radial continuity of seismic velocity in the whole mantle in some places where slabs are present, and for two superplumes in central Pacific and Africa. All these issues are still subject of vigorous debate and call for more reliable tomographic models in the transition zone. However, there is a good consensus that the mantle cannot be divided into independent convecting cells but is characterized by imbricated convection where different scales coexist and where exchange of matter is possible.

4.4.2 Other Applications in Geochemistry, Tectonics and Petrology

Earth sciences are by essence multidisciplinary, and progress in one field can benefit other fields. For example, there were some attempts to make a quantitative comparison between the major-element chemistry of basalts erupted at mid-ocean ridges (MORB) and upper mantle shear-wave velocity derived from seismic tomography [117]. The main advantage of this approach is that it can provide a way for locating at depth the reservoirs displayed by geochemists. For instance, a strong correlation between basalt chemistry and seismic velocity has been found at depths 100–170 km, for lateral wavelengths of 1000–2400 km, supporting a common thermal origin for the two types of signal. This kind of simple approach can be easily generalized to other types of geochemical parameters such as isotopic elements.

Seismic profiles have long been used to infer the mineralogy of the mantle [118]. The competing petrological models for the upper mantle and transition zone are pyrolite [35] and piclogite [38,39]. So far, the isotropic seismic velocities can be explained down to 400 km by a pyrolite model, but in the transition zone, neither the pyrolite nor the piclogite model is better. *Montagner* and *Anderson* [37] investigated the correlations between anisotropic parameters for realistic mineralogical and petrological models of the upper mantle. They show that the anisotropic parameters involved in a radiallly anisotropic medium, A, C, F, L and N, are strongly correlated but that the 8 other anisotropic parameters, $B_c, B_s, G_c, G_s, H_c, H_s, E_c$ and E_s, involved in azimuthal anisotropy are less correlated. A complete exploitation of the 3-dimensional anisotropic tomographic models has not yet been done. This kind of approach might provide some important constraints on mineralogy in the deep mantle in the future.

The strain field near the surface is probably different from that at depth and could also be related to the strain field prevailing during the setting of materials. This shallow anisotropy could be very useful for understanding the strain field responsible for surficial tectonics. For example, seismic anisotropy could be used for explaining geological observations, such as mountain-range building or more generally continental deformation. Such an application has been attempted by *Vinnik* et al. [48,49,50], *Silver* and *Chan* [55,119] and *Silver* [57] by using anisotropy derived from SKS splitting. The poor lateral resolution of large-scale anisotropic tomography can be considered as a strong limitation in continental areas. This technique can only be efficiently applied to areas where large-scale strain and stress fields are implied. By applying this technique to Central Asia, *Griot* et al. [68,69] were able to discriminate between two extreme models of deformation, the heterogeneous model of *Avouac* and *Tapponnier* [120] and the homogeneous model of *England* and *Houseman* [121]. They show that the heterogeneous model is in better agreement with observations in the first uppermost 200 km, whereas the homogeneous model better fits the deep anisotropy below 200 km. This result provides a rough estimate of the thickness of the continental lithosphere.

These simple examples show that seismic anisotropy tomography is still largely unexploited by the community of Earth scientists and will constitute an invaluable source of information and inspiration.

5 Conclusions

We have presented in this paper some basic first-order asymptotic theories which make it possible to derive seismic anisotropy tomographic models. By applying the anisotropic technique to seismic data, seismologists make use of the full potential of the three-component seismograms. In addition, they are able not only to image temperature anomalies but also to map the flow of matter in the Earth's mantle.

The next steps will consist of taking simultaneous account of the phase and amplitude of seismic waves. By using new theoretical developments (see, for example, *Clévédé* and *Lognonné* [122]), it will be possible to calculate synthetic seismograms in complex a priori laterally heterogeneous media and to make a direct comparison with seismic waveforms in the time domain. In addition, it will be possible to correctly assess the effect of scattering on seismograms. Notwithstanding these future theoretical improvements, it is desirable to increase the lateral resolution of tomographic models in order to obtain images in a broad spatial scale range. This will be done, firstly, by installing broadband networks at smaller scale (wavelengths smaller than 1000 km), and secondly, by implementing an ocean seismic network, which will provide a better coverage of the whole Earth by seismic waves.

Acknowledgements

CNRS UMR 7580 — Institut Universitaire de France — Institut de Physique du Globe. This is IPGP contribution no. 1780.

References

1. L. Knopoff, Observation and inversion of surface wave inversion, Tectonophysics **13**, 497–519 (1972)
2. J. H. Woodhouse, A. M. Dziewonski, Mapping the upper mantle: Three dimensional modelling of Earth structure by inversion of seismic waveforms, J. Geophys. Res. **89**, 5953–5986 (1984)

3. A. M. Dziewonski, Mapping the lower mantle: Determination of lateral heterogeneity in P velocity up to degree and order 6, J. Geophys. Res. **89**, 5929–5952 (1984)
4. I. Nakanishi, D. L. Anderson, Measurement of mantle wave velocities and inversion for lateral heterogeneity and anisotropy, II. Analysis by the single station method, Geophys. J. R. Astron. Soc. **78**, 573–618 (1984)
5. H.-C. Nataf, I. Nakanishi, D. L. Anderson, Anisotropy and shear velocity heterogeneities in the upper mantle, Geophys. Res. Lett. **11**, 109–112 (1984)
6. H.-C. Nataf, I. Nakanishi, D. L. Anderson, Measurement of mantle wave velocities and inversion for lateral heterogeneity and anisotropy, III. Inversion, J. Geophys. Res. **91**, 7261–7307 (1986)
7. D. Agnew, J. Berger, R. Buland, W. Farrell, F. Gilbert, International deployment of accelerometers: A network of very long period seismology, EOS, Trans. Am. Geophys. Union **57**, 180–188 (1976)
8. J. Peterson, H. M. Butler, L. G. Holcomb, C. R. Hutt, The Seismic Research Observatory, Bull. Seism. Soc. Am. **66**, 2049–2068 (1977)
9. B. Romanowicz, M. Cara, J. F. Fels, D. Rouland, GEOSCOPE: a French initiative in long period, three component, global seismic networks, EOS, Trans. Am. Geophys. Union **65**, 753–754 (1984)
10. S. W. Smith, IRIS, a program for the next decade, EOS, Trans. Am. Geophys. Union **67**, 213–219 (1986)
11. J.-P. Montagner, Surface waves on a global scale — Influence of anisotropy and anelasticity, In *Seismic Modeling of the Earth's Structure*, ed. by E. Boschi, G. Ekström, A. Morelli, Summer School of Erice, (Bologna 1996) p. 81–148
12. J.-P. Montagner, First results on the three dimensional structure of the Indian Ocean inferred from long period surface waves, Geophys. Res. Lett. **13**, 315–318, (1986)
13. J.-P. Montagner, T. Tanimoto, Global anisotropy in the upper mantle inferred from the regionalization of phase velocities, J. Geophys. Res. **95**, 4797–4819 (1990)
14. J.-P. Montagner, T. Tanimoto, Global upper mantle tomography of seismic velocities and anisotropies, J. Geophys. Res. **96**, 20 337–20 351 (1991)
15. T. Tanimoto, Waveform inversion of Love waves: The Born Seismogram approach, Geophys. J. R. Astron. **78**, 641–660 (1984)
16. G. Roult, B. Romanowicz, J.-P. Montagner, 3D upper mantle shear velocity and attenuation from fundamental mode free oscillation data, Geophys. J. Int. **101**, 61–80 (1990)
17. B. Romanowicz, The upper mantle degree two: Constraints and inferences from global mantle wave attenuation measurements, J. Geophys. Res. **95**, 11051–11071 (1990)
18. T. Tanimoto, Waveform inversion of mantle Love waves: The Born seismogram approach, Geophys. J. R. Astron. Soc. **78**, 641–660 (1984)
19. J. K. Wong, Upper mantle heterogeneity from phase and amplitude data of mantle waves, PhD Thesis, Harvard University, Cambridge MA (1989)
20. R. Snieder, Large-scale waveform inversions of surface waves for lateral heterogeneity, 1. Theory and numerical examples, J. Geophys. Res. **93**, 12 055–12 066 (1988)
21. R. Snieder, Large-scale waveform inversions of surface waves for lateral heterogeneity, 2. Application to surface waves in Europe and the Mediterranean, J. Geophys. Res. **93**, 12 067–12 080 (1988)

22. E. Wielandt, G. Streickeisen, The leaf-spring seismometer: design and performances, Bull. Seism. Soc. Am. **72**, 2349–2367 (1982)
23. S. Cacho, Etude et Réalisation d'un sismomètre très large bande, 3 axes, qualifié spatial, Thèse de l'Université Paris VII (1996)
24. P. Lognonné, J. Gagnepain-Beyneix, W. B. Banerdt, S. Cacho, J.-F. Karczewski, An ultra-broadband seismometer in InterMarsnet, Planet. Space Sci. **44**, 1237–1249 (1996)
25. J.-P. Montagner, P. Lognonné, R. Beauduin, G. Roult, J.-F. Karczewski, E. Stutzmann, Towards multiscale and multiparameter networks for the next century: The French efforts, Phys. Earth Planet. Int. **108**, 155–174 (1998)
26. J. Peterson, Observation and modeling of background seismic noise, I. S. Geol. Surv. Open-file report 93–222, Albuquerque (1993)
27. J.-P. Montagner, B. Romanowicz, J. F. Karczewski, A first step towards an Oceanic Geophysical observatory, EOS, Trans. Am. Geophys. Union **75**, 150–154 (1994)
28. K. Suyehiro, T. Kanazawa, N. Hirata, M. Shinohara, H. Kinoshita, Broadband downhole digital seismometer experiment at site 794: a technical paper, Proc. ODP, Sc. Results, (1992) p. 127–128
29. J. H. Woodhouse, F. A. Dahlen, The effect of a general aspherical perturbation on the free oscillations of the Earth, Geophys. J. R. Astron. Soc. **53**, 335–354 (1978)
30. A. R. Edmonds, *Angular Momentum and Quantum Mechanics* (Priceton University Press, Priceton NJ, 1960)
31. J.-P. Montagner, Where can seismic anisotropy be detected in the Earth's mantle? In boundary layers ..., Pure Appl. Geophys. **151**, 223–256 (1998)
32. L. Peselnick, A. Nicolas, P. R. Stevenson, Velocity anisotropy in a mantle peridotite from Ivrea zone: Application to upper mantle anisotropy, J. Geophys. Res. **79**, 1175–1182 (1974)
33. D. L. Anderson, *Theory of the Earth* (Blackwell, Oxford 1989)
34. V. Babuska, M. Cara, *Seismic Anisotropy in the Earth* (Kluwer Academic, Dordrecht 1991)
35. A. E. Ringwood, *Composition and petrology of the Earth's mantle* (McGraw-Hill, New York 1975) pp. 618
36. N. I. Christensen, S. Lundquist, Pyroxene orientation within the upper mantle, Bull. Geol. Soc. Am. **93**, 279–288 (1982)
37. J.-P. Montagner, D. L. Anderson, Constraints on elastic combinations inferred from petrological models, Phys. Earth Planet. Int. **54**, 82–105 (1989)
38. D. L. Anderson, J. D. Bass, Mineralogy and composition of the upper mantle, Geophys. Res. Lett. **11**, 637–640 (1984)
39. D. L. Anderson, J. D. Bass, Transition region of the Earth's upper mantle, Nature **320**, 321–328 (1986)
40. L. Peselnick, A. Nicolas, Seismic anisotropy in an ophiolite peridotite. Application to oceanic upper mantle, J. Geophys. Res. **83**, 1227–1235 (1978)
41. A. Nicolas, Why fast polarization directions of SKS seismic waves are parallel to mountain belts? Phys. Earth Planet. Int. **78**, 337–342 (1993)
42. A. Vauchez, A. Nicolas, Mountain building: strike-parallel motion and mantle anisotropy, Tectonophysics **185**, 183–191 (1991)
43. N. M. Ribe, Seismic anisotropy and mantle flow, J. Geophys. Res. **94**, 4213–4223 (1989)

44. D. L. Anderson, Elastic wave propagation in layered anisotropic media, J. Geophys. Res. **66**, 2953–2963 (1961)
45. K. Aki, K. Kaminuma, Phase velocity of Love waves in Japan (part 1): Love waves from the Aleutian shock of March 1957, Bull. Earthq. Res. Inst. **41**, 243–259 (1963)
46. H. Hess, Seismic anisotropy of the uppermost mantle under the oceans, Nature **203**, 629–631 (1964)
47. L. P. Vinnik, G. L. Kosarev, L. I. Makeyeva, Anisotropiya litosfery po nablyudeniyam voln SKS and SKKS, Dokl. Akad. Nauk USSR **278**, 1335–1339 (1984)
48. L. P. Vinnik, R. Kind, G. L. Kosarev, L. I. Makeyeva, Azimuthal Anisotropy in the lithosphere from observations of long–period S–waves, Geophys. J. Int. **99**, 549–559 (1989)
49. L. P. Vinnik, V. Farra, B. Romanowicz, Azimuthal anisotropy in the earth from observations of SKS at GEOSCOPE and NARS broadband stations, Bull. Seism. Soc. Am. **79**, 1542–1558 (1989)
50. L. Vinnik, L. I. Makayeva, A. Milev, A. Y. Usenko, Global patterns of azimuthal anisotropy and deformations in the continental mantle, Geophys. J. Int. **111**, 433–447 (1992)
51. M. Ando, ScS polarization anisotropy around the Pacific Ocean, J. Phys. Earth **32**, 179–196 (1984)
52. J. Fukao, Evidencec from Core-reflected Shear waves for Anisotropy in the Earth's mantle, Nature **309**, 695–698 (1984)
53. M. Ando, Y. Ishikawa, F. Yamazaki, Shear wave polarization anisotropy in the upper mantle beneath Honshu, Japan, J. Geophys. Res. **88**, 5850–5864 (1983)
54. J. R. Bowman, M. Ando, Shear-wave splitting in the upper mantle wedge above the Tonga subduction zone, Geophys. J. R. Astron. Soc. **88**, 25–41 (1987)
55. P. G. Silver, W. W. Chan, Implications for continental structure and evolution from seismic anisotropy, Nature **335**, 34–39 (1988)
56. V. Ansel, H.C Nataf, Anisotropy beneath 9 stations of the Geoscope broadband network as deduced from shear wave splitting, Geophys. Res. Lett. **16**, 409–412 (1989)
57. P. G. Silver, Seismic anisotropy beneath the continents: Probing the depths of geology, Annu. Rev. Earth Planet. Sci. **24**, 385–432 (1996)
58. A. Levshin, L. Ratnikova, Apparent anisotropy in inhomogeneous media, Geophys. J. R. Astron. Soc. **76**, 65–69 (1984)
59. D. W. Forsyth, The early structural evolution and anisotropy of the oceanic upper mantle, Geophys. J. R. Astron. Soc. **43**, 103–162 (1975)
60. B. J. Mitchell, G.-K. Yu, Surface wave dispersion, regionalized velocity models and anisotropy of the Pacific crust and upper mantle, Geophys. J. R. Astron. Soc. **63**, 497–514 (1980)
61. J.-P. Montagner, Seismic anisotropy of the Pacific Ocean inferred from long-period surface wave dispersion, Phys. Earth Planet. Int. **38**, 28–50 (1985)
62. T. Tanimoto, D. L. Anderson, Lateral heterogeneity and azimuthal anisotropy of the upper mantle: Love and Rayleigh waves 100–250 s, J. Geophys. Res. **90**, 1842–1858 (1985)
63. E. Debayle, J.-J. Lévêque, Upper mantle heterogeneities in the Indian Ocean froml waveform inversion, Geophys. Res. Lett. **24**, 245–248 (1997)

64. D. Suetsugu, I. Nakanishi, Regional and azimuthal dependence of phase velocities of mantle Rayleigh waves in the Pacific Ocean, Phys. Earth Planet. Int. **47**, 230–245 (1987)
65. C. E. Nishimura, D. W. Forsyth, The anisotropic structure of the upper mantle in the Pacific, Geophys. J. **96**, 203–229 (1989)
66. G. Silveira, E. Stutzmann, J.-P. Montagner and L. Mendes-Victor, Anisotropic tomography of the Atlantic Ocean from Rayleigh surface waves, Phys. Earth Planet. Int. **106**, 259–275 (1998)
67. O. Hadiouche, N. Jobert and J. P. Montagner, Anisotropy of the African continent inferred from surface waves, Phys. Earth Planet. Int. **58**, 61–81 (1989)
68. D.-A. Griot, J.-P. Montagner, P. Tapponnier, Surface wave phase velocity and azimuthal anisotropy in Central Asia, J. Geophys. Res. **103**, 21215–21232 (1998)
69. D.-A. Griot, J.-P. Montagner, P. Tapponnier, Heterogeneous versus homogeneous strain in Central Asia, Geophys. Res. Lett. **25**, 1447–1450 (1998)
70. J. J. Lévêque, M. Cara, Inversion of multimode surface wave data: evidence for sub–lithospheric anisotropy, Geophys. J. R. Astron. Soc. **83**, 753–773 (1985)
71. M. Cara, J.-J. Lévêque, Anisotropy of the asthenosphere: The higher mode data of the Pacific revisited, Geophys. Res. Lett. **15**, 205–208 (1988)
72. J.-P. Montagner, H.-C. Nataf, On the inversion of the azimuthal anisotropy of surface waves, J. Geophys. Res. **91**, 511–520 (1986)
73. S. Crampin, An introduction to wave propagation in anisotropic media, Geophys. J. R. Astron. Soc. **76**, 17–28 (1984)
74. J.-P. Montagner, N. Jobert, Investigation of upper mantle structure under young regions of the Sout-East Pacific using long-period Rayleigh waves, Phys. Earth Planet. Int. **27**, 206–222 (1981)
75. M. L. Smith, F. A. Dahlen, Correction to 'The azimuthal dependence of Love and Rayleigh wave propagation in a slightly anisotropic medium', J. Geophys. Res. **80**, 1923 (1975)
76. G. Laske, G. Masters, Surface-wave polarization data and global anisotropic structure, Geophys. J. Int. **132**, 508–520 (1998)
77. E. W. Larson, J. Tromp, G. Ekström, Effects of slight anisotropy on surface waves, Geophys. J. Int. **132**, 654–666 (1998)
78. J.-P. Montagner, D.-A. Griot, J. Lavé, How to relate body wave and surface wave anisotropies?, J. Geophys. Res. **105**, 19 015–19 027 (2000)
79. J.-P. Montagner, Surface waves on a global scale — Influence of anisotropy and anelasticity, In *Seismic Modeling of the Earth's Structure*, ed. by E. Boschi, G. Ekström, A. Morelli, Summer School of Erice, (Bologna 1996) p. 81–148
80. J. Bass, D. L. Anderson, Composition of the upper mantle: Geophysical tests of two petrological models, Geophys. Res. Lett. **11**, 237–240 (1984)
81. J.-P. Montagner, D. L. Anderson, Constrained reference mantle model, Phys. Earth Planet. Int. **58**, 205–227 (1989)
82. J. W. Schlue, L. Knopoff, Shearwave Polarization in the Pacific Ocean, Geophys. J. R. Astron. Soc. **49**, 145–165, (1977)
83. A. M. Dziewonski, D. L. Anderson, Preliminary Reference Earth Model, Phys. Earth Planet. Int. **25**, 297–356 (1981)
84. D. L. Anderson, A. M. Dziewonski, Upper mantle anisotropy: Evidence from fre oscillations, Geophys. J. R. Astron. Soc. **69**, 383–404 (1982)

85. G. Nolet, Higher Rayleigh modes in Western Europe, Geophys. Res. Lett. **2**, 60–62 (1975)
86. M. Cara, Regional variations of Rayleigh-mode velocities: a spatial filtering method, Geophys. J. R. Astron. Soc. **57**, 649–670 (1978)
87. E. Okal, B.-G. Jo, stacking investigation of higher-order mantle Rayleigh waves, Geophys. Res. Lett. **12**, 421–424 (1985)
88. A. L. Lerner-Lam, T. H. Jordan, Earth structure from fundamental and higher-mode waveform analysis, Geophys. J. R. Astron. Soc. **75**, 759–797 (1983)
89. G. Nolet, Partitioned waveform inversion and two-dimensional structure under the network of autonomously recording seismographs, J. Geophys. Res. **95**, 8499–8512 (1990)
90. J. J. Lévêque, M. Cara, D. Rouland, Waveform inversion of surface-wave data: a new tool for systematic investigation of upper mantle structures, Geophys. J. Int. **104**, 565–581 (1991)
91. E. Stutzmann, J. P. Montagner, Tomography of the transition zone from the inversion of higher-mode surface waves, Phys. Earth Planet. Int. **86**, 99–116 (1994)
92. H. J. Van Heijst, J. Woodhouse, Measuring surface-wave overtone phase velocities using a mode-branch stripping technique, Geophys. J. Int. **131**, 209–230 (1997)
93. R. Snieder, Surface wave inversions on a regional scale, In *Seismic Modeling of Earth Structure*, ed. by E. Boschi, G. Ekström, A. Morelli, Summer School of Erice, (Bologna 1996) p. 149–182
94. G. E. Backus, F. Gilbert, Numerical applications of a formalism for geophysical inverse problems, Geophys. J. R. Astron. Soc. **13**, 247–276 (1967)
95. G. E. Backus, J. F. Gilbert, The resolving power of gross earth data, Geophys. J. R. Astron. Soc. **16**, 169–205 (1968)
96. G. E. Backus, F. Gilbert, Uniqueness in the inversion of inaccurate gross earth data, Philos. Trans. R. Soc. Lond. Ser. A **266**, 123–192 (1970)
97. A. Tarantola, B. Valette, Generalized non-linear inverse problems solved using the least squares criterion, Rev. Geophys. Space Phys. **20**, 219–232 (1982)
98. T. Tanimoto, The Backus-Gilbert approach to the three-dimensional structure in the upper mantle, 1. Lateral variation of surface wave phase velocity with its error and resolution, Geophys. J. R. Astron. Soc. **82**, 105–123 (1985)
99. A. M. Dziewonski, J. H. Woodhouse, Global images of the Earth's interior, Science **236**, 37–48 (1987)
100. R. A. Fisher, Dispersion on a sphere, Proc. R. Soc. London A **217**, 295 (1953)
101. J.-P. Montagner, Regional three-dimensional structures using long-period surface waves, Ann. Geophys. **4**, B3, 283–294 (1986)
102. P. Ho-Liu, J.-P. Montagner, H. Kanamori, Comparison of iterative back-projection inversion and generalized inversion without blocks: Case studies in Attenuation tomography, Geophys. J. **97**, 19–29 (1989)
103. J.-P. Montagner, N. Jobert, Vectorial Tomography. II: Application to the Indian Ocean, Geophys. J. R. Astron. Soc. **94**, 309–344 (1988)
104. J.-P. Montagner, H.-C. Nataf, Vectorial Tomography. I: Theory, Geophys. J. R. Astron. Soc. **94**, 295–307 (1988)
105. T. Tanimoto, Long-wavelength *S*-wave velocity structure throughout the mantle, Geophys. J. Int. **100**, 327–336 (1990)

106. B. H. Hager, R. W. Clayton, M. A. Richards, R. P. Comer, A. M. Dziewonski, Lower mantle heterogeneity, dynamic topography and the geoid, Nature **313**, 541–545 (1985)
107. J.-P. Montagner, What can seismology tell us about mantle convection? Rev. Geophys. **32**, 115–137 (1994)
108. J. B. Minster, T. H. Jordan, Present-day plate motions, J. Geophys. Res. **83**, 5331–5354 (1978)
109. V. Babuska, J.-P. Montagner, J. Plomerova, N. Girardin, Age-dependent large-scale fabric of the mantle lithosphere as derived from surface-wave velocity anisotropy, Pure Appl. Geophys. **151**, 257–280 (1998)
110. G. Masters, T. H. Jordan, P. G. Silver, F. Gilbert, Aspherical Earth structure from fundamental spheroidal- mode data, Nature **298**, 609–613 (1982)
111. J.-P. Montagner, B. Romanowicz, Degrees 2, 4, 6 inferred from seismic tomography, Geophys. Res. Lett. **20**, 631–634 (1993)
112. A. Cazenave, A. Souriau, K. Dominh, Global coupling of Earth surface topography with hotspots geoid and mantle heterogeneities, Nature **340**, 54–57 (1989)
113. M. A. Richards, B. H. Hager, The Earth's geoid and the large scale structure of mantle convection, In *The Physics of the Planets*, ed. by S. J. Runcorn (Wiley, New York 1988) p. 247–271
114. L. Vinnik, J.-P. Montagner, Shear wave splitting in the mantle from Ps phases, Geophys. Res. Lett. **23**, 2449–2452 (1996)
115. Y. Capdeville, E. Stutzmann, J.-P. Montagner, Effect of a plume on long period surface waves computed with normal mode coupling, Phys. Earth Planet. Int. **119**, 57–74 (2000)
116. J. Ying, H. C. Nataf, Detection of mantle plume in the lower mantle by deffrection tomography, Earth Planet. Sci. Lett. **159**, 87–98 (1998)
117. E. Hunter, J. L. Thirst, J.-P. Montagner, Global correlations of ocean ridge basalt chemistry with seismic tomographic images, Nature **364**, 225–228 (1993)
118. F. Birch, Elasticity and constitution of the Earth's interior, J. Geophys. Res. **57**, 227–28, (1952)
119. P. G. Silver, W. W. Chan, Shear wave splitting and subcontinental mantle deformation, J. Geophys. Res. **96**, 16429–16454 (1991)
120. J.-P. Avouac, P. Tapponnier, Kinematic model of active deformation in central Asia, Geophys. Res. Lett. **20**, 895–898 (1993)
121. P. England, G. Houseman, Finite strain calculations of continental deformation, 2. comparison with the India-Asia collision zone, J. Geophys. Res. **91**, 3664–3676 (1986)
122. E. Clévédé, P. Lognonné, Fréchet derivatives of coupled seismograms with to an anelastic rotating Earth, Geophys. J. Int. **124**, 456–482 (1996)
123. L. Peselnick, A. Nicolas, P. R. Stevenson, Velocity anisotropy in a mantle peridotite from Ivrea zone: Application to upper mantle anisotropy, J. Geophys. Res. **79**, 1175–1182 (1974)
124. M. L. Smith, F. A. Dahlen, The azimuthal dependence of Love and Rayleigh wave propagation in a slightly anisotropic medium, J. Geophys. Res. **78**, 3321–3333 (1973)
125. Y. Yu, J. Park, Anisotropy and coupled long-period surface waves, Geophys. J. Int. **114**, 473–489 (1993)

126. H. Takeuchi, M. Saito, Seismic surface waves, Methods Comput. Phys. **11**, 217–295 (1972)
127. R. D. Van der Hilst, H. Karason, Compositional heterogeneity in the bottom 1000 km of the Earth's mantle: Toward a hybrid convection model, Science **283**, 1885–1888 (1999)
128. G. Barruol, D. Mainprice, A quantitative evaluation of the contribution of crustal rocks to the shear-wave splitting of teleseismic SKS waves, Phys. Earth Planet. Int. **78**, 281–300 (1993)
129. B. Dost, Upper mantle structure under western Europe from fundamental and higher mode surface waves using the NARS array, Geophys. J. R. Astron. Soc. **100**, 131–151 (1990)
130. G. Ekström, A. M. Dziewonski, The unique anisotorpy of the Pacific upper mantle, Nature **394**, 168–172 (1998)
131. T. Lay, T. C. Wallace, *Modern Global Seismology* (Academic, San Diego, Calif. 1995)
132. H.-C. Nataf, Y. Ricard, 3-SMAC: An a priori tomographic model of the upper mantle based on geophysical modeling, Phys. Earth Planet. Int. **95**, 101–122 (1996)
133. R. Snieder, B. Romanowicz, A new formalism for the effect of lateral heterogeneity on normal modes and surface waves, I: Isotropic perturbations, perturbations of interfaces and gravitational perturbations, Geophys. J. R. Astron. Soc. **92**, 207–222 (1988)

Elastic-Wave Propagation in Random Polycrystals: Fundamentals and Application to Nondestructive Evaluation

Bruce R. Thompson

Center for Nondestructive Evaluation and Ames Laboratory, Departments of Materials Science and Engineering Mechanics, Iowa State University, Ames, Iowa 50011, USA
cnde@cnde.iastate.edu

Abstract. The fundamental principles that govern the propagation of elastic waves in metal polycrystals are discussed in the context of their influence on nondestructive evaluation. The major influence of the polycrystalline microstructure is to determine the velocity, attenuation and backscattering of the elastic waves. For randomly oriented, equi-axed polycrystals, these effects are reasonably well understood. Waves travel at the same velocity in all directions and are exponentially attenuated at a rate controlled by the frequency and grain size. Signals backscattered from the grains, also controlled by the wavelength and grain size, produce a background noise that competes with flaw signals. The same basic phenomena exist in more complex materials. However, the understanding of these phenomena is not as well understood. Recent progress towards the development of such an understanding is discussed within this chapter. Examples include cases in which the grains have preferred crystallographic orientation, elongation in one or more dimension, or correlations in orientation from crystallite to crystallite. The latter case is particularly rich, in that the two dimensions scales of the media, associated with the grain size and the correlation length, can lead to a number of unusual phenomena such as highly anisotropic backscattering and phase modulations of an elastic beam. These modulations make the measurement, and even definition, of attenuation problematic. The current status of experimental observation and theoretical description of these phenomena is discussed. The chapter concludes with a discussion of the implications of these effects on the imaging of flaws in complex media.

1 Introduction

Ultrasonic energy, in the form of elastic waves, is used extensively in assuring the quality of structural components, either during their manufacture or service life. Motivations include the detection, characterization, and sizing of discrete flaws that might lead to failure (e.g., inclusion or cracks), the characterization of degradation in materials during their service lives (e.g., fatigue of aircraft components or embrittlement of materials in the pressure vessels of nuclear power plants), and monitoring the changes in material structure that occur during manufacturing (e.g., grain size and porosity) to provide information for control of the manufacturing process.

Flaw detection and characterization is often done in the pulse–echo (monostatic) mode, in which a single transducer (which may be a focused transducer or an array) is used to examine the material. Modes of operation include the formation of images using a focused transducer or the inference of the properties of the flaw from the dependence of its reflectivity on various parameters such as frequency or position of the transducer. Some of these procedures are discussed in Chap. 7. Further discussions may also be found in handbooks [1,2] and in review articles that have been prepared by the author [3,4].

Consideration of the simple case of the formation of an image illustrates the role played by the complexity of the medium. The resolution of a simple lens is limited by its diffraction-limited spot size, which is proportional to $\lambda F/D$, where λ is the wavelength, D is the diameter of the transducer, and F is its focal length. In order to improve resolution, one must either decrease λ or F or increase D. Attempts to do any of these face limits. The latter two strategies are limited by technology. Decreasing F can lead to aberrations associated with refraction at the interface between water (in which the transducer is often placed to provide a coupling medium) and the solid under inspection, and increasing D can run into problems associated with transducer fabrication. In both of these strategies, array technology offers the potential for improvements. Decreasing λ, on the other hand, is limited by fundamental issues associated with the inhomogeneity of the material, as is conceptually illustrated in Fig. 1. Here, the ultrasonic signals are shown, as they would appear when a planar transducer is placed on the surface of a solid with parallel faces. If the material were homogenous, one would see a train of equally spaced echoes of equal amplitude (neglecting beam spread due to diffraction). Inhomogeneities within the material modify this by scattering energy from the beam as it propagates, leading to an exponential decay of the echo train (attenuation) and the appearance of noise between the echoes (backscattering). In addition, the interval between echoes, controlled by the ultrasonic velocity, might be modified. These scattering phenomena limit the ability to decrease the wavelength, since attenuation and backscattering become more significant at higher frequencies. Thus flaw signals become more attenuated, and the competing noise tends to obscure them.

Many structural materials are made of metal polycrystals. In this case the inhomogeneities are primarily associated with the individual grains or crystallites. Since these are elastically anisotropic, orientation changes from grain to grain lead to a continuously changing set of elastic properties. An understanding of the detailed relationship between the structure of these grains, known as the microstructure of the material, and the ultrasonic attenuation and backscattering is then central to understanding and improving the ability to detect flaws with ultrasound. This problem is made richer by

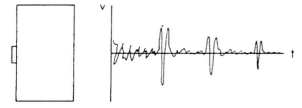

Fig. 1. Pulse–echo response in a sample with a distribution of microstructural inhomogeneities

the fact that some structural metals, with titanium being an important example, have microstructures in which there can be long range correlations between the orientation of crystallites. In the context of the subject matter of this book, an alternate title for the present chapter might be "Imaging in random media with distributed clutter having both short and long range correlations".

The importance of understanding the interaction of elastic waves with microstructure is not limited to its effects on flaw detection and characterization. As noted previously, other functions of ultrasonic nondestructive evaluation include the characterization of the material as its microstructure is developed during processing/manufacturing or degraded during service. Central to material characterization techniques is the understanding of the ultrasound–microstructure interactions.

This paper will have two major sections, respectively dealing with the effects of simple and complex microstuctures on ultrasonic propagation. A brief summary of the implications of these effects on imaging will then be presented, followed by concluding remarks. Extensive references are given to prior work of the author and colleagues, much of which has been only published in conference proceedings. No attempt is made at preparing a comprehensive review, and apologies are extended to other researchers in this field whose work could not be cited because of time and/or space constraints.

2 Simple Polycrystals

2.1 Background

Figure 2 schematically illustrates the microstructure of a simple polycrystal, which can be considered to consist of an aggregate of grains or crystallites, each defined by its size, shape and crystallographic orientation. For the purpose of this paper, a simple polycrystal will be defined as one in which all of the crystallites are of the same phase and in which the crystallographic orientations of separate grains are independent of one another. The microstructure is then defined by the distributions of crystallite size, shape, and orientation.

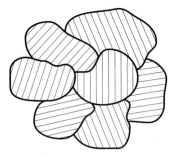

Fig. 2. Schematic microstructure of a simple polycrystal, with parallel lines representing crystallographic orientation

2.2 Theory

One can intuitively view the interaction of an elastic wave with a polycrystal as a superposition of scattering processes. When the wavelength is sufficiently large with respect to the grain size, the consequences of scattering are small and the attenuation and backscattering have low values. The velocity V is determined by a relationship of the form

$$V = \sqrt{C/\rho}, \tag{1}$$

where C is the appropriate component of the macroscopic elastic stiffness tensor and ρ is the density. Within the Voigt approximation [5], the macroscopic elastic stiffness is determined from the stiffnesses of the individual crystallites by the relation

$$\overleftrightarrow{C}_v = \langle \overleftrightarrow{C} \rangle, \tag{2}$$

where $\langle \ldots \rangle$ denotes an average over an ensemble of macroscopically equivalent polycrystals. The double arrow indicates that the elastic stiffness is a fourth-rank tensor. The element of this tensor that appears in (1) depends on the direction of wave propagation and polarization.

As the wavelength becomes shorter, scattering from the crystallites must be explicitly considered. The basic concepts are illustrated in Fig. 3a, showing the scattering from an inclusion in an elastic medium. Assuming an incident plane wave, the scattering amplitude A describes the spherically spreading scattered wave. The scattering amplitude is a function of the inclusion size, shape and composition, as well as the frequency and the angles of incidence and scattering. In the Born approximation (weak-scattering limit), the backscattered scattering amplitude for incidence in the 3-direction is given by the expression [6]

$$A = \frac{1}{(4\pi\rho V^2)} \left(\delta\rho\omega^2 - \delta C_{3333} k^2 \right) \int d^3 x \exp(2\mathrm{i}kz), \tag{3}$$

(a)

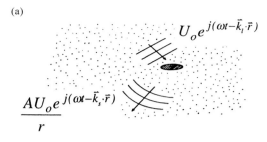

(b)

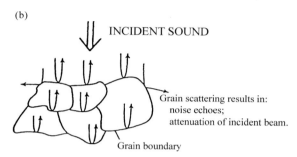

Fig. 3. Scattering processes. (**a**) Scattering from a single discontinuity, illustrating the definition of the scattering amplitude A. (**b**) Scattering from grain boundaries

where ω is the angular frequency, k is the wavevector (ω/V), $\delta\rho$ is the deviation of the density with respect to the host medium and δC is the deviation of the elastic stiffness from the host medium. The integral is evaluated over the volume of the inclusion.

As an elastic wave propagates through a polycrystal, its response may be considered as the superposition of the scattering from the individual grains, as illustrated in Fig. 3b. The fields that are scattered back towards the transducer contribute to the backscattered signals. However, the attenuation is controlled by the energy removed from the beam by scattering in all directions.

The backscattering coefficient, η, is a fundamental quantity characterizing the degree to which the microstructure generates backscattered noise. It is formally defined as the differential scattering cross-section (in the backward direction) per unit volume. Under the assumptions that the elastic anisotropy of the crystallites is not too great (so that the Born approximation can be utilized) and that only single scattering need be considered, the relationship between the backscattering coefficient and the microstructure has been developed in a series of papers [7,8,9,10]. For the case of incidence along the 3-direction the result is the expression

$$\eta(\omega) = \left(\frac{\omega^2}{4\pi\rho V^4}\right) \int d^3\,(\boldsymbol{r}-\boldsymbol{r}')\, \langle \delta C_{3333}(\boldsymbol{r})\delta C_{3333}\boldsymbol{r}'\rangle e^{2ik(\boldsymbol{r}-\boldsymbol{r}')\hat{k}}, \qquad (4)$$

where $\delta C_{3333}(\boldsymbol{r}) = C_{3333}(\boldsymbol{r}) - C^0_{3333}$ is the local deviation of the elastic stiffness from its average value and $\hat{k}$ is a unit vector in the direction of wave propagation.

Space does not allow the presentation of a full derivation of this result. However, it should be noted that the key idea is that the differential scattering cross-section is proportional to $|A|^2$. Since the density is the same in all grains, this leads to the factor $\langle \delta C_{3333}(\boldsymbol{r})\delta C_{3333}(\boldsymbol{r}')\rangle$ appearing in the integrand. This is the key ingredient in the theory, since this is the factor that relates the ultrasonic backscattering to the mictrostructure. It is known as the two-point correlation of the elastic stiffnesses.

Figure 4 illustrates the concept of the two-point correlation of elastic stiffnesses. In the evaluation of the integral, one needs to determine $\langle \delta C_{3333}(\boldsymbol{r})\delta C_{3333}(\boldsymbol{r}')\rangle$ as a function of $\boldsymbol{r}$ and $\boldsymbol{r}'$. Consider any value of these points. The crystallographic orientation of the grain in which the point falls will determine the value of δC_{3333}. When the two points are close to one another, there is a high probability that they will fall in the same grain. In such cases, the value of δC_{3333} will be the same at the two points and $\delta C_{3333}(\boldsymbol{r})\delta C_{3333}(\boldsymbol{r}')$ will be a positive quantity. In the limit that the two points approach one another, evaluation of the ensemble average is equivalent to averaging over all possible orientations of the grain, leading to a well-defined value for $\langle \delta C_{3333}(\boldsymbol{r})\delta C_{3333}(\boldsymbol{r}')\rangle$ which can be expressed in terms of the anisotropy of the single crystal elastic constants. On the other hand, when the two points are far apart, there is a high probability that they will fall in different crystallites. Because the orientation of the crystallites are assumed to be independent of one another, the quantity $\langle \delta C_{3333}(\boldsymbol{r})\delta C_{3333}(\boldsymbol{r}')\rangle$ will be found to vanish. If we also assume that the material is macroscopically homogeneous, then these results will only depend on the difference in the positions of $\boldsymbol{r}$ and $\boldsymbol{r}'$ which we will denote by $\boldsymbol{s}$. Under these conditions, the two-point correlations of elastic stiffnesses factors into the product

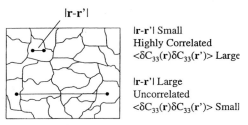

Fig. 4. Schematic illustrating the two-point correlation of elastic constants

$\langle \delta C_{3333}\delta C_{3333}\rangle W(\boldsymbol{s})$, where $\langle \delta C_{3333}\delta C_{3333}\rangle$ is the average over possible crystallite orientations (not necessarily random) and $W(\boldsymbol{s})$ is the probability that two points, separated by a distance $\boldsymbol{s}$, will fall in the same crystallite. Equation (4) then simplifies to

$$\eta(\omega) = \left(\frac{\omega^2}{4\pi\rho V^4}\right)^2 \langle \delta C_{3333}\delta C_{3333}\rangle \int \mathrm{d}^3 \boldsymbol{s}\, W(\boldsymbol{s}) \mathrm{e}^{2ik\boldsymbol{s}\hat{k}}. \tag{5}$$

This is a particularly useful form, since the function $W(\boldsymbol{s})$ is also used by materials scientists in the analysis of micrographs. *Stanke* [11] has examined this function in some detail, noting that, when cord lengths (the distance between intersections with grain boundaries of lines drawn randomly in a micrograph) follow Poisson statistics, $W(\boldsymbol{s})$ has the exponential form

$$W(\boldsymbol{s}) = \exp\left(-|\boldsymbol{s}|/A_\mathrm{g}\right), \tag{6}$$

where A_g is the correlation distance, equal to $1/2$ of the effective linear dimension of the grains.

The backscattering coefficient, η, was originally defined in the context of medical applications of ultrasonics. In nondestructive evaluation, the term FOM (figure-of-merit) has been used to define the quantity $\eta^{1/2}$. The FOM is of interest since it controls such parameters as the rms noise level that will be observed in an inspection.

In the theory of attenuation, a variety of approaches have been considered. Here, we follow the work of *Stanke* and *Kino* [12], motivated by successful comparisons to experiment [13] and the fact that the controlling microstructural feature is the same two-point correlation of elastic stiffnesses. As interpreted by *Ahmed* and *Thompson* [14], the governing equation is

$$\left(\Gamma_{ik} - \rho\frac{\omega^2}{k^2}\delta_{ik}\right)\hat{u}_k = 0, \tag{7}$$

where

$$\Gamma_{ik} = \left\{ C^0_{ijkl} + \langle \delta C_{ijkl}\rangle + (\langle \delta C_{ij\alpha\beta}\delta C_{ijkl}\rangle - \langle \delta C_{ij\alpha\beta}\rangle \right.$$
$$\left. \times \langle \delta C_{\gamma\delta kl}\rangle) \int_v G_{\alpha\gamma}(\boldsymbol{s}) \left[W(\boldsymbol{s})\mathrm{e}^{ik\boldsymbol{s}\cdot\hat{k}}\right]_{\beta\delta} \mathrm{d}v \right\} \hat{k}_j \hat{k}_l \tag{8}$$

and G is a Green's function.

Despite its rather formidable appearance, this result has a relatively simple interpretation. Consider first the case in which the material is homogeneous, e.g., when the grains have no elastic anisotropy. Then the matrix on the left-hand side of (7) would be diagonal. Its three eigenvalues would give the velocities of a longitudinal wave and a transverse wave (a double root) in the material. The eigenvectors would simply give the polarizations of these waves. If the material were a polycrystal, the same result would be obtained if there were no single-crystal anisotropy or if the grains were randomly oriented and sufficiently small that the last two terms in (8) could be neglected.

Next, suppose that there is a preferred orientation but that the grains are still quite small. Then the term $\langle \delta C_{ijkl} \rangle$ in (8) must be considered, leading to off-diagonal elements in (7). This is just the Christoffel matrix used to describe waves in anisotropic media [15]. It is well known that the eigenvalues give the wave speeds and the eigenvectors the polarizations of the three "quasi-pure" waves that can propagate in the medium.

Finally, suppose that the grain size is not small, so that the last term in (8) must be considered. This describes the scattering contributions. Because this term has an imaginary part, the eigenvalues of (7) also have an imaginary part and thus describe the attenuation as well as the velocity. A number of studies have used this equation to study the effects of preferred grain orientation and grain elongation, as will be discussed in subsequent sections. The key point to be noted at this time is that (7) and (8) depend on exactly the same microstructural feature, the two-point correlation of elastic stiffnesses, as the theory for backscattering. This two-point correlation thus plays a central role in understanding the interaction of ultrasound with material structure.

In the following subsections, the current status of using the above theory to interpret experimental results for simple polycrystals with various degrees of preferred grain orientation and grain elongation will be reviewed.

2.3 Randomly Oriented, Equi-axed Polycrystals

For this case, the ultrasonic velocity is isotropic and not of particular interest.

Equations (5) and (6) can be combined and evaluated analytically with the result that the backscattering coefficient is given by

$$\eta = \left(\frac{\omega^2}{4\pi\rho V^4}\right)^2 \langle \delta C_{33}^2 \rangle \left[8\pi A_g^3/(1+4k^2 A_g^2)^2\right], \qquad (9)$$

where the reduced (matrix) notation is used for the elastic constant. For hexagonal crystallites,

$$\begin{aligned}\langle \delta C_{33}^2 \rangle = \big(&192C_{11}^2 - 128C_{11}C_{13} + 48C_{13}^2 - 256C_{11}C_{33} \\ &+ 32C_{13}C_{33} + 112C_{33}^2 - 256C_{11}C_{44} + 192C_{13}C_{44} \\ &+ 64C_{33}C_{44} + 192C_{44}^2\big)/1575,\end{aligned} \qquad (10)$$

where the values of the elastic constants on the right-hand side are for single crystals. This result has been compared to experiment, with an absolute agreement better than a factor of 2 [16].

The attenuation can be found by evaluation of (8) analytically for this case, with the rather complex expressions that result being found in [12]. The result is a prediction of attenuation that grows initially as the fourth power of frequency in the Rayleigh regime, passes through a stochastic regime with

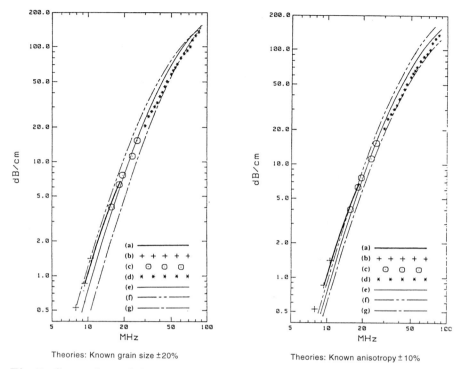

Fig. 5. Comparison of theory and experiment for the attenuation of longitudinal waves propagating in copper (after [13])

a lower frequency dependence and then approaches a constant value in the geometrical regime when the wavelength is small with respect to the grain size. Figure 5 presents a direct comparison to experiment for the case of copper [13]. The experiment is compared to the theory using best estimates of the grain size and the single crystal elastic constants, and the effects of variations of these quantities on the predictions are also examined.

2.4 Equi-axed Polycrystals with Preferred Orientation

The introduction of preferred grain orientation (texture) will effect the velocity, attenuation and backscattering. The former effect has received the greatest attention since it has led to an important materials characterization application, the characterization of sheet-metal-forming parameters [17]. Figure 6 presents, as an example, the angular variation of the speeds of guided ultrasonic modes in the plane of sheets of aluminum. Sheet 2 had what is known as a rolling texture, while sheets 1 and 3 had recrystallization textures

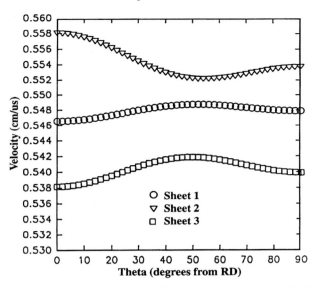

Fig. 6. Influence of recrystallization on anisotropy of ultrasonic velocity in an aluminum sheet. Sheet 2 has a rolling texture, and sheets 1 and 3 have varying degrees of recrystallization textures. RD denotes the rolling direction. (After [19])

of varying degrees. The ability of the ultrasonic measurements to distinguish between these two metallurgical conditions is clearly evident. It has been further shown that one can extract from such measurements quantitative information about the texture, in particular the lower order coefficients in an expansion of the orientation distribution function in terms of generalized spherical harmonics for both cubic [18,19] and hexagonal [20] crystallites. Not so obvious is the fact that this information correlates strongly with the plastic properties, a result that has stimulated much interest in the sheet-metal-forming community, including the evaluation of online measurement systems [17,21].

Limited theoretical studies of the effects of texture on backscattering and attenuation have also been conducted. *Yalda* et al. [22] examined the modifications that occur in (10) when texture is present. Many additional terms occur, each of which is a product of a factor that is a function of the single crystal elastic constants and a factor that is a coefficient in a spherical harmonic expansion of the orientation distribution function. *Ahmed* and *Thompson* [23] considered the case of the influence of a very strong texture, one in which the [001] axes of cubic crystallites were fully aligned along a preferred direction, upon the attenuation. This is a configuration of some practical interest, since it is commonly found in solidification microstructures such as those found in welds. A significant consequence was anisotropy in the attenuation.

2.5 Randomly Oriented Polycrystals with Grain Elongation

Elongation of grains might also be expected to produce anisotropy in the attenuation and backscattering, a phenomena that was studied experimentally be *Guo* et al. [24]. Figure 7 presents a rather surprising result of the study. The measurements were made on rolled rod that had been prepared for use in ultrasonic standards. Figure 7a shows the microstructure that resulted, consisting of grains that were highly elongated along the rod axis. It would be intuitively expected that the backscattering would be highest when the waves illuminated the grains from the "broadside" direction. This was confirmed by experiment, as seen in Fig. 7b. It might also be imagined that the attenuation in this direction would be the greatest, but this was not the case, as seen in Fig. 7c. Theoretical analyses to explain these results, based on (5) and (7) [10,25,26] are in progress. Initial results, neglecting the effects of texture are in semi-quantitative agreement with the data. It can be concluded that, although attenuation and backscattering are consequences of the scattering of elastic waves at grain boundaries, they depend quite differently on grain shape.

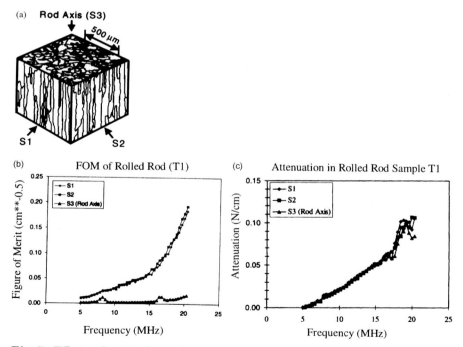

Fig. 7. Effects of grain elongation on anisotropy of ultrasonic scattering. (**a**) Microstructure, (**b**) backscattering in terms of FOM, (**c**) attenuation (after [24])

2.6 Polycrystals with Both Preferred Orientation and Grain Elongation

This general case is the true state of many metal polycrystals produced by forging processes. Limited theoretical analyses have been conducted, including calculations of attenuation in microstructures typical of rolled aluminum metal sheet [27] and the columnar microstructures (elongated grains with aligned [001] axes) develop in steel under certain solidification conditions [28].

3 Complex Microstructures

As defined in this paper, complex microstructures occur when multiple phases are present and/or when there is a correlation between the orientations of different grains. Such phenomena occur in some materials that have very high technological importance, including steel and titanium alloys. In this chapter, we will concentrate on the latter.

3.1 Background

Titanium can exhibit a high level of backscattered noise and attenuation. Because of the extensive use of titanium alloys in the rotating components of aircraft engines and the undesirable consequences of in-flight ruptures, considerable effort has been devoted to understanding and mitigating these high noise and attenuation levels since they influence flaw detection.

Titanium alloys often exhibit two-phase microstructures with long-range orientation correlations. For example, in the extensively used Ti–6Al–4V alloy, the presence of vanadium stabilizes the cubic beta phase, which occurs in addition to the primary, hexagonal alpha phase. The beta phase occurs in a variety of morphologies and locations, depending on the processing. Moreover, during processing, the material is often heated to such a temperature that it is entirely in the beta phase. Upon cooling, solid-state transformations occur in which the high-temperature beta phase converts to the alpha phase. However, the alpha crystallites do not grow with random orientations but rather have a fixed crystallographic relationship to the prior beta phase. In particular, when a prior beta crystallite transforms, the resulting alpha crystallite tends to adopt one of 12 possible variants in which there is a correspondence between high-density planes and directions of the initial and transformed crystallites.

Figure 8 illustrates the consequences of this effect by showing metallographic images produced at different magnifications under different etching and illumination conditions [29]. The top and bottom rows represent samples that were taken from two different regions of the same billet of a titanium alloy (Ti–6Al–4V). The top sample was taken from a region that exhibited relatively high ultrasonic noise (when examined in the radial direction), while the lower sample was taken from a region that exhibited relatively low noise.

High Noise (OD)

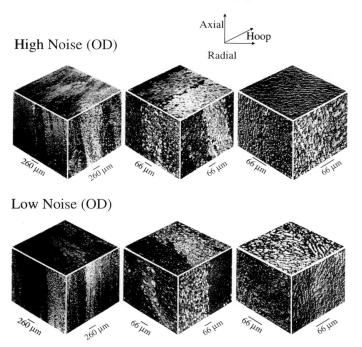

Low Noise (OD)

Fig. 8. Micrographs and macrographs of high- and low-noise regions of a titanium alloy billet (after [29])

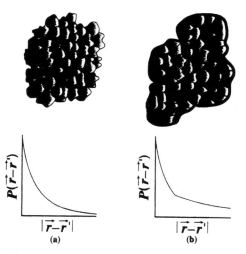

Fig. 9. Typical form of two-point correlations of elastic stiffnesses: (**a**) single phase equiaxial microstructure, (**b**) duplex microstructure of alpha releases in prior beta grains

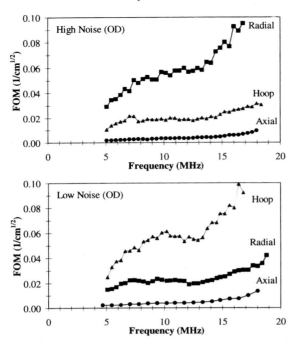

Fig. 10. Ultrasonic backscatter for samples taken from high-noise (*top*) and low-noise (*bottom*) regions of a titanium alloy billet (after [29])

For each case, the images on the right show the hexagonal alpha phase crystallites, with a small amount of beta phase present. There is no evident anisotropy. The images on the left, produced at a lower magnification and examined under crossed polarizers, show considerable banding. These structures so revealed are called macrograins and are believed to each represent a region that was part of the same prior beta grain. There are clearly two dimension scales present in what is called a duplex microstructure. It is important to understand which aspect of this duplex microstructure controls the ultrasonic attenuation and backscattering.

Han and *Thompson* [10] took a first step towards answering this question in a theoretical study applying (3) to this problem. In their work, it was assumed that, during the transformation, each of the possible variants occurred with equal probability. The implications of this assumption are illustrated in Fig. 9. The quantity $P(s)$ shown here is closely related to the $W(s)$ discussed previously. It is defined as the probability that the two points fall within crystallites of the same orientation, and reduces to $W(s)$ for a simple polycrystal. For a single-phased, equi-axed microstructure, this is expected to follow the exponential behavior shown in Fig. 9a, as previously described by (6). Figure 9b shows how this behavior would be modified in a duplex microstructure. The essential feature is that there is an extended "tail" on

$P(s)$, representing the fact that, when s is much greater that the size of an individual crystallite but still small enough to be in the same prior beta grain, there will be a finite probability that the two points will fall within crystallites of the same orientation. This tail is thus a measure of the correlation in orientation of different grains that arises because of the prior beta phase structure. The analysis made many interesting predictions regarding the relative importance of the crystallites and macrograins in determining the backscattering, all of which were in qualitative agreement with experiment.

Panetta et al. subsequently conducted a quantitative test of the model [29]. Figure 10 shows the ultrasonic backscattering, as a function of frequency, for each of the three propagation directions in the two samples whose microstructures were illustrated in Fig. 8. The relative values of the noise in the various directions is strongly correlated with the shapes of the macrograins shown on the left-hand side of Fig. 8. To quantitatively test the model, orientation imaging microscopy (OIM) was used to determine the orientation of individual crystallites. Assuming that the single crystal elastic constants were the same as those for pure titanium, the elastic constants as seen by the ultrasonic wave were then calculated. Figure 11 presents some typical results in which the elastic consequences of the correlation in crystallographic orientation is clearly evident. From such data, the two-point correlation of elastic constants can be determined and used as an input for the prediction of the backscattering coefficient from (3). Figure 12 presents the results, in remark-

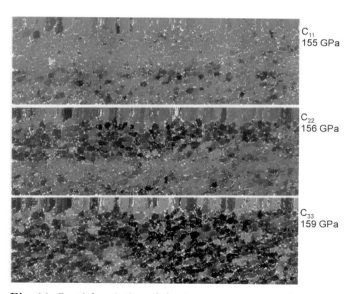

Fig. 11. Spatial variation of elastic constants on a radial face of a sample taken from the low-noise region of a titanium alloy billet (after [29]). The elongated features are aligned with the billet axis

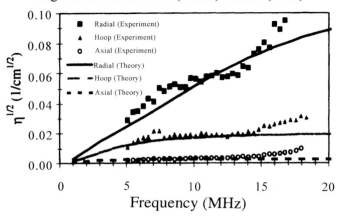

Fig. 12. Comparison of experiment with model predictions (based on OIM data) for backscattering (after [29])

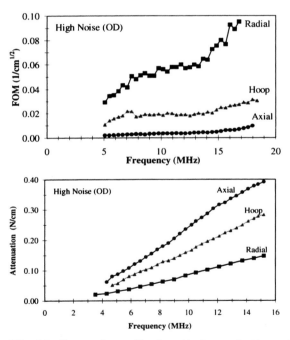

Fig. 13. Comparison of backscattering and attenuation in a high-noise region of a titanium alloy billet (after [29])

able agreement with the experimental observations. It can be concluded that, at least to the first order, a good understanding of the noise is in hand.

A rich set of phenomena is also observed in the attenuation measurements. For example, Fig. 13 compares the "attenuation" to the backscattering for the "high-noise" sample [29] discussed above. Counterintuitively, it is observed that directions of high noise correspond to low "attenuation" and vice versa, a matter that has been the subject of considerable experimental and theoretical investigation. The word "attenuation" is placed in quotations because, in these materials, different classical methods of measuring attenuation are

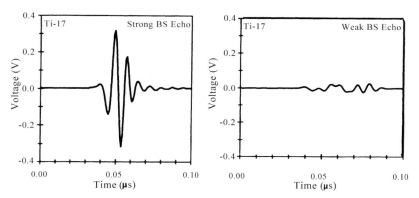

Fig. 14. Fluctuations in backsurface echo strength for ultrasonic wave propagating in the axial direction of a titanium alloy billet (after [31])

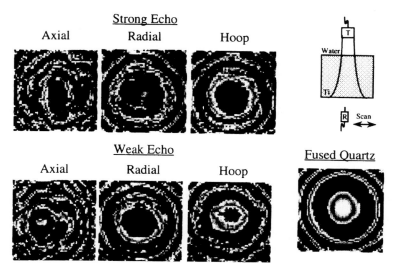

Fig. 15. Maps of the 9.2-MHz component of phase of an ultrasonic wave propagating in a sample from a titanium alloy billet in regions of high and low backsurface echo (after [31])

found to give different results [30]. For example, the data in Fig. 13 were obtained in a pulse–echo mode. However, since there were considerable fluctuations in the back surface echo strength as the transducer was translated over the sample, as illustrated in Fig. 14, the values of the "attenuation" were measured at several locations and averaged, leading to the data plotted in Fig. 13.

These observations suggested that the beam was developing significant phase and amplitude fluctuations, a hypothesis that was confirmed by experimentally mapping the fields, as illustrated in Figs. 15 and 16. In each case, a point probe was used to map out the phase and amplitude, for both strong and weak echo positions, in each of three propagation directions. For comparison, the results of similar scans are shown for fused quartz. It can be

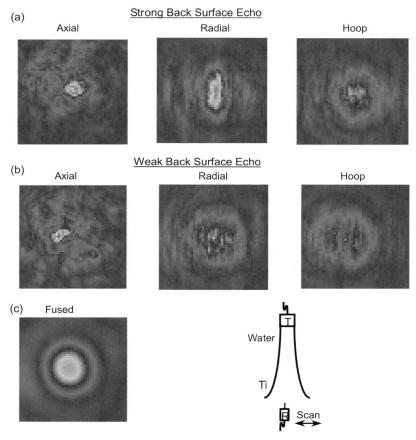

Fig. 16a–c. Maps of the 9.2-MHz component of amplitude of an ultrasonic wave propagating in a sample from a titanium alloy billet in regions of high and low backsurface echo (after [31])

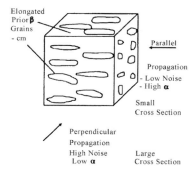

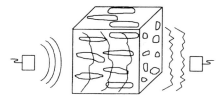

Fig. 17. A physical picture for effects of macrostructure on noise and attenuation (after [32])

seen that significant phase and amplitude distortion does exist, that the distortion is most severe when the wave propagates in the direction of highest "attenuation", and that there is no obvious difference between the distortions observed at the strong and weak echo positions. Further studies showed that there was little difference in the energy (determined by summing the squares of the signals at each observation point) passing through the sample at the strong and weak echo positions, and it was concluded that the phase fluctuations are the major contribution to the anomalous "attenuation" observations. The attenuation inferred from the energy was much lower than that observed in the pulse–echo mode.

A series of diagnostic experiments [30] and preliminary modeling efforts [32] strongly support the idea that sound-velocity inhomogeneities associated with the macrostructure are responsible for these effects, as conceptually illustrated in Fig. 17.

4 Effects on Imaging

The phenomena discussed in the previous two sections can have a profound effect on the ability to use ultrasound to image defects in solids. Backscattering and attenuation can obscure flaw signals, as is illustrated in Fig. 18 [33]. In this experiment, a focused transducer was used to examine a set of 16 nominally identical flat-bottom holes in a titanium alloy. The presence of the

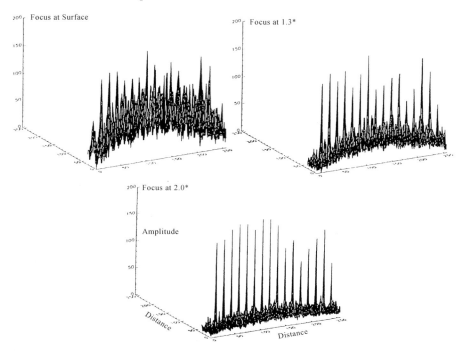

Fig. 18. Dependence of echo strengths and backscattered noise on the position of the focal plane. The reflectors are flat-bottom holes of 1/64 in diameter located at a depth of 2 in in titanium (after [33])

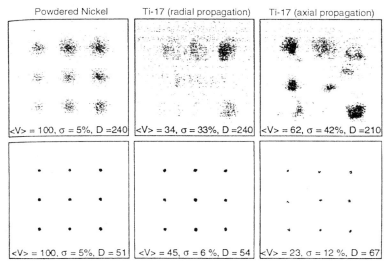

Fig. 19. Effects of material on C-scan images in fine-grained nickel and titanium alloy samples. *Top:* planar probe. *Bottom:* focused probe (after [30])

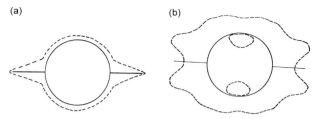

Fig. 20. Reconstruction of a "saturn ring" defect consisting of a spherical cavity with a larger crack at the equator. The material is assumed to have a 2% velocity anisotropy. (**a**) Inverse Born sizing utilizing low-frequency phase information to compensate for anisotropy. (**b**) Synthetic aperture imaging neglecting anisotropy. *Solid lines* are actual boundaries of the defect, and *dashed lines* are the results of the reconstruction

clutter and the fact that the ratio of flaw signal to clutter can be changed by adjusting the position of the focal point can be clearly seen.

Beam distortion can also significantly degrade flaw image, as shown in Fig. 19 [30]. The degree of this distortion also depends critically on the position of the flaw with respect to the beam. Here one can see that the degradation of the flaw image is much greater for planar probes than for focused probes and that the severity of the effect depends on propagation direction. The distortion is greatest for propagation in the axial direction, which is along the elongation of the macrograins.

Finally, anisotropy can have a significant effect on the formation of images. For example, if one tries to reconstruct a defect based on a synthetic aperture technique neglecting material anisotropy, significant distortions can occur if anisotropy is present. Consider a defect consisting of a spherical cavity with a large crack at its equator. Figure 20 compares two approaches for reconstruction [34]. The right image is a simulation of a synthetic aperture reconstruction, assuming that defect is embedded in a material with 2% velocity anisotropy. The left image shows the results of an algorithm known as the Inverse Born Approximation (IBA), which uses low-frequency phase information to stabilize the inversion. The improvement is clearly evident.

5 Conclusions

Complex microstuctures can have a significant influence on the velocity, attenuation and beam patterns of propagating elastic waves and can produce backscattered signals that obscure flaw echoes. There are a wide range of intriguing questions associated with wave propagation in these materials, and their solution requires a combination of the perspectives of elastodynamics and materials science. Theoretical approaches are being developed which provide a good understanding of many of the phenomena associated with backscattering, and the two-point correlation of elastic constants plays a central role. Phase fluctuations have been identified as playing a central role in the understanding of attenuation, and good theoretical progress is being made. However, more work is required to complete our understanding. Least well understood are the contributions of phase and amplitude fluctuations of beams to the fluctuation of signals reflected from small flaws.

These effects can have important implications on ultrasonic nondestructive evaluation. Improved scientific understanding of the phenomena can provide the basis for microstru ctural design and inspection system design to mitigate the effects. In addition, the rich phenomena that have been observed contain much information about the material and its microstructure. These phenomena have the potential to serve as a foundation for improved tools for materials characterization.

Note in Proof

In reading this manuscript, a colleague pointed out a semantic problem with the use of the word "random" in the title. To the community studying preferred orientation of polycrystals, "random" refers to the case in which all crystallographic orientations occur with equal probability. We have used the term more generally here, to describe the case in which the orientation is described by a random variable, not necessarily uniformly distributed.

References

1. *Ultrasonics, Nondestructive Testing Handbook*, Vol. 7 (American Society for Nondestructive Testing, Columbus, OH 1991)
2. *Metals Handbook*, Vol. 17 (ASM International, Metals Park, Ohio 1989)
3. R. B. Thompson, Quantitative ultrasonic nondestructive evaluation methods, J. Appl. Mech **50**, 1191–1201 (1983)
4. R. B. Thompson, D. O. Thompson, Ultrasonics in nondestructive evaluation, Proc. IEEE, **73**, 1716–1755 (1985)
5. W. Voigt, *Lehrbuch der Kristallphysik* (Tauber, Leipzig 1928)
6. J. E. Gubernatis, E. Domany, J. A. Krumhansl, M. Huberman, The Born approximation in the theory of the scattering of elastic waves from flaws, J. Appl. Phys. **50**, 4046 (1979)
7. J. H. Rose, Ultrasonic backscattering from polycrystalline aggregates using time-domain linear response theory, Rev. Prog. Quant. Nondestr. Eval. B **10**, 1715–1720 (1991)
8. J. H. Rose, Ultrasonic backscattering from microstructure, Rev. Prog. Quant. Nondestruct. Eval. B **11**, 1677–1684 (1992)
9. J. H. Rose, Theory of ultrasonic backscatter from multiphase polycrystalline solids, Rev. Prog. Quant. Nondestruct. Eval. B **12**, 1719–1729 (1993)
10. Y. K. Han, R. B. Thompson, Ultrasonic backscattering in duplex microstructures: Theory and application to titanium alloys, Metal. Trans. A **28**, 91–104 (1997)

11. F. E. Stanke, Spatial autocorrelation functions for calculations of effective propagation constants in polycrystalline materials, J. Acoust. Soc. Am. **80**, 1479 (1986)
12. F. E. Stanke, G. S. Kino, A unified theory for elastic wave propagation on polycrystalline materials, J. Acoust. Soc. Am. **75**, 665 (1984)
13. F. E. Stanke, Inversion of attenuation measurements in terms of parameterized autocorrelation function, In *NDE for Microstructure for Process Control*, ed. by H. N. G. Wadley (ASM, Metals Park, Ohio 1985) p. 55
14. S. Ahmed, R. B. Thompson, propagation of elastic waves in equiaxed stainless steel polycrystals with aligned [001] axes, J. Acoust. Soc. Am. **99**, 2086–2096 (1996)
15. M. J. P. Musgrave, *Crystal Acoustics* (Holden-Day, San Francisco 1970)
16. F. J. Margetan, R. B. Thompson, I. Yalda-Mooshabad, Backscattered microstructural noise in ultrasonic toneburst measurements, J. Nondestr. Eval. **13**, 111–136 (1994)
17. R. B. Thompson, Determination of texture and grain size in metals: An example of materials characterization, In *Sensing for Materials Characterization, Processing, and Manufacturing*, ed. by G. Birnbaum, B. A. Auld (ASNT, Columbus, Ohio 1998) p. 23–45
18. R. B. Thompson, J. F. Smith, S. S. Lee, G. C. Johnson, A comparison of ultrasonic and X-ray determinations of texture in thin Cu and Al plates, Metal. Trans. A **20**, 2431–2447 (1989)
19. A. Anderson, R. B. Thompson, R. Bolingbroke, J. Root, Ultrasonic characterization of rolling and recrystallization textures in hot rolled aluminum sheet, Textures Microstruct. **26–27**, 39–58 (1996)
20. A. J. Anderson, R. B. Thompson, C. S. Cook, Ultrasonic measurements of the kearns texture factors in zircaloy, zirconium, and titanium, Metal. Trans. A **30**, 1981–1988 (1999)
21. R. B. Thompson, E. P. Papadakis, D. D. Bluhm, G. A. Alers, K. Forouraghi, H. D. Shank, S. J. Wormley, Measurement of texture and formability parameter with a fully automated ultrasonic instrument, J. Nondestr. Eval. **12**, 45–62 (1993)
22. I. Yalda-Mooshabad, R. B. Thompson, Influence of texture and grain morphology on the two-point correlation of elastic constraints: Theory and implications on ultrasonic attenuation and backscattering, Rev. Prog. Quant. Nondestr. Eval. B **14**, 1939–1946 (1995)
23. S. Ahmed, R. B. Thompson, Propagation of elastic waves in equiaxed iron polycrystalline with aligned [001] axes, Rev. Prog. Quant. Nondestr. Eval. B **10**, 1999–2005 (1991)
24. Y. Guo, R. B. Thompson, D. K. Rehbein, F. J. Margetan, M. Warchol, The effects of microstructure on the response of aluminum E-127 calibration standards, Rev. Prog. Quant. Nondestr. Eval. B **18**, 2337–2344 (1999)
25. S. Ahmed, R. B. Thompson, Influence of columnar microstructure on ultrasonic backscattering, Rev. Prog. Quant. Nondestr. Eval. B **14**, 1617–1624 (1995)
26. P. D. Panetta, unpublished results
27. S. Ahmed, R. B. Thompson, Attenuation and dispersion of ultrasonic waves in rolled aluminum, Rev. Prog. Quant. Nondestr. Eval. B **17**, 1649–1655 (1998)
28. S. Ahmed, R. B. Thompson, Effect of preferred grain orientation and grain elongation on ultrasonic wave propagation in stainless steel, Rev. Prog. Quant. Nondestr. Eval. B **11**, 1999–2006 (1992)

29. P. D. Panetta, R. B. Thompson, F. J. Margetan, Use of electron backscatter diffraction in understanding texture and the mechanisms of backscattered noise generation in titanium alloys, Rev. Prog. Quant. Nondestr. Eval. A **17**, 89–96 (1998)
30. F. J. Margetan, P. D. Panetta, R. B. Thompson, Ultrasonic signal attenuation in engine titanium alloys, Rev. Prog. Quant. Nondestr. Eval. B **17**, 1469–1476 (1998)
31. P. D. Panetta, F. J. Margetan, I. Yalda, R. B. Thompson, Ultrasonic attenuation measurements in jet engine titanium alloys, Rev. Prog. Quant. Nondestr. Eval. B **15**, 1525–1532 (1996)
32. P. D. Panetta, R. B. Thompson, Ultrasonic attenuation in duplex titanium alloys, Rev. Prog. Quant. Nondestr. Eval. B **18**, 1717–1724 (1999)
33. E. J. Nieters, R. S. Gilmore, R. C. Trzaskos, J. D. Young, D. C. Copley, P. J. Howard, M. E. Keller, W. J. Leach, A multizone technique for billet inspection, Rev. Prog. Quant. Nondestr. Eval. B **14**, 2137–2144 (1995)
34. R. B. Thompson, K. M. Lakin, J. H. Rose, A comparison of the inverse born and imaging techniques for reconstructing flaw shapes, In *1981 Ultrasonics Symposium Proceedings*, Vol. 2 (IEEE, New York 1981) p. 930–93

Imaging the Viscoelastic Properties of Tissue

Mostafa Fatemi and James F. Greenleaf

Department of Physiology and Biophysics, Mayo Clinic and Mayo Foundation, Rochester, Minn. 55905, USA
{fatemi.mostafa,jfg}@mayo.edu

Abstract. Elasticity and viscosity of soft tissues are often related to pathology. These parameters, along with other mechanical parameters, determine the dynamic response of tissue to a force. Tissue mechanical response, therefore, may be used for diagnosis. Measuring and imaging of the mechanical properties of tissues is the aim of a class of techniques generally called elasticity imaging or elastography. The general approach is to measure tissue motion caused by a force or displacement and use it to reconstruct the elastic parameters of the tissue. The excitation stress can be either static or dynamic (vibration). Dynamic excitation is of particular interest because it provides more comprehensive information about tissue properties in a spectrum of frequencies. In one approach an external stress field must pass through the superficial portion of the object before reaching the region of interest within the interior. An alternative strategy is to apply a localized stress directly in the region of interest. One way to accomplish this task is to use the radiation force of ultrasound. This approach offers several benefits, including: (a) safety – acoustic energy is a noninvasive means of exerting force; (b) adaptability – existing ultrasound technology and devices can be readily modified for this purpose; (c) remoteness – radiation force can be generated remotely inside tissue without disturbing its superficial layers; (d) localization – the radiation stress field can be highly localized, thus allowing for precise positioning of the excitation point; and (e) a wide frequency spectrum. Several methods have been developed for tissue probing using the dynamic radiation force of ultrasound, including: (a) transient methods which are based on impulsive radiation force; (b) shear-wave methods which are based on generation of shear-waves; and (c) vibro-acoustography, recently developed by the authors, where a localized oscillating radiation force is applied to the tissue and the acoustic response of the tissue is detected by a hydrophone. Here, we focus on vibro-acoustography and present a detailed description of the theory and the experimental results. We conclude with the capabilities and limitations of these radiation-force methods.

1 Introduction

It is well known that changes in elasticity of soft tissues are often related to pathology. Traditionally, physicians use palpation as a simple method for estimating mechanical properties of tissue. In palpation, a static force is applied and a crude estimation of tissue elasticity is obtained through the sense of touch. The force is applied on the body surface and the result is a collective response of all the tissues below. Clinicians can sense abnormalities if the response to palpation of the suspicious tissue is sufficiently different from that of normal tissue. However, if the abnormality lies deep in the body, or if it is too small to be resolved by touch, then the palpation method fails. The dynamic response of soft tissue to a force is also valuable in medical diagnosis. For instance, rebound of tissue upon sudden release of pressure exerted by the physician's finger on the skin provides useful diagnostic information about the tissue.

Quantitative measurement of the mechanical properties of tissues and their display in a raster format is the aim of a class of techniques generally called elasticity imaging or elastography [1]. The general approach is to measure tissue motion caused by an external (or, in some methods, internal) force or displacement and use it to reconstruct the elastic parameters of the tissue. The excitation stress can be either static or dynamic (vibration). The dynamic excitation is of particular interest because it provides more comprehensive information about tissue properties in a spectrum of frequencies, or alternatively, the transient behavior of the tissue could be deduced from the measurements. In most elasticity imaging methods ultrasound is used to detect the motion or displacement resulting from the applied stress. Magnetic resonance elastography is a recently developed method [2] that employs a mechanical actuator to vibrate the body surface and then measures the strain waves with phase-sensitive magnetic resonance imaging (MRI). The majority of elasticity imaging methods are based on an external source of force, resulting in a spatially wide stress field distribution. This requires the stress field to pass through the superficial portion of the object before reaching the region of interest within the interior. This requirement can complicate the estimation of stiffness because the stress-field patterns change at different depths. Also, because the stress field is widely distributed, the response of the object is an integral sum of the stress field. An alternative strategy is to apply a localized stress directly in the region of interest. One way to accomplish this is to use the radiation pressure of ultrasound.

Acoustic radiation force is the time-average force exerted by an acoustic field on an object. This force is produced by a change in the energy density of an incident acoustic field [3], for example, due to absorption or reflection. Several benefits may result from using ultrasound radiation force for evaluating tissue properties, including: (a) acoustic (ultrasound) energy is a noninvasive means of exerting force; (b) existing ultrasound technology and devices can be readily modified for this purpose, thus eliminating the need for developing a new technology; (c) radiation force can be generated remotely inside tissue without disturbing its superficial layers; (d) the radiation stress field can be highly localized, thus allowing for precise positioning of the excitation point; and (e) radiation force can be produced in a wide range of frequencies or temporal shapes. These features make radiation-force methods more attractive than other, mostly mechanical, excitation methods used in elasticity imaging.

Tissue probing with the radiation force produced by ultrasound can be accomplished in a variety of techniques, depending on the excitation and detection methods used. We may categorize these methods as follows: (a) transient methods, where an impulsive radiation force is used and the transient response of the tissue is detected by Doppler ultrasound [4]; (b) shear-wave methods, where an oscillating radiation is applied to the tissue and the resulting shear wave is detected by ultrasound or other methods [5,6,7]; and (c) vibro-acoustography, recently developed by the authors, where a localized oscillating radiation force is applied to the tissue and the acoustic response of the tissue is detected by a hydrophone [8,9]. Figure 1 illustrates the relationship between various methods that have been developed for evaluating or imaging the elastic properties of tissue.

Dynamic radiation-force methods seem to be evolving rapidly as a new field in tissue characterization and imaging. The purpose of this chapter is to systematically discuss the features and capabilities that are offered in this group of elasticity measurement techniques. Here, we describe the general approach used in the transient, shear-wave, and vibro-acoustography methods. We pay particular attention to the last method, present its theory, and discuss its features. Finally we discuss the capabilities and limitations of all three methods.

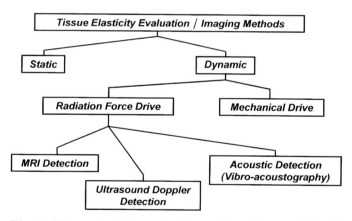

Fig. 1. Tissue elasticity evaluation and imaging methods. These methods can be divided into two general groups. The static group employs a steady force to deform the object. The dynamic group use either a momentary or an oscillating force to vibrate the object and can be divided into two subgroups. Mechanical-drive methods use a mechanical actuator, whereas the radiation-force-drive methods use the radiation force of ultrasound to excite the object. The latter subgroup is divided into three subsubgroups based on the detection method used. Vibro-acoustography uses a hydrophone to detect object response

2 Theory of the Radiation Force

The study of radiation force and radiation pressure dates back nearly one century, to the time of *Rayleigh* [10]. Since then, this subject has been under continuous investigation. A historical review of radiation force and radiation pressure is presented in [11], and a critical review of the subject can be found in [3]. Some recent analyses of radiation force/pressure in an attenuating medium, which may be applicable to biological tissues, are presented in [12] and [13].

The acoustic radiation force is an example of a universal phenomenon in any wave motion that introduces some type of unidirectional force on absorbing or reflecting targets in the wave path. Radiation force in fluids is often studied in the context of radiation pressure. Depending on the boundary conditions, radiation pressure can be defined differently. Simple explanations of these definitions can be presented by considering a sound traveling inside, and along the axis of, a cylindrical container toward the opposite wall [11]. Rayleigh radiation pressure is the excess pressure produced on the opposite wall when the container's side wall is confining the fluid inside. Langevin radiation pressure is the excess pressure on the opposite wall when we remove the confining side wall, so that the fluid is free to move (Fig. 2). Here we will focus on Langevin radiation pressure because the conditions for which this pressure is defined apply to our experimental situation. It can be shown that the Langevin radiation pressure of a plane wave impinging normally on a perfectly absorbing wall is equal to the total energy density $\langle E \rangle$, where $\langle \ldots \rangle$ represents the time average. If the wall is partially reflecting, this pressure would be equal to $(1+R)\langle E \rangle$, where R is the power reflection coefficient [11]. Thus, in general, we can write the radiation force of a normally impinging sound beam on a wall as

$$F = d_\mathrm{r} S \langle E \rangle, \tag{1}$$

where F is the force in the beam direction, and d_r is the "radiation-force function" or the drag coefficient. This dimensionless coefficient is defined per unit incident energy density and unit projected area. For a planar object, d_r is numerically equal to the force on the object. Physically, the drag coefficient represents the scattering and absorbing properties of the object [14]. For a perfectly absorbing object $d_\mathrm{r} = 1$, and for a perfectly reflecting object $d_\mathrm{r} = 2$. In the case of oblique incidence, the radiation force will have a normal as well as a transverse component. A more detailed description of d_r is presented in [14].

To produce a time-varying radiation force, the intensity of the incident beam can be modulated in various ways. For example, a short ultrasound pulse can produce a transient pulsed radiation force, and a sinusoidally modulated beam can result in a sinusoidally varying force.

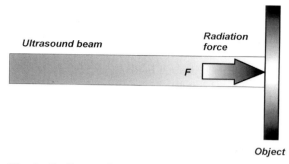

Fig. 2. Radiation force produced by projection of ultrasound on an object. Interaction of an ultrasound beam with an object that scatters and/or absorbs results in a force on the object in the beam direction. The magnitude of this force is proportional to the time-average energy density of the incident beam and a factor that represents the scattering and absorbing properties of the object

3 Radiation-Force Methods

In transient methods the radiation force of ultrasound is used to make a minute deformation in the tissue. The transient recoil of the tissue resulting from this deformation is measured and used for evaluation of tissue elastic properties. A method for measuring tissue hardness, presented by *Sugimoto* et al. [4], uses the radiation force of a single focused ultrasound beam. Ideally, hardness may be represented by the spring constant of the object, which is the ratio of the applied force to the displacement. Principles of the transient methods are presented in Fig. 3.

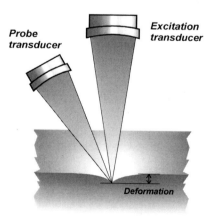

Fig. 3. Principle of the transient method. The excitation transducer produces a momentary radiation force which induces a minute deformation in the object. Doppler methods are used to measure this deformation using the probe transducer

3.1 Transient Method

In this method an ultrasound pulse is used to generate a short-duration radiation force which produces localized deformation of the tissue. Immediately after the force pulse, the resulting transient deformation of the tissue is measured as a function of time with Doppler ultrasound using a separate transducer. The deformation includes an initial rapid squeeze of the tissue, followed by a relaxation and possibly a rebound. The deformation is a function of tissue viscoelastic parameters, as well as the applied force. The authors argue that, because quantitatively measuring the internal radiation force is difficult, it is advantageous to derive a relative quantity that is representative of tissue hardness. To derive a single relative quantity from the deformation data, the relaxation part of the function is approximated by a sum of several exponential curves, and the sum of the first-order derivatives of such exponentials is calculated. *Sugimoto* et al. [4] show that this quantity is correlated to the spring constant of the tissue; thus it may be used as a measure of tissue hardness.

3.2 Shear-Wave Methods

The shear modulus is related to the hardness or elasticity of the material. It is known that the shear moduli of various soft tissues range over several orders of magnitude, while the bulk modulus, a parameter that is associated with the conventional pulse–echo ultrasound compressional wave speed, varies significantly less than an order of magnitude [15,16]. These features indicate that the shear modulus may be a better parameter for tissue characterization than bulk modulus. Tissue attenuation of the shear wave is very large, even at low kHz frequencies. One way to induce localized shear waves inside tissue is to use the radiation force of focused ultrasound [5]. In a method called shear-wave elasticity imaging (SWEI) [6], an amplitude-modulated, single-focused ultrasound beam is used to induce a localized radiation stress inside the soft tissue. Localization of the stress field is critical to the success of the method. To achieve a high degree of localization, the method uses a focused ultrasound beam. It is shown that the radiation stress exerted within a dissipative medium peaks about the focal region of the highly focused transducer. Also, it has been suggested that localization can be improved by designing the transducer and selecting the beam parameters such that a nonlinear shock wave is produced in the focal region, increasing the magnitude of the stress field in the vicinity of the focal region, thus augmenting localization of the stress field.

Modulation of the ultrasound beam can be in the form of an oscillating wave or short pulse. The resulting radiation force elicits a shear wave propagating in the radial direction with respect to the beam axis, with particle motion parallel to the beam axis. Shear waves in soft tissue travel at very low speed, typically around a few meters per second; thus the corresponding wavelength is much shorter than that of the compressional waves for the same frequency. Shear waves are also highly attenuated in soft tissue, with an attenuation coefficient two or three orders of magnitude higher than that of the compressional waves. Because of high attenuation of shear waves, it is possible to induce them in a very limited region in the vicinity of the focal point of the ultrasound beam, hence avoiding the influence of tissue boundaries. Shear parameters of the tissue, such as shear modulus and shear viscosity, can be calculated by measuring the amplitude and the temporal characteristics of this wave. For example, the time required for the wavefront to propagate from one point to another can be used to calculate the shear-wave speed, and consequently the shear-wave modulus μ as

$$\mu = \rho c_t^2 \tag{2}$$

where ρ and c_t are the density and the shear-wave velocity, respectively.

Shear waves may be detected optically. In this method a laser source and a photo detector are used to detect the displacement of particles due to the shear wave in a transparent phantom [6]. Because this method requires a transparent medium, its application in vivo is difficult. Phase-sensitive MRI [2,6] is an alternative method that can be used to measure the 3-dimensional distribution of particle displacement in a given direction versus time in a material. Figure 4 illustrates a simplified system for shear-wave elasticity measurement using MRI. In an experiment presented in [6], an ultrasound pulse of 3.6-ms duration produced by a 70-mm diameter transducer focused at 100 mm was transmitted within a cylindrical rubber phantom. The displacement was measured by 2.0-T MRI system at two different times after the acoustic pulse was applied. The position of the peak displacement at these time points was used to estimate the shear velocity, which was shown to be consistent with the independently measured value. Shear waves can also be detected by Doppler ultrasound [5]. It has been shown [7] that to achieve the appreciable displacement needed for Doppler detection, most soft tissues require high ultrasound intensities which might be beyond the safe limit.

3.3 Vibro-Acoustography

This technique produces a map of the mechanical response of an object to a dynamic force applied at each point. The method utilizes an ultrasound radiation force to remotely exert a localized oscillating stress field at a desired frequency within (or on the surface of) an object, and records the resultant acoustic response [8,9]. Figure 5 illustrates the principle of this method. This acoustic response, which is normally in the low kHz range, is a function of the viscoelastic properties of the object and can be used to produce an image of the object.

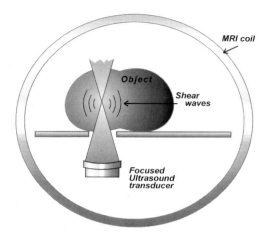

Fig. 4. Principle of shear-wave elastography. The focused ultrasound beam is absorbed by the object, resulting a localized radiation stress, which squeezes the elastic object. This deformation propagates in the form of shear waves through the object. The phase-sensitive MRI machine detects the spatial distribution of the resulting displacement in the object

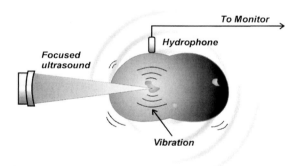

Fig. 5. Principle of vibro-acoustography. The ultrasound beam is used to produce a localized oscillatory radiation stress to vibrate the object. Vibration of the object produces an acoustic emission field in the medium. This field, which is a function of the viscoelastic properties of the object, is detected by a sensitive hydrophone

3.3.1 Method

This method operates on an oscillating radiation force to vibrate the tissue and consequently produce an acoustic emission from the object. To produce an oscillating radiation force the intensity of the incident ultrasound must be amplitude modulated at the desired low frequency. Using a single amplitude-modulated beam seems to be the simplest means to attain this purpose.

However, such a beam could exert a radiation force on any object that is present along the beam path, producing undesirable acoustic emission. To confine the radiation stress to the desired region, we use two unmodulated cw beams at slightly different frequency, propagating along separate paths. The beams are positioned to cross each other at their respective foci, and thus produce a modulated field at a confined, small, cross-sectional region.

3.3.2 Theory

In the beamforming method to be described, the amplitude-modulated field is obtained by the interference of two unmodulated ultrasound beams. The main advantage of this approach is that the interference volume size can be limited to a small region. Radially symmetric interfering beams are obtained when two coaxial, confocal transducers are used [9]. For this purpose, elements of a two-element spherically focused annular array (consisting of a central disc with radius a_1 and an outer ring with the inner radius of a'_2 and outer radius of a_2) are excited by separate cw signals at frequencies $\omega_1 = \omega_0 - \Delta\omega/2$ and $\omega_2 = \omega_0 + \Delta\omega/2$, respectively. We assume that the beams are propagating in the $+z$-direction with the joint focal point at $z = 0$, as shown in Fig. 6. The resultant field on the $z = 0$ plane may be written as follows:

$$p(t) = P_1(r)\cos\left[\omega_1 t + \psi_1(r)\right] + P_2(r)\cos\left[\omega_2 t + \psi_2(r)\right], \tag{3}$$

where $r = \sqrt{x^2 + y^2}$ is the radial distance. The amplitude functions are [17]

$$P_1(r) = \rho c U_{01}\frac{\pi a_1^2}{\lambda_1 z_0}\mathrm{jinc}\left(\frac{ra_1}{\lambda_1 z_0}\right), \tag{4}$$

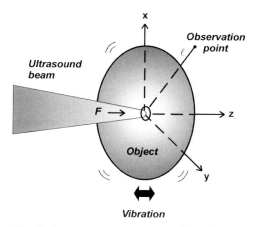

Fig. 6. The coordinate system describing the position of the ultrasound beam and the observation point relative to the object. The beam travels in the z-direction

and
$$P_2(r) = \rho c U_{02} \frac{\pi}{\lambda_2 z_0} \left[a_2^2 \text{jinc}\left(\frac{ra_2}{\lambda_2 z_0}\right) - a_2'^2 \text{jinc}\left(\frac{ra_2'}{\lambda_2 z_0}\right) \right], \tag{5}$$

where U_{0i} is the velocity amplitude at the ith transducer element surface, and $\lambda_i = 2\pi/\omega_i$ for $i = 1, 2$ are the ultrasound wavelengths. The phase functions,

$$\psi_i(r) = -\frac{\pi r^2}{\lambda_i z_0}, \tag{6}$$

for $i = 1, 2$, are conveniently set to be zero at the origin. Also, $\text{jinc}(X) = J_1(2\pi X)/\pi X$, where $J_1(\ldots)$ is the first-order Bessel function of the first kind.

The instantaneous energy is $E = p^2(t)/\rho c^2$. Replacing $p(t)$ from (3), this energy will have a time-independent component, a component at the difference frequency $\Delta\omega = \omega_2 - \omega_1$ which results from the cross-product of the two pressure fields, and high-frequency components at ω_1 and ω_2 and their harmonics. The energy component at the difference frequency is

$$e_{\Delta\omega}(t) = \frac{P_1(r_0)P_2(r_0)}{4\rho c^2} \cos\left[\Delta\omega t - \Delta\psi(r_0)\right], \tag{7}$$

where $\Delta\psi(r_0) = \psi_2(r_0) - \psi_1(r_0)$.

Now, we define a *unit point target* with an area of $dxdy$ at position (x_0, y_0) on the focal plane, and with a drag coefficient $d_r(x_0, y_0)$ such that $d_r(x_0, y_0)dxdy = 1$ on the target and zero elsewhere. This equation is merely used as a mathematical model. In this case, the projected area can be considered to be $S = dxdy$. Therefore, if the projected area is unity, then $d_r(x, y) = 1$, which corresponds to a totally absorptive object.

Referring to (1), and replacing $d_r S$ with unity and $\langle E \rangle$ with $e_{\Delta\omega}(t)$ of (7), we can write the low-frequency component of the radiation force on the unit point target as

$$f_{\Delta\omega}(x_0, y_0; t) = \frac{1}{\rho c^2} P_1(r_0)P_2(r_0) \cos\left[\Delta\omega t + \Delta\psi(r_0)\right], \tag{8}$$

where arguments x_0 and y_0 are added to denote the position of the point target, and $r_0 = \sqrt{x_0^2 + y_0^2}$. Referring to (4) and (5), the complex amplitude of the stress field can be found as

$$F_{\Delta\omega}(x_0, y_0) = \rho U_{01} U_{02} \frac{\pi a_1^2}{\lambda_1 z_0} \text{jinc}\left(\frac{r_0 a_1}{\lambda_1 z_0}\right)$$
$$\times \left[\frac{\pi a_2^2}{\lambda_2 z_0} \text{jinc}\left(\frac{r_0 a_2}{\lambda_2 z_0}\right) - \frac{\pi a_2'^2}{\lambda_2 z_0} \text{jinc}\left(\frac{r_0 a_2'}{\lambda_2 z_0}\right)\right]$$
$$\times \exp\left(-j\frac{r_0^2}{2\Delta\lambda z_0}\right), \tag{9}$$

where $\Delta\lambda = 2\pi c/\Delta\omega$ is the wavelength associated with $\Delta\omega$. The above equation indicates that the radiation stress is concentrated at the focal point and decays quickly as jinc$(\ldots)^2$. It should be noted that because $f_{\Delta\omega}(x_0, y_0; t)$ and $F_{\Delta\omega}(x_0, y_0)$ are defined per unit point target, they have the dimension of force per unit area, or equivalently the unit of stress. As an example, we consider a confocal transducer with dimensions $a_1 = 14.8\,\text{mm}$, $a_2' = 16.8\,\text{mm}$, $a_2 = 22.5\,\text{mm}$, and the focal length 70 mm. Also, we assume that the center frequency is 3 MHz, and the difference frequency is 7.3 kHz. The radiation stress at the focal plane of this transducer is plotted in Fig. 7.

In medical applications the maximum ultrasonic intensity is regulated for safety reasons. It is therefore useful to write the stress field in terms of the peak ultrasonic intensity at the focal point. The long-term average of ultrasonic intensity at the focal point can be written as

$$I(0) = \frac{P_1^2(0) + P_2^2(0)}{2\rho c}, \tag{10}$$

where $P_1(0)$ and $P_2(0)$ can be found from (4) and (5) as

$$I(0) = \frac{\rho c}{2} \left[U_{01}^2 \left(\frac{\pi a_1^2}{\lambda_1 z_0}\right)^2 + U_{02}^2 \left(\frac{\pi}{\lambda_2 z_0}\right)^2 \left(a_2^2 - a_2'^2\right)^2 \right]. \tag{11}$$

Assuming $U_{01} = U_{02} = U_0$, we can write the focal plane stress field in terms of $I(0)$:

$$F_{\Delta\omega}(x_0, y_0) = \frac{2I(0)}{c} \times \frac{\frac{\pi a_1^2}{\lambda_1 z_0}}{\left(\frac{\pi a_1^2}{\lambda_1 z_0}\right)^2 + \left(\frac{\pi}{\lambda_2 z_0}\right)^2 \left(a_2^2 - a_2'^2\right)^2}$$

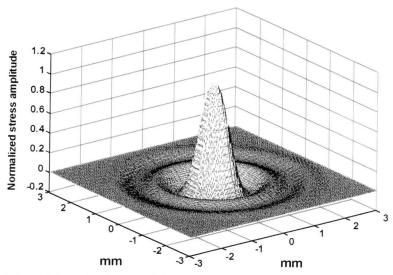

Fig. 7. Normalized stress field of the vibro-acoustography system with a 3-MHz confocal transducer

$$\times \text{jinc}\left(\frac{r_0 a_1}{\lambda_1 z_0}\right)\left[\frac{\pi a_2^2}{\lambda_2 z_0}\text{jinc}\left(\frac{r_0 a_2}{\lambda_2 z_0}\right) - \frac{\pi a_2'^2}{\lambda_2 z_0}\text{jinc}\left(\frac{r_0 a_2'}{\lambda_2 z_0}\right)\right]$$
$$\times \exp\left(-j\frac{r_0^2}{2\Delta\lambda z_0}\right). \tag{12}$$

The first fraction in the above equation represents the static radiation force produced by the two beams on a total absorber, the second fraction is a constant factor, and the rest represents the spatial distribution of the stress field on the focal plane. If $\Delta\omega \ll \omega_2$, then we may replace λ_1 and λ_2 with image spatial λ_0, and simplify the expression for the stress field. Under these conditions, the stress at the focal point is

$$F_{\Delta\omega}(0,0) = \frac{2I(0)}{c}\frac{a_1^2\left(a_2^2 - a_2'^2\right)}{a_1^4 + \left(a_2^2 - a_2'^2\right)^2}. \tag{13}$$

The fraction on the right represents the effect of transducer dimension on the stress field. For the transducer used in the previous example, the value of this fraction is 0.4999. [This fraction can be also calculated for a single-element transducer of the same diameter excited by an amplitude-modulated signal. For this purpose, we may let $a_1 = a_2$ and $a_2' = 0$. For these values, the last fraction in (13) is 0.5.] The resulting radiation stress (assuming $c = 1500\,\text{m/s}$ for water) is $F_{\Delta\omega}(0,0) = 6.67\times 10^{-4}I(0)\text{N/m}^2$. Now, letting $I(0) = 7200\,\text{W/m}^2$, which is the intensity suggested by the FDA for safe in vivo applications, the resulting stress at the focal point is $F_{\Delta\omega}(0,0) = 4.80\,\text{N/m}^2$. Referring to Fig. 7, we note that this stress field is applied only in a small region around the focal point to the object.

Acoustic Emission To explain the acoustic emission we consider an "object" within an homogeneous infinite medium. This model allows us to separate the roles played by the parameters of the object and the surrounding medium. Also, this model can be fitted to various applications. When an oscillating stress field is applied to the object, the object vibrates at the frequency of the stress field. Vibrational energy of the object is partly transferred to the surrounding medium, resulting in an acoustic emission field. Here, we calculate the acoustic emission of an object subjected to cyclic radiation stress.

To explain the physics in an analytical form, we consider a flat plate facing the beam. Here, we assume that the vibrating object has a circular cross-section of radius b and uniformly vibrates back and forth like a piston. We also consider an area $S \leq \pi b^2$ of the piston surface to be projected normally to the beam. We can always return to our elementary point object by reducing the area of this disk to $dxdy$. Similar solutions can be carried out for objects of other forms. The theory can be also extended to include arbitrary vibrating-part shapes and nonuniform displacement of the object. The total radiation force on this object, $\overline{F_{\Delta\omega}}$, can be found by integrating the radiation stress over the area of the object. This force vibrates the target object at frequency $\Delta\omega$. The steady-state normal velocity amplitude of a piston at frequency $\Delta\omega$, $U_{\Delta\omega}$, due to a harmonic force $\overline{F_{\Delta\omega}}$, can be written as

$$U_{\Delta\omega} = \frac{\overline{F_{\Delta\omega}}}{Z_{\Delta\omega}}, \tag{14}$$

where $Z_{\Delta\omega}$ is the mechanical impedance of the object at $\Delta\omega$. The mechanical impedance of the object has two components, one resulting from the inertia, friction, and the elasticity of the object itself, and the other resulting from the loading effect of the surrounding medium on the vibrating object. The mechanical impedance can be interpreted as a measure of object rigidity and how much it yields to the applied force. For example, for a rigid object, $Z_{\Delta\omega}$ is high and hence resisting the force.

Knowing $U_{\Delta\omega}$, we can calculate the pressure field it produces in the medium. We assume that the acoustic emission signal propagates in a free and homogenous medium. The far-field acoustic pressure due to a piston source of radius b set in a planar boundary of infinite extent is given by [17]:

$$P_{\Delta\omega} =$$
$$-j\Delta\omega\rho \frac{\exp\left(j\Delta\omega l/c\right)}{4\pi l} \left(\frac{2J_1\left(\frac{\Delta\omega b}{c}\sin\vartheta\right)}{\frac{\Delta\omega b}{c}\sin\vartheta} \times \frac{\cos\vartheta}{\cos\vartheta + \beta_\mathrm{B}} \right) \left(2\pi b^2 U_{\Delta\omega}\right), \tag{15}$$

where l is the distance from the observation point to the center of the piston, ϑ is the angle between this line and the piston axis, and β_B is the specific acoustic admittance of the boundary surface (the specific acoustic admittance is $\beta_\mathrm{B} = \frac{\rho c}{Z_\mathrm{B}}$, where Z_B, the acoustic impedance of the boundary, represents the ratio between the pressure and normal fluid velocity at a point on the object). The factor two comes from the presence of the boundary wall. It would be replaced by unity if the boundary wall were not present [17].

The acoustic emission field resulting from object vibration can be written in terms of object mechanical impedance by combining (14) and (15) as

$$P_{\Delta\omega} = \rho c^2 \left[j\frac{\Delta\omega}{c^2} \times \frac{\exp(j\Delta\omega l/c)}{4\pi l} \left(\frac{2J_1\left(\frac{\Delta\omega b}{c}\sin\vartheta\right)}{\frac{\Delta\omega b}{c}\sin\vartheta} \times \frac{\cos\vartheta}{\cos\vartheta + \beta_\mathrm{B}} \right) \right]$$
$$\times \frac{2\pi b^2}{Z_m} \overline{F_{\Delta\omega}}. \tag{16}$$

For wavelengths long compared to the object size, i.e., when $b\Delta\omega/c \to 0$, the term in the paranthesis approaches a constant, hence we may consider the contents of the brackets to be an object-independent function (the specific acoustic admittance β_B relates to the surrounding boundary surface). Under these conditions, the bracket contents in the above equation represent the effect of the medium on the acoustic emission field, which we may call the *medium transfer function*, which my be denoted by

$$H_{\Delta\omega}(l) = j\frac{\Delta\omega}{c^2} \times \frac{\exp\left(j\Delta\omega l/c\right)}{4\pi l} \times \frac{\cos\vartheta}{\cos\vartheta + \beta_\mathrm{B}}. \tag{17}$$

The last fraction in (16) includes $\frac{1}{Z_{\Delta\omega}}$, which is the mechanical admittance of the object at the frequency of the acoustic emission ($\Delta\omega$), and we denote this by $Y_{\Delta\omega}$. It is convenient to combine this term with the next term ($2\pi b^2$) in (16), as $Q_{\Delta\omega} = 2\pi b^2 Y_{\Delta\omega} = 2\pi b^2/Z_{\Delta\omega}$, which is the total acoustic outflow by the object per unit force (acoustic outflow is the volume of the medium, e.g., the fluid, in front of the object surface that is displaced per unit time due to object motion). Function $Q_{\Delta\omega}$ represents the object characteristics at the acoustic frequency. We may thus rewrite (16) in a more compact form as

$$P_{\Delta\omega} = \rho c^2 H_{\Delta\omega}(l) Q_{\Delta\omega} \overline{F_{\Delta\omega}}. \tag{18}$$

Equation (18) indicates that the acoustic emission pressure is proportional to the following: (a) the radiation force, which itself is proportional to the square of ultrasound pressure and the ultrasound characteristics of the object, d_r, in the projected area S; (b) the acoustic outflow by this object, $Q_{\Delta\omega}$, representing the object size b and its mechanical admittance at the acoustic frequency, $Y_{\Delta\omega}$; and (c) the transfer function of the medium at the acoustic frequency, $H_{\Delta\omega}(l)$. Note the difference between the projection area S and the vibrating area πb^2. The projection area determines the extent of the force applied to the object (1). The vibrating area, however, influences the total acoustic outflow in the medium caused by object vibration.

3.3.3 Applications

Evaluating the characteristics of an object (or a medium) by listening to its sound is a traditional approach that has been used for many purposes. Qualitative evaluation of a crystal glass by tapping on it is a simple example. Vibro-acoustography implements the same approach but in microscale, and in a way that could be applied to tissues.

Equation (18) presents a general relationship between the mechanical parameters of the objects, surrounding medium, and the acoustic emission field resulting from object vibration. By measuring the acoustic emission field, it would be possible, in principle, to estimate some of the object or medium parameters, either in an absolute or in a relative sense. For example, one can use vibro-acoustography to measure the resonance frequency of an object, and from that information it is possible to estimate some viscoelastic parameters of the object or the medium (Fig. 8). This method has been used to measure the Young's modulus of a metallic rod [18]. In another experiment, by measuring the resonance frequency of a known resonator in a liquid medium, the viscosity of the fluid has been estimated with good accuracy [19]. These applications are not necessarily medical, but the principle could be used for evaluation of soft tissues, blood, bone, etc.

In medical applications, one can use vibro-acoustography to obtain images of the human body for diagnostic purposes. In such applications, the image may not represent a physical quantity, rather it provides a means to visualize

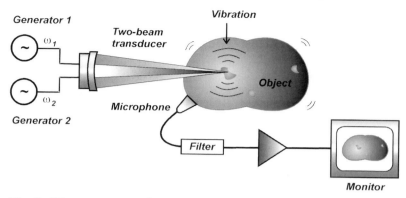

Fig. 8. Vibro-acoustography system. The object is vibrated by the radiation force of the confocal, two-beam transducer. Resulting vibrations are detected by the microphone (or hydrophone) and mapped into the image

object details. To demonstrate the capability of vibro-acoustography, we used this method to image a specimen of the human iliac artery that included calcifications. A 3-MHz confocal transducer with dimensions $a_1 = 14.8$ mm, $a'_2 = 16.8$ mm, and $a_2 = 22.5$ mm and a focal length of 70 mm, was used for this experiment. The difference frequency was set at 44 kHz. Figure 9 shows a photo of the artery that is cut open. The areas with calcification

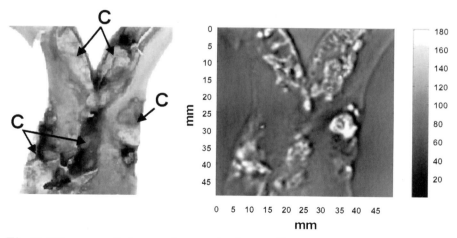

Fig. 9. Vibro-acoustic image of an ex-vivo human iliac artery near the bifurcation. On the left is a photo of the cut-open artery specimen. Calcified areas are marked with letter "C". Vibro-acoustic image of the specimen is shown on the right. Calcified areas stand out as bright spots. This image was obtained with a 3-MHz confocal transducer as described in the text. The difference frequency was set at 44 kHz. The gray-level scale is in arbitrary units

are marked by the letter "C". The vibro-acoustic image of the same artery is shown on the right. Calcium deposits, which are formed in large (a few mm to a few cm long) and stiff (compare to the arterial wall) plates, are efficient acoustic radiators, producing strong acoustic emission when they are exposed to the radiation force of the incident ultrasound. Hence, they stand out in the vibro-acoustography image. The vibro-acoustic image has high resolution, no speckle, and a high signal-to-noise ratio. These features allow vibro-acoustography to delineate calcium deposits with high definition and contrast.

4 Capabilities and Limitations

Although the methods described above tend to estimate the mechanical properties of the object, the quantities they measure are different. The quantity measured in the transient method is somewhat arbitrary, and not a direct measure of any physical quantity. The merit of this method is that tissue hardness is measured from relative displacement values, without the need to measure the applied force. The shear-wave method directly measures a physical parameter of the material in an absolute sense. The acoustic emission field measured in vibro-acoustography is a function of several physical parameters that relate to the object and the surrounding medium. It is possible to measure a single parameter only if one has enough knowledge about other parameters involved. For example, one can measure the shear viscosity of a liquid using a known resonator as an object in the liquid [19], or, for a given the geometry, the Young's modulus of a metallic object can be measured [18] by vibro-acoustography.

Detection sensitivity is critical to the success of any of the above methods. Greater ultrasound intensity would be needed if the detection sensitivity is poor. As a result, the peak ultrasound intensity in the focal area may be limited by the safe in vivo limit. Ultrasound Doppler methods, at conventional ultrasound frequencies, probably can detect displacements on the order of a few micrometers. The sensitivity of the MRI technique is on the order of 100 nm [2]. Vibro-acoustography has been shown to detect motions as small as a few nanometers [8]. The high sensitivity of this method is a result of the fact that small motions of the object can produce an acoustic emission pressure field that is easily detectable by a sensitive hydrophone.

Image spatial resolution is an important parameter when assessing the capability of an imaging method in delineating fine structures in the object. Vibro-acoustography can be used for high-quality imaging. The spatial resolution is proportional to the width of the stress-field main lobe. For example, for the 3-MHz ultrasound transducer presented in Fig. 7, the spatial resolution is about 700μm. Imaging complex objects using the transient or the shear-wave methods has not been fully explored in the literature.

Performance of each method is partly influenced by viscoelastic properties of the material. Both the transient and shear-wave methods work best if the material under test supports the shear wave, has enough attenuation to allow the build-up of enough radiation force, and is compliant enough to allow appreciable displacement. Application of these methods for hard materials, such as bone and calcifications, would be difficult or impractical because (a) stiff materials have low compliance, hence their displacement in response to the force is relatively small and difficult to detect and (b) the shear wave travels at high speed in such materials. In vibro-acoustography hard materials produce structural vibrations which are often stronger than those of soft tissue. Such vibrations result in strong and easily detectable acoustic emissions. The bright appearance of the arterial calcifications in the experiment presented in [8] is a demonstration of this phenomenon. Vibro-acoustography can be used to detect particles in materials that do not support shear waves, for example, detecting gas bubbles in liquids. The acoustic emission resulting from particle vibration includes a compressional-wave component that can travel in the surrounding medium including the liquid. Tissue attenuation for compressional waves at low frequencies is small; hence such waves can be detected by the hydrophone from a distance. Shear waves in a liquid medium decay very rapidly and are difficult to detect by most methods.

The ultimate goal of viscoelastic parameter imaging methods is to measure these parameters in the human body. In vivo application of the transient method is not clear because it requires high ultrasound power to produce a detectable displacement. Application of the shear-wave method in the human body largely depends on the detection method used. Optical methods are not likely to be usable in vivo. MRI detection methods are generally complex, especially when the technique is to be linked to the ultrasound system in the magnetic field. Ultrasound Doppler is a more practical detection method for shear-wave detection; however its sensitivity may not be sufficient for ultrasound intensities within the safe limit [7]. Vibro-acoustography uses a hydrophone for detection which is simple to operate in clinical settings. It however requires an acoustically quiet environment for proper detection. The biological noise spectrum seems to be below 1 kHz, which can be easily filtered out if the operation frequency is above this limit [8].

Before any of the methods discussed here are used in the human body, one must be certain that the ultrasound exposure does not harm the subject. Vibro-acoustography uses either cw or tone-burst ultrasound. Therefore, the continuous ultrasound intensity (spatial peak temporal average intensity) limit must be observed for the safety reasons. This limit according to the FDA is $720\,\mathrm{mW/cm^2}$. However, because of the high sensitivity of hydrophone detectors, it is possible to detect very small levels of the acoustic emission while using low transmittance power. For example, it is shown that a submillimeter object can be detected at ultrasound intensities far below the FDA limit [9]. The shear-wave method presented in [6] uses high-intensity pulsed ultrasound for excitation and either the optical or MRI methods for detection. It is argued that, although the intensity is high, the total exposure is within the FDA limits. However, in vivo use of such detection methods would be difficult. Using Doppler ultrasound for detection is less sensitive; hence it requires higher peak ultrasound intensity to produce enough radiation force on the tissue. This intensity level may exceed the safe spatial peak pulse average intensity limit.

5 Summary

The radiation force of ultrasound is a noninvasive means for introducing a highly localized vibrating force inside tissue. The three methods described here utilize this tool to evaluate the dynamic viscoelastic properties of tissue. The transient method measures minute transient tissue deformation versus time. The shear-wave method measures the amplitude and velocity of the shear waves resulting from excitation by a radiation force. Finally, vibroacoustography measures the acoustic field resulting from object vibration at a specified frequency.

Acknowledgements

This work was supported in part by grant BC971878 from the Army Medical Research and Materiel Command and grant HL 61451 from the National Institutes of Health.

References

1. L. Gao, K. J. Parker, R. M. Lerner, S. F. Levinson, Imaging of the elastic properties of tissue – a review, Ultrasound Med. Biol. **22**, 959–977 (1996)
2. R. Muthupillai, D. J. Lomas, P. J. Rossman, J. F. Greenleaf, A. Manduca, R. L. Ehman, Magnetic resonance elastography by direct visualization of propagating acoustic strain waves, Science **269**, 1854–1857 (1995)
3. B. T. Chu, R. E. Apfel, Acoustic radiation pressure produced by a beam of sound. J. Acoust. Soc. Am. **72**, 1673–1687 (1982)
4. T. Sugimoto, S. Ueha, K. Itoh, Tissue hardness measurement using the radiation force of focused ultrasound, IEEE Ultrasonics Symp. Proc. **3**, 1377–1380 (1990)
5. V. Andreev, V. Dmitriev, O. V. Rudenko, A. Sarvazyan, A remote generation of shear-wave in soft tissue by pulsed radiation pressure, J. Acoust. Soc. Am. **102**, 3155 (1997)
6. A. P. Sarvazyan, O. V. Rudenko, S. D. Swanson, J. B. Fowlkes, S. Y. Emelianov, Shear wave elasticity imaging: a new ultrasonic technology of medical diagnostics, Ultrasound Med. Biol. **24**, 1419–1435 (1998)
7. W. F. Walker, Internal deformation of a uniform elastic solid by acoustic radiation, J. Acoust. Soc. Am. **105**, 2508–2518 (1999)

8. M. Fatemi, J. F. Greenleaf, Ultrasound-stimulated vibro-acoustic spectrography, Science **280**, 82–85 (1998)
9. M. Fatemi, J. F. Greenleaf, Vibro-acoustography: An imaging modality based on ultrasound-stimulated acoustic emission, Proc. Natl. Acad. Sci. USA **96**, 6603–6608 (1999)
10. Lord Raleigh, Philos. Mag. **3**, 338–346 (1902); see also: Lord Raleigh, Philos. Mag. **10**, 364–374 (1905)
11. R. T. Beyer, Radiation pressure – the history of a mislabeled tensor, J. Acoust. Soc. Am. **63**, 1025–1030 (1978)
12. O. V. Rudenko, A. P. Sarvazian, S. Y. Emelionov, Acoustic radiation force and streaming induced by focused nonlinear ultrasound in a dissipative medium, J. Acoust. Soc. Am. **99**, 1–8 (1996)
13. Jiang Z-Y, J. F. Greenleaf, Acoustic radiation pressure in a three-dimensional lossy medium, J. Acoust. Soc. Am. **100**, 741–747 (1996)
14. P. J. Westervelt, The theory of steady force caused by sound waves, J. Acoust. Soc. Am. **23**, 312–315 (1951)
15. S. A. Goss, R. L. Johnston, F. Dunn, Comprehensive compilation of empirical ultrasonic properties of mammalian tissues, J. Acoust. Soc. Am. **64**, 423–457 (1978)
16. L. A. Frizzell, E. L. Carstensen, Shear properties of mammalian tissues at low megahertz frequencies, J. Acoust. Soc. Am. **60**, 1409–1411 (1976)
17. P. M. Morse, K. U. Ingard (Eds.) *Theoretical Acoustics* (McGraw-Hill, New York 1968)
18. M. Fatemi, J. F. Greenleaf, Application of radiation force in noncontact measurement of the elastic parameters, Ultrasonic Imaging **21**, 147–154 (1999)
19. M. Fatemi, J. F. Greenleaf, Remote measurement of shear viscosity with ultrasound-stimulated vibro-acoustic spectrography, Acta Phys. Sin. **8**, S27–S32 (1999)

Estimation of Complex-Valued Stiffness Using Acoustic Waves Measured with Magnetic Resonance

Travis E. Oliphant, Richard L. Ehman, and James F. Greenleaf

Mayo Foundation, 200 First St. SW, Rochester, Minn. 55905, USA
olipt@mayo.edu

Abstract. Tissue stiffness can be a useful indicator of diseased tissue. Noninvasive quantitation of the mechanical properties of tissue could improve early detection of such pathology. A method for detecting displacement from propagating shear waves using a phase-contrast MRI technique has been developed previously. In this chapter the principles behind the measurement technique are reviewed, and the mechanical properties that can be determined from the displacement data are investigated for isotropic materials. An algebraic inversion approach useful for piecewise homogeneous materials is described in detail for the general isotropic case, which is then specialized to incompressible materials as a model for tissue. Results of the inversion approach are presented for an experimental phantom and in-vivo breast tumor. These results show that the technique can be used to obtain shear-wave speed and attenuation in regions where there is sufficient signal-to-noise ratio in the displacement and its second spatial derivatives. The sensitivity to noise is higher in the attenuation estimates than in the shear-wave speed estimates.

1 Introduction

In medical imaging, tissue stiffness is often more correlated with pathology than conventionally imaged parameters such as $T1, T2$, and X-ray density. As a result, elastography, which seeks to develop images based on stiffness, has been the source of significant study [1]. Material hardness is an important parameter of interest in other areas as well. Historically, much elastography research has been conducted using displacement measured with ultrasound [2,3,4,5,6]. With the advent of a phase-contrast technique for measuring displacement using magnetic resonance (MR), interest in magnetic resonance elastography (MRE) has been increasing [7,8,9,10,11]. One reason for the interest in using MR in elastography is that the phase-contrast method used in MRE can provide a 4-dimensional vector-displacement data set throughout the region of interest. While a similar data set can be obtained with ultrasound physics [12,13], there are limited acoustic windows into certain regions of the body (most notably the brain) that hamper the use of current ultrasonic methods in those regions.

Obtaining stiffness from displacement data requires a reconstruction process that is related to but unique from inverse scattering. Typical inverse scattering problems seek reconstruction of material parameters using measurements of the scattered field outside the region of interest coupled with knowledge of the incident field. The reconstruction problem presented when the entire displacement field can be measured seeks determination of the material parameters using measurements of the field inside the region of interest. There have been two general approaches to stiffness estimation using such measurements of displacement: quasi static [7] and dynamic [8,11]. In quasi static reconstruction, displacement measurements are obtained at two times separated by the application of a known stress load. The tissue strain estimated from displacement measurements is used to estimate relative tissue stiffness. Estimating absolute tissue stiffness requires knowledge of the stiffness at the boundary using this quasi static technique. Dynamic elastography uses knowledge of displacement collected over time to reconstruct absolute displacement directly. Absolute estimates of elastic parameters can be obtained with only local knowledge of dynamic displacement.

In this review we present a method for the reconstruction of complex-valued stiffness based on dynamic displacement measurements using an algebraic inversion of the differential equation and demonstrate the method on MR phase-contrast data. As a guide to the reader unfamiliar with elastography and MR measurement of displacement, we first present our model for acoustic motion and review MR measurement theory to show how the displacement is related to a reconstructed phase image. Following this review of the measurement model, we remark on the inversion problem in general and discuss some of the methods that have been presented in the literature before closing with a discussion and demonstration of our straightforward inversion method.

2 Measurement Model

A classic description of continuum mechanics can be given in either Eulerian or Lagrangian coordinates. Eulerian coordinates are fixed in space, while Lagrangian coordinates are fixed on the material and move as the material moves. A description in Lagrangian coordinates keeps track of material and where it is going, while a description in Eulerian coordinates keeps track of space and what is coming through it. When using MR to image displacement, of primary concern is the motion of collections of material nuclei with intrinsic spin, called isochromats, or "spin-regions." This suggests that a description of motion in a Lagrangian coordinate system connected with the material is most applicable.

2.1 Acoustic Model

In Lagrangian coordinates, conservation of momentum and mass gives the partial differential equation satisfied by the material. Consider a continuous region of moving material. Its motion can be characterized by a vector-valued function describing displacement:

$$q(a, t) = a + \xi(a, t),$$

where, for convenience, $\xi(a, 0) = 0$ so that $q(a, t)$ gives the position at time t of the point-mass that was at position a (the Lagrangian coordinate) at time $t = 0$. Conservation of mass gives the following density relation:

$$\rho^{L}(a, t) J(a, t) = \rho^{L}(a, 0) \equiv \rho_0(a),$$

where

$$J(a, t) = \left|\frac{\partial q}{\partial a}\right| = \left|I + \frac{\partial \xi}{\partial a}\right|.$$

$\rho^{L}(a, t) \equiv \rho(q(a, t), t)$, and $\rho(q, t)$ is the (Eulerian) density of whatever is at position q at time t. Force-balance considerations provide the dynamic relation between material properties and displacement. The force-balance equation is written here in summation notation so that repeated indices implies a sum over the index from 1 to 3 and commas imply differentiation with respect to the indices coming after the comma.

$$\rho_0(a) \frac{\partial^2}{\partial t^2} \xi_i(a, t) = \left[\sigma_{ij}^{L}(a, t) G_{jk}(a, t)\right]_{,k} + F_i, \tag{1}$$

where

$$G_{jk} = \text{Cofactor}_{jk}\left(I + \frac{\partial \xi}{\partial a}\right)$$

and $\sigma_{ij}^{L}(a, t) \equiv \sigma_{ij}(q(a, t), t)$; $\sigma_{ij}(q, t)$ is the ijth element of the stress tensor describing internal forces that arise due to the strain of the material, and F_i is the ith component of the body force (such as gravity). Note that the gradient matrix is defined here so that $(\partial \xi / \partial a)_{ij} = \partial \xi_i / \partial a_j$, and the equations are nonlinear even for a "linear material," which can be well modeled with a linear relationship between stress and strain.

In the next section we will show how with MR we can obtain a time-filtered measurement of the vector field, $\xi(a, t)$. With this measurement it is possible to characterize $\sigma_{ij}^{L}(a, t)$ for the sample in question using (1). Such a characterization could be quite arbitrary, including effects such as loss, nonlinearity, and hysteresis. In this presentation, however, we will give only the details of the characterization and estimation of its parameters for materials that are well modeled as linear and isotropic.

We define a linear material as one with a linear relationship between stress and the symmetric part of the displacement gradient (strain). The coefficients of this general linear relationship can be expressed [14] with a fourth-order stiffness tensor, $c_{ijkl}(\boldsymbol{a}, t)$, and a fourth-order viscosity tensor, $c^v_{ijkl}(\boldsymbol{a}, t)$, such that if $s_{kl}(\boldsymbol{a}, t) = (\xi_{k,l} + \xi_{l,k})/2$ and $\star$ denotes linear convolution, then

$$\sigma^{\mathrm{L}}_{ij}(\boldsymbol{a}, t) = c_{ijkl}(\boldsymbol{a}, t) \star s_{kl}(\boldsymbol{a}, t) + c^v_{ijkl}(\boldsymbol{a}, t) \star \frac{\partial s_{kl}(\boldsymbol{a}, t)}{\partial t}.$$

Convolution is used in this expression because we expect a frequency dependence in both stiffness and viscosity. In the frequency domain, this relationship is a product:

$$\Sigma_{ij}(\boldsymbol{a}, f) = C_{ijkl}(\boldsymbol{a}, f) S_{kl}(\boldsymbol{a}, f),$$

where capital letters denote temporal Fourier transforms, except for C_{ijkl} which is defined as

$$C_{ijkl}(\boldsymbol{a}, f) = F_t \{c_{ijkl}(\boldsymbol{a}, t)\} + \mathrm{i} 2\pi f F_t \{c^v_{ijkl}(\boldsymbol{a}, t)\},$$

where F_t denotes the temporal Fourier transform operator. Since $S_{kl} = S_{lk}$, it is also true that $C_{ijkl} = C_{ijlk}$. Consequently,

$$2 C_{ijkl} S_{kl} = C_{ijkl}(\Xi_{k,l} + \Xi_{l,k}) = C_{ijkl} \Xi_{k,l} + C_{ijlk} \Xi_{l,k} = 2 C_{ijkl} \Xi_{k,l}.$$

In other words we can write the relationship between stress and displacement as

$$\Sigma_{ij}(\boldsymbol{a}, f) = C_{ijkl}(\boldsymbol{a}, f) \Xi_{k,l}(\boldsymbol{a}, f).$$

We expect this linear relationship between stress and strain only when strain is small, so we also assume in the equation of motion that $G_{jk} = \delta_{jk}$, where δ_{jk} is the Kronecker delta which is one only if $j = k$ and zero otherwise. Under this assumption and ignoring body forces, the temporal Fourier transform of both sides of the force-balance equation gives

$$[C_{ijkl}(\boldsymbol{a}, f) \Xi_{k,l}(\boldsymbol{a}, f)]_{,j} = -\rho_0(\boldsymbol{a})(2\pi f)^2 \Xi_i(\boldsymbol{a}, f).$$

In principle we can use this equation to find $C_{ijkl}(\boldsymbol{a}, f)$ if we know $\Xi_i(\boldsymbol{a}, f)$ and the spatial derivatives of $\Xi_i(\boldsymbol{a}, f)$.

For isotropic materials the number of independent parameters of C_{ijkl} is reduced to 2. Using complex Lamé parameters to represent these unknowns, we can write the stiffness tensor as

$$C_{ijkl} = \Lambda \delta_{ij} \delta_{kl} + M(\delta_{ik}\delta_{jl} + \delta_{il}\delta_{jk}).$$

This reduces the equation of motion to

$$[\Lambda(\boldsymbol{a}, f) \Xi_{j,j}(\boldsymbol{a}, f)]_{,i} + [M(\boldsymbol{a}, f)(\Xi_{i,j}(\boldsymbol{a}, f) + \Xi_{j,i}(\boldsymbol{a}, f))]_{,j} = -\rho_0(\boldsymbol{a})(2\pi f)^2 \Xi_i(\boldsymbol{a}, f). \qquad (2)$$

2.2 Displacement Measurement with Magnetic Resonance

To expose the relationship between the MR measurement and the motion of the (excited) spins, and to emphasize the assumptions made in developing it, we review MR measurement theory. We adopt a classical continuum perspective of material spins and consider the magnetization density, $\boldsymbol{m}(\boldsymbol{a},t)$, the gyromagnetic ratio, $\gamma(\boldsymbol{a})$, the spin-lattice relaxation constant, $T1(\boldsymbol{a})$, the spin–spin relaxation constant, $T2(\boldsymbol{a})$, and the static magnetic density, $m0(\boldsymbol{a})$, all in a Lagrangian coordinate system. The components of $\boldsymbol{m}$ are given in a reference frame which rotates around the static magnetic field at the Larmor frequency, $\omega_L = \gamma_0 B_0$, where γ_0 is the gyromagnetic ratio for the dominant nuclei and B_0 is a large, static magnetic field whose direction we define as the z-axis. It is natural to consider the magnetic field, $\mathbf{B}(\boldsymbol{x},t)$, in an Eulerian coordinate system; however, the Bloch equations, which describe the evolution of spin magnetization density, are typically given in a Lagrangian coordinate system. The magnetic field can be written in Lagrangian coordinates by noting that $\boldsymbol{x} = \boldsymbol{a} + \boldsymbol{\xi}(\boldsymbol{a},t)$ so that if $B_{\text{ih}}(\boldsymbol{x})$ represents static field inhomogeneities, $\mathbf{G}(t)$ represents applied gradients to the static field, and $\mathbf{B}_1(t)$ represents an applied radio-frequency (RF) pulse, then

$$\mathbf{B}^L(\boldsymbol{a},t) \equiv \mathbf{B}\left(\boldsymbol{a} + \boldsymbol{\xi}(\boldsymbol{a},t), t\right)$$
$$= \{B_0 + B_{\text{ih}}\left[\boldsymbol{a} + \boldsymbol{\xi}(\boldsymbol{a},t)\right] + \mathbf{G}(t)\left[\boldsymbol{a} + \boldsymbol{\xi}(\boldsymbol{a},t)\right]\}\hat{\boldsymbol{z}} + \mathbf{B}_1(t). \quad (3)$$

This model ignores effects due to the physical constraints on the magnetic field requiring it to be curl- and divergence-free, as the effects of these can be corrected [15]. For a circularly polarized RF pulse at a frequency of $\omega_L + \Delta\omega$, the Bloch equations can be written as

$$\frac{\mathrm{d}\boldsymbol{m}(\boldsymbol{a},t)}{\mathrm{d}t} = \mathbf{A}(\boldsymbol{a},t)\boldsymbol{m}(\boldsymbol{a},t) + \begin{bmatrix} 0 \\ 0 \\ \frac{m0(\boldsymbol{a})}{T1(\boldsymbol{a})} \end{bmatrix}, \quad (4)$$

where

$$\mathbf{A}(\boldsymbol{a},t) =$$
$$\begin{bmatrix} -1/T2(\boldsymbol{a}) & \gamma(\boldsymbol{a})B_z(\boldsymbol{a},t) - \omega_L & \gamma(\boldsymbol{a})B_1(t)\sin(\Delta\omega t) \\ -\gamma(\boldsymbol{a})B_z(\boldsymbol{a},t) + \omega_L & -1/T2(\boldsymbol{a}) & \gamma(\boldsymbol{a})B_1(t)\cos(\Delta\omega t) \\ -\gamma(\boldsymbol{a})B_1(t)\sin(\Delta\omega t) & -\gamma(\boldsymbol{a})B_1(t)\cos(\Delta\omega t) & -1/T1(\boldsymbol{a}) \end{bmatrix}.$$

A classical description models MR as a collection of spins with initial magnetization density $\boldsymbol{m}(\boldsymbol{a},t) = m0(\boldsymbol{a})\hat{\boldsymbol{z}}$, individually obeying the Bloch equations with the measurement consisting of a volume integral of the transverse magnetization. To obtain an analytic solution of this measurement, we solve the Bloch equations for $B_1(t) = 0$. Define $m_T(\boldsymbol{a},t) = m_x(\boldsymbol{a},t) + im_y(\boldsymbol{a},t)$ to be the complex transverse magnetization density and t_0 to be the time directly after application of the excitation electromagnetic pulse. These definitions reduce the first two rows of the Bloch equations to a complex first-order differential equation:

$$\frac{dm_T(\boldsymbol{a},t)}{dt} = -\left(\frac{1}{T2(\boldsymbol{a})} + i\gamma(\boldsymbol{a})\{B_0 + B_{ih}[\boldsymbol{a}+\boldsymbol{\xi}(\boldsymbol{a},t)]\right.$$
$$\left. + \mathbf{G}(t)\cdot[\boldsymbol{a}+\boldsymbol{\xi}(\boldsymbol{a},t)]\} - i\gamma_H B_0\right)m_T(\boldsymbol{a},t).$$

The solution to this equation, assuming an initial condition of $m_T(\boldsymbol{a},t_0)$, can be found by integration:

$$m_T(\boldsymbol{a},t) = |m_T(\boldsymbol{a},t)|\,e^{i\theta(\boldsymbol{a},t)-i\varphi(\boldsymbol{a},t)-i2\pi\mathbf{k}(t)\cdot\boldsymbol{a}},$$

$$|m_T(\boldsymbol{a},t)| = |m_T(\boldsymbol{a},t_0)|\exp\left(-\frac{(t-t_0)}{T2(\boldsymbol{a})}\right),$$

$$\theta(\boldsymbol{a},t) = \arg m_T(\boldsymbol{a},t_0)$$
$$- B_0 t[\gamma(\boldsymbol{a})-\gamma_H] - \gamma(\boldsymbol{a})\int_{t_0}^{t} B_{ih}[\boldsymbol{a}+\boldsymbol{\xi}(\boldsymbol{a},\tau)]\,d\tau,$$

$$\varphi(\boldsymbol{a},t) = \gamma(\boldsymbol{a})\int_{t_0}^{t}\mathbf{G}(\tau)\cdot\boldsymbol{\xi}(\boldsymbol{a},\tau)\,d\tau - [\gamma(\boldsymbol{a})-\gamma_0]\boldsymbol{a}\cdot\int_{t_0}^{t}\mathbf{G}(\tau)\,d\tau,$$

$$\mathbf{k}(t) = \frac{\gamma_0}{2\pi}\int_{t_0}^{t}\mathbf{G}(\tau)\,d\tau. \tag{5}$$

The terms in this expression can be interpreted as follows: $|m_T(\boldsymbol{a},t)|$ represents the contrast in a standard magnitude MR image, $\mathbf{k}(t)$ represents the path in spatial Fourier coordinates traversed by the gradients, and $\theta(\boldsymbol{a},t)$ is phase error that necessitates taking the difference between two phase images to obtain the phase signal, $\varphi(\boldsymbol{a},t)$. Reconstruction of an MR image usually requires several "views," which are repeated experiments with an altered $k(t)$ path. We model the (demodulated) measured signal as the weighted integral over space of (5), where the (complex) weight, $A(\boldsymbol{a})e^{i\vartheta(\boldsymbol{a})}$, accounts for receiver sensitivity and the proportionality relationship between magnetization and voltage on a coil. For the nth view, the signal as a function of view-time is

$$s_n(t) = \int_V A(\boldsymbol{a})e^{i\vartheta(\boldsymbol{a})}m_{nT}(\boldsymbol{a},t)\,d^3\boldsymbol{a},$$

where V defines the volume containing excited spins and $m_{nT}(\boldsymbol{a},t)$ denotes the transverse magnetization (in the rotating reference frame) for the nth view. Written in this way, it is evident that the demodulated (complex) signal at view n and time t represents the Fourier transform of

$$I_n(\boldsymbol{a},t) = A(\boldsymbol{a})|m_{nT}(\boldsymbol{a},t)|\,e^{i\vartheta(\boldsymbol{a})+i\theta_n(\boldsymbol{a},t)-i\varphi_n(\boldsymbol{a},t)}$$

at Fourier position $\boldsymbol{k}_n(t)$ where the pulse-sequence employed specifies the Fourier-space path $\boldsymbol{k}_n(t)$.

To obtain a single spatial function using the inverse Fourier transform, it is necessary to make the assumption that $I_n(\boldsymbol{a},t)$ is constant between views (inter-view) and during read-out time (intra-view). In other words we assume $I_n(\boldsymbol{a},t) \approx I_\nu(\boldsymbol{a},TE)$, where TE is the center of the read-out time and ν is the middle view. Then,

$$s_n(t) = \int_V I_\nu(\boldsymbol{a},TE) e^{-\mathrm{i}2\pi \boldsymbol{k}_n(t)\cdot \boldsymbol{a}} \, \mathrm{d}^3\boldsymbol{a}\,.$$

Image reconstruction normally takes place in a computer. As a result, aliasing and "partial-volume" effects occur. We normally assume that spatial aliasing is minimal so that only the first term of the aliasing sum contributes, so the output image consists of samples of the function $P(\boldsymbol{a}) \star I_\nu(\boldsymbol{a},TE)$. The function $P(\boldsymbol{a})$ is the inverse Fourier transform of the k-space window function present due to the finite sampling of Fourier space, and $\star$ denotes spatial convolution. Note that the presence of the blurring function in the digital reconstruction implies that the reconstructed magnitude image can potentially contain information that is strictly in the phase of $I_\nu(\boldsymbol{a},TE)$. For displacement-field imaging we assume that $P(\boldsymbol{a})$ has small enough spatial support so that

$$\arg\left(P(\boldsymbol{a}) \star I_\nu(\boldsymbol{a},TE)\right) \approx \arg I_\nu(\boldsymbol{a},TE)\,.$$

To eliminate contributions to the phase signal that are not due to spin motion, it is common to take two MR images with opposite polarity on the motion-encoding gradients. If we include the possibility of shifting the phase of the motion by T with respect to the imaging sequence, then we can obtain the phase-difference time-series

$$\phi(\boldsymbol{a},t) = \varphi_1(\boldsymbol{a},TE) - \varphi_2(\boldsymbol{a},TE) = 2\gamma(\boldsymbol{a}) \int_{t_0}^{TE} \bar{\mathbf{G}}(\tau) \cdot \boldsymbol{\xi}(\boldsymbol{a},\tau+T)\,\mathrm{d}\tau\,,$$

where $\bar{\mathbf{G}}(t)$ represents just the motion-encoding gradient, which is assumed to have a null zeroth moment so that the second term of $\varphi(\boldsymbol{a},TE)$ is zero. If the gyromagnetic ratio of each excited spin is equal to the dominant spin, then

$$\phi(\boldsymbol{a},t) = 2\gamma_0 \int_{t_0}^{TE} \bar{\mathbf{G}}(\tau) \cdot \boldsymbol{\xi}(\boldsymbol{a},\tau+T) \mathrm{d}\tau\,. \tag{6}$$

Viewing this relationship in the temporal Fourier domain by taking the Fourier transform with respect to T and writing Fourier transforms as capital letters gives

$$\Phi(\boldsymbol{a},f) = \mathbf{H}(f) \cdot \boldsymbol{\Xi}(\boldsymbol{a},f)\,, \tag{7}$$

where

$$\mathbf{H}(f) = 2\gamma_0 \left[\boldsymbol{\Gamma}^*(f) \star W^*(f) \right] .$$

Note that $\boldsymbol{\Gamma}(f)$ is the Fourier transform of the motion-encoding gradients (extended to infinite support if desired) and $W(f)$ is the Fourier transform of a window function which provides the restriction on the support of the true motion-encoding gradients. Equations (6) and (7) show that the phase difference data (which must be free of phase-wrapping) and the displacement field are related via a linear projection filter whose characteristics depend only on the motion-encoding gradients. In summary, it is worthwhile reviewing the assumptions that led to (6) and (7):

- the spins obey the classical Bloch equations (e.g. no diffusion or spin-coupling terms),
- uniform RF pulse and spatially linear magnetic field (concomitant magnetic fields can be ignored or corrected),
- the spins giving signal have a common gyromagnetic ratio equal to γ_0,
- the inter- and intra-view changes to the spins have minimal influence on the phase-difference data,
- the spatial bandwidth of the displacement is low enough so that partial-volume effects are unimportant and there is no aliasing.

Using MR it is thus possible to obtain a full, vector-field measurement of the internal displacement of any object for which an MR signal can be obtained. The frequencies at which the displacement can be measured are limited by the available frequency content of the motion-sensitizing gradients. This limits the frequencies which can be measured in vivo due to the effects on tissue of high-frequency gradients. The amplitude that can be reliably measured is a function of the motion-sensitizing gradient amplitude and the signal-to-noise ratio in the received transverse magnetization signal.

3 Estimating Material Properties

With MR displacement imaging it is feasible to obtain the complete (time-filtered) vector field as a function of space and time. This data set presents new opportunities for obtaining material property distributions distinct from the inverse scattering approach, where measurement is made only outside the region of interest. To date, several algorithms for obtaining material stiffness parameters from displacement measurements have been proposed. In this section we present our straightforward inversion method based on algebraically inverting (2) for homogeneous materials and show examples of the results. Afterwards we briefly explain some of the other methods that have been proposed.

3.1 Algebraic Inversion of the Differential Equation (AIDE)

Because the MR phase-difference data is proportional to displacement in the temporal frequency domain we can collect a phase data set that satisfies the same differential equation of motion as the displacement. Let Φ_i be the phase-difference data set collected when the direction of the motion-sensitizing gradient, $\bar{\mathbf{G}}$, is aligned with the ith coordinate axis. Then, Φ_i is related to Ξ_i through a linear filter and thus satisfies the same linear differential equation of motion as Ξ:

$$[\Lambda(\boldsymbol{a},f)\Phi_{j,j}(\boldsymbol{a},f)]_{,i} + [M(\boldsymbol{a},f)(\Phi_{i,j}(\boldsymbol{a},f) + \Phi_{j,i}(\boldsymbol{a},f))]_{,j} = -\rho_0(\boldsymbol{a})(2\pi f)^2 \Phi_i(\boldsymbol{a},f). \tag{8}$$

3.1.1 Full Inversion

In order to algebraically invert this equation of motion to estimate the shear modulus, we assume local homogeneity of the Lamé coefficients, which allows (8) to be written as a matrix equation applicable at each spatial point, $\boldsymbol{a}$, and frequency, f.

$$\mathbf{A} \begin{bmatrix} \Lambda(\boldsymbol{a},f) + M(\boldsymbol{a},f) \\ M(\boldsymbol{a},f) \end{bmatrix} = -4\pi^2 f^2 \rho_0(\boldsymbol{a}) \begin{bmatrix} \Phi_1(\boldsymbol{a},f) \\ \Phi_2(\boldsymbol{a},f) \\ \Phi_3(\boldsymbol{a},f) \end{bmatrix},$$

where

$$\mathbf{A} \equiv \begin{bmatrix} A_{11} & A_{12} \\ A_{21} & A_{22} \\ A_{31} & A_{32} \end{bmatrix} = \begin{bmatrix} \Phi_{j,j1}(\boldsymbol{a},f) & \Phi_{1,jj}(\boldsymbol{a},f) \\ \Phi_{j,j2}(\boldsymbol{a},f) & \Phi_{2,jj}(\boldsymbol{a},f) \\ \Phi_{j,j3}(\boldsymbol{a},f) & \Phi_{3,jj}(\boldsymbol{a},f) \end{bmatrix}.$$

This expression can be inverted to obtain an estimate of the complex Lamé coefficients for each position and temporal frequency contained in the phase-difference data.

$$\begin{bmatrix} \Lambda(\boldsymbol{a},f) + M(\boldsymbol{a},f) \\ M(\boldsymbol{a},f) \end{bmatrix} = -4\pi^2 f^2 \rho_0(\boldsymbol{a}) \left(\mathbf{A}^H \mathbf{A}\right)^{-1} \mathbf{A}^H \begin{bmatrix} \Phi_1(\boldsymbol{a},f) \\ \Phi_2(\boldsymbol{a},f) \\ \Phi_3(\boldsymbol{a},f) \end{bmatrix}. \tag{9}$$

In this expression, $\mathbf{A}^H$ denotes conjugate transpose of the matrix $\mathbf{A}$. Sometimes it is useful to assume only 2-dimensional data; therefore, we note that to obtain a 2-dimensional version of (9) all derivatives in the z-direction ($i=3$) should be zero. This implies that $A_{31} = 0$ and sums run from 1 to 2 in the remaining elements of $\mathbf{A}$. Under this assumption, the equations decouple into two independent, 2-dimensional modes. Each mode can be used to find $M(\boldsymbol{a},f)$ independently. Alternatively, both modes may be used in an effort to improve estimate quality. The simplest inversion scheme possible occurs when we measure the shear mode (out-of-plane motion) for a purely 2-dimensional problem. In this case

$$M(\boldsymbol{a},f) = \frac{-4\pi^2 f^2 \rho_0(\boldsymbol{a})\Phi_3(\boldsymbol{a},f)}{\Phi_{3,11}(\boldsymbol{a},f) + \Phi_{3,22}(\boldsymbol{a},f)}. \tag{10}$$

3.1.2 Helmholtz Inversion

The inversion method presented here provides estimates of both $\Lambda(\boldsymbol{a}, f)$ and $M(\boldsymbol{a}, f)$, which is appropriate for materials with similar magnitudes of these values. In elastography of soft tissues, however, $|M|/|\Lambda|$ is very small (≈ 0.0001), which creates difficulties in trying to reliably estimate both Lamé coefficients on the same scale using (9). As a result, we must either estimate Λ as a "nuisance" parameter and ignore it, guess a value of Λ to use, or obtain an expression that relates Φ_i to M alone. To accomplish the latter we can assume incompressibility of the material of interest, i.e., $\Phi_{j,j} = 0$. This assumption (along with the homogeneity assumption) decouples the equations of motion into three independent equations and allows measurement of a single sensitization direction to produce an estimate for M using

$$M(\boldsymbol{a}, f) = -4\pi^2 f^2 \rho_0(\boldsymbol{a}) \frac{\Phi_i(\boldsymbol{a}, f)}{\Phi_{i,11}(\boldsymbol{a}, f) + \Phi_{i,22}(\boldsymbol{a}, f) + \Phi_{i,33}(\boldsymbol{a}, f)}. \quad (11)$$

This will be called Helmholtz, or direct, inversion. If the incompressibility assumption is made, we obtain (10) for all polarizations of motion for 2-dimensional data by ignoring out-of-plane derivatives.

3.1.3 Numerical Details

To find the shear modulus, (9) is solved for each point for which Φ_i is available and for which spatial derivatives can be calculated using data in a local window. How best to calculate derivatives from sampled, noisy data to give the best results for this application is an open question. As a first implementation we have calculated derivatives using the best least-squares fit to a local polynomial. This fitting procedure can be implemented as a separable finite impulse response (FIR) filter on the data. Prior to point-by-point division, the original data are also smoothed by using the value of the fitted polynomial at that point instead of the measured value. This is also accomplished with a separable FIR filter.

Finding the complex Lamé coefficients using this direct division approach ultimately results in the division of two numbers, which can cause numerical difficulties if the denominator is nearly zero in (9) or (11). We describe as "poorly interrogated" those regions whose displacement field results in a denominator which is nearly zero. In the presence of noise, the effect of poorly interrogated regions is amplified, and we must take steps to control the magnitude of the computed results in these problem areas so that contrast is not lost in the resulting image. In the simplest case where we apply Helmholtz inversion it can be seen that poorly-interrogated regions correspond to regions where $\nabla^2 \Phi_i(\boldsymbol{a}, f) \approx 0$. In order to regularize the solution in these regions without introducing restrictions on the spatial frequency content of the result and eliminating the fast point-by-point method developed here, we apply two independent methods: (a) project the real part of $M(\boldsymbol{a}, f)$ onto non-negative numbers (zero-out negative results) and (b) apply a median filter to the output.

Once the complex-valued shear modulus, $M = M_\mathrm{R} + \mathrm{i}M_\mathrm{I}$, is determined for each position, the shear speed and attenuation coefficients may be computed from the well-known formulas [obtained from noting that $k = \mathrm{i}2\pi f/c_\mathrm{s} + \alpha$ and $k^2 = (2\pi f)^2 \rho_0/M$]:

$$c_\mathrm{s}(f) = \sqrt{\frac{2\left(M_\mathrm{R}^2 + M_\mathrm{I}^2\right)}{\rho_0\left(M_\mathrm{R} + \sqrt{M_\mathrm{R}^2 + M_\mathrm{I}^2}\right)}},$$

$$\alpha(f) = \sqrt{\frac{\rho_0(2\pi f)^2\left(\sqrt{M_\mathrm{R}^2 + M_\mathrm{I}^2} - M_\mathrm{R}\right)}{2\left(M_\mathrm{R}^2 + M_\mathrm{I}^2\right)}}.$$

Note that AIDE can be carried out for each frequency with significant amplitude in the data set to obtain frequency-dependent estimates of the shear-wave speed and loss parameter. It is useful to point out for some applications that the longitudinal wave speed and attenuation may be obtained using these same two equations by replacing M with $\Lambda + 2M$.

3.2 Other Inversion Methods

The first general approach suggested for estimating stiffness from internal displacement data was to make use of an algorithm to generate a local frequency estimate (LFE) such as the one described in [16] and demonstrated on MRE data in [9].

3.2.1 Local Frequency Estimate

The LFE returns a "direction-independent" estimate of the local phase gradient using a sequence of log-normal filters with the property that the output of the next filter in the sequence is proportional to the spatial derivative of the output of the previous filter. The LFE assumes that a local (radial) frequency can be well defined, which in this case means that the motion satisfies the lossless Helmholtz equation. This equation is relevant whenever the material is sufficiently incompressible, lossless, and homogeneous. The algebraic approach presented here differs from the LFE in that it can be applied to more general cases such as compressible and lossy materials, although it still makes the homogeneity assumption. Even in the incompressible, lossy case in which both methods should apply, AIDE uses finite-impulse-response filters so the homogeneity assumption can be more directly controlled by the size of the window used.

3.2.2 Variational Method

The reconstruction technique described in this paper is a direct solution of the strong form (derivative form) of the dynamic differential equation relating modulus to displacement. Previously, other authors have suggested using the weak (variational, integral) form of the differential equation and appropriately chosen "test functions" to estimate both Lamé coefficients [11]. Their method, which we have termed the variational method (VM), could be adapted to find only the shear modulus using the methods suggested here. It is also a straightforward extension of their method to estimate attenuation in a manner similar to what is demonstrated here.

The variational method "avoids derivative calculations" by taking analytic derivatives of test functions and integrating them over local windows in product with the data. The more direct interpretation presented here, however, ultimately calculates derivatives in a very similar way (i.e., by filtering with the derivative of a smooth polynomial). An additional feature of the variational interpretation allows selection of test functions which are zero on the boundary of a local region in order to eliminate the inhomogeneity from the problem. However, the modulus/density ratio is also assumed to be constant over this local-volume region which is typically on the order of a wavelength. This is very similar to the assumption of quasihomogeneity given here that results in the Helmholtz relation (11). Thus, the variational derivation, while providing valuable alternative insights into the problem, results in a very similar algorithm that can be viewed from the perspective of our technique as using a particular method of taking local derivatives of the data.

3.2.3 Overlapping Subzone Technique

Another technique that has been proposed to invert the full partial differential equation uses Newton's method to minimize the difference between measured displacement and displacement calculated through finite elements from an iteratively updated Young's modulus estimate with an assumed Poisson ratio [8]. This method follows the more traditional approach of other inverse problems in establishing a measurement model and iteratively varying the parameters of the model seeking a "best fit" between the measured data and the model-computed results. So far, this method has only been studied for 2-dimensional problems without attenuation and has shown good results on synthetic data sets. One way to see the difference between AIDE and this overlapping subzone technique (OST) is that AIDE is a direct technique while OST is an indirect one. As a result AIDE is significantly faster and has already demonstrated 3-dimensional reconstructions of both shear speed and attenuation [17]. On the other hand, OST inverts a partial differential equation with inhomogeneous material parameters, which may more realistically model complicated domains. In addition, though the technique could be straightforwardly adapted, the OST as presented in [8] requires measurement of all polarizations, as no provision is described for assumptions that could lead to using only data from one polarization.

The method described in this review and the last two methods described all hold the possibility of reconstructing quantitative shear-wave speeds and attenuations over a 3-dimensional region of interest. One aspect of the reconstruction methods we have described that should be discussed is that they do not model the motion as a plane wave and as a result they are not restricted to single plane-wave shear illumination. This is a common approach to obtaining displacement data with MRE. The only restriction of the methods is the presence of motion in the region of interest. These methods could all benefit from increased complexity and amplitude of motion in the regions of interest.

4 Examples

To illustrate the MR technique for measuring displacement and to demonstrate the AIDE method for estimating material properties from those measurements, two examples will be presented in this section. The first example uses 3-dimensional data from a phantom constructed with two concentrations of gelatin. The second example uses 2-dimensional data from a breast cross-section of a person with breast cancer.

4.1 Experimental Phantom

A phantom was constructed from 1.5% marine-algae gel and with a 3.0% gel inclusion. A 400-Hz shear vibration was applied for each view of the experiment to one side of the phantom with an offset between the motion and experiment start time. The phantom was scanned using an inter-view time of 100 ms (TR) and an echo time of 20 ms (TE). Each read-out period collected a single horizontal line of k-space defining the x-direction. The number of views taken to generate each MR data set was 1024 ($y \times z = 64 \times 16$). The data was reconstructed onto a $256 \times 256 \times 32$ grid corresponding to a field of view of $8 \times 8 \times 4.8$ cm. A motion-sensitizing gradient consisting of 1 cycle of a 400-Hz square wave was used and applied according to the following tetrahedral strategy that required four independent phantom data sets. The first was a reference data set with the motion-sensitizing gradient applied to all three gradient directions. The next three data sets each reversed the polarity of the motion-sensitizing gradient for one of the directions. When the reconstructed phase from one of these last data sets is subtracted from the reference data set, a filtered-displacement, phase-difference data set is obtained for one of the directions of motion. In this way data corresponding to all three polarizations of the motion were measured. Eight time offsets covering one period of the applied vibration were obtained. The entire data set required $1024 \times 4 \times 8 = 32\,768$ read-out times, which at 100 ms per readout required 54min 36.8 s. Note that if the material can be well modeled as incompressible, so that all polarizations of motion are not required for reconstruction, then this time would be cut in half.

A region of the full 3-dimensional magnitude data (averaged over all 32 views) is represented with two orthogonal cross-sections in Fig. 1a. The cross-sections are connected along the black horizontal lines shown in that figure. The lines run through the 3.0% gel, which is just visible in this image. A phase-difference image at one temporal offset is shown in Fig. 1b, where by visual inspection we see that the wavelength in the inclusion is longer, indicating a stiffer material. From this image it is possible to estimate the wavelength in the background material to be about 1 cm, which gives a rough estimate for the speed at 400 Hz as 4 cm/s. The wavelength in the inclusion looks roughly twice as large, giving a speed of sound estimate of 8 cm/s for the inclusion.

The first harmonic of the eight temporal offsets was calculated using the fast fourier transform (FFT) at each point in the 3-dimensional data set. This was used as $\Phi_i(\boldsymbol{a}, f)$ and fed as input into code written to implement full AIDE inversion. For computing derivatives and smoothing of the data, filter lengths of $13 \times 13 \times 3$ were used, and a final $3 \times 3 \times 3$ median filter was applied to eliminate outliers caused by "poorly-interrogated" regions. As speed is a principle advantage of this direct method, it is useful to note for reference that inversion took approximately 50 s on a single-processor 350-MHz Pentium II.

Results of the inversion are shown in Fig. 2. Figure 2a represents the reconstructed shear-speed data with two connected, orthogonal slices of the data, while Fig. 2b shows the reconstructed attenuation constant for the same two

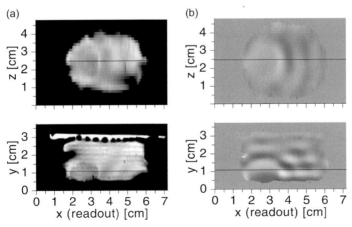

Fig. 1. Magnetic resonance data from the gel phantom: (**a**) average magnitude, (**b**) phase-difference image for the first offset

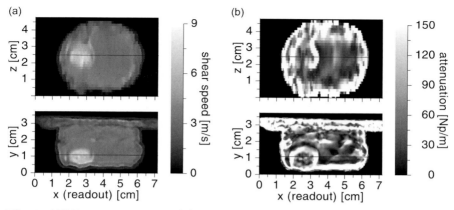

Fig. 2. AIDE inversion results: (**a**) shear-wave speed, (**b**) shear-wave attenuation

slices of data. The inclusion is clearly visible in the shear-speed reconstruction, although it is smaller due to edge effects, which are present because of the window used to estimate spatial derivatives. In addition, the shear speed in the phantom's background gel shows good uniformity, except in "poorly-interrogated" regions, where the amplitude of the motion is insufficient (from inspection of the magnitude of the first harmonic of the motion).

The attenuation images demonstrate that reconstructing attenuation is much more sensitive to noise. This is expected because the signal variation due to attenuation in the wave image is much less than signal variation due to wave speed, and the technique for estimating them both relies on spatial windows of fixed length. The results suggest in fact that separate window sizes be used to estimate shear speed and attenuation and that a decreased spatial resolution be attempted for the attenuation image. One feature of the attenuation data is its emphasis of edges. For example, the edge of the phantom is clearly visible in the attenuation images as are the edges of the "poorly-interrogated" regions.

The second example we present is a 2-dimensional axial slice collected from a person with a large tumor located in the center of the image. A 2-dimensional pulse sequence was used to collect the data as described in [9] with a field of view of 16 × 16 cm reconstructed on a 256 × 256 grid. The frequency of vibration and the applied motion-sensitizing square wave was again 100 Hz. The motion-sensitizing gradient was applied in the out-of-plane direction, and six phase-difference images were collected witht emporal offsets between the vibration and the imaging sequence equally spaced over one period of the 100-Hz vibration (10 ms).

A magnitude image from one of the collected views is shown in Fig. 3a. The tumor appears in the magnitude image as a large dark mass in the center of the breast. One temporal offset from of the phase-difference data is shown in Fig. 3b. This figure shows both that the amplitude of motion in the tumor

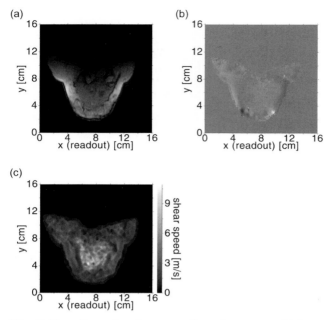

Fig. 3. Two-dimensional cross-section of a person with breast cancer: (**a**) magnitude image, (**b**) phase-difference image, (**c**) reconstructed shear-speed image

is small and that the wavelength is large. The first temporal harmonic of the time-series was found using the FFT, and this was used with 2-dimensional Helmholtz inversion using a filter-window size of 15×15 and a median filter size of 5×5. The shear speed results are shown in Fig. 3c, where the tumor can be seen as the stiffer material. The homogeneity of the reconstruction is only fair, as the low signal-to-noise ratio of the displacement inside the tumor makes reconstruction difficult. This points out that, while the reconstruction technique is not inherently limited by the wavelength of the illuminating shear wave, it is noise-limited. As a result, for a fixed window size, the errors in derivative estimation increase in the presence of noise as the wavelength becomes larger. To keep the same error in the derivative estimate for a given amount of noise, the window size should increase as the wavelength increases. The filter window size affects the resolution of the image. Consequently, the resolution of this imaging technique is a function of both the signal-to-noise ratio and the applied shear wavelength.

5 Conclusion

Measuring the mechanical properties of materials is useful in many areas. For medical applications, differences in tissue stiffness is often relevant to pathology. As a result, interest in elastography, which seeks to produce an image of tissue stiffness, is growing. In this chapter we have shown how quantitative images of the complex-valued Lamé constants can be estimated from the difference between the phase of two magnetic resonance data sets collected with special motion-sensitizing gradients. These complex-valued Lamé constants can be used to generate images of wave speed and attenuation. For tissues, an incompressibility assumption allows the images to be reconstructed using the ratio of the outputs of two linear filters (a derivative filter and a low-pass filter), where the input is phase-difference data corresponding to a single polarization of motion. The resolution of these images depends only on the noise in the image and how it affects estimates of spatial derivatives.

References

1. L. Gao, K. J. Parker, R. M. Lerner, S. F. Levinson, Imaging of the elastic properities of tissue, Ultrasound Med. Biol. **22**, 959–977 (1996)
2. S. Catheline, F. Wu, M. Fink, A solution to diffraction biases in sonoelasticity: The acoustic impulse technique, J. Acoust. Soc. Am. **105**, 2941–2950 (1999)
3. M. Fatemi, J. F. Greenleaf, Ultrasound-stimulated vibro-acoustic spectrography, Science **280**, 82–85 (1998)
4. A. P. Sarvazyan, O. V. Rudenko, S. D. Swanson, J. B. Fowlkes, S. Y. Emelianov, Shear wave elasticity imaging: a new ultrasonic technology of medical diagnosis, Ultrasound Med. Biol. **24**, 1419–1435 (1998)
5. C. Sumi, K. Nakayama, A robust numerical solution to reconstruct a globally relative shear modulus distribution from strain measurements, IEEE Trans. Med. Imag. **17**, 419–428 (1998)
6. C. Sumi, A. Suzuki, K. Nakayama, Estimation of shear modulus distribution in soft tissue from strain distribtution, IEEE Trans. Biomed. Eng. **42**, 193–202 (1995)
7. T. L. Chenevert, A. R. Skovoroda, M. O'Donnell, S. Y. Emelianov, Elasticity reconstructive imaging via stimulated echo MRI, Magn. Res. Med. **39**, 482–490 (1998)
8. E. E. W. Van Houten, K. D. Paulsen, M. I. Miga, F. E. Kennedy, J. B. Weaver, An overlapping subzone technique for MR-based elastic property reconstruction, Magn. Res. Med. **42**, 779–786 (1999)
9. R. Muthupillai, D. J. Lomas, P. J. Rossman, J. F. Greenleaf, A. Manduca, R. L. Ehman, Magnetic resonance elastography by direct visualization of propagating acoustic strain waves, Science **269**, 1854–1857 (1995)
10. R. Muthupillai, P. J. Rossman, D. J. Lomas, J. F. Greenleaf, S. J. Riederer, R. L. Ehman, Magnetic resonance imaging of transverse acoustic strain waves, Magn. Res. Med. **36**, 266–274 (1996)
11. A. J. Romano, J. J. Shirron, J. A. Bucaro, On the noninvasive determination of material parameters from a knowledge of elastic displacements: Theory and simulation, IEEE Trans. Ultrason. Ferroelectr. Freq. Control **45**, 751–759 (1998)
12. V. Dutt, R. R. Kinnick, R. Muthupillai, T. E. Oliphant, R. L. Ehman, J. F. Greenleaf, Acoustic shear-wave imaging using echo ultrasound compared to magnetic resonance elastography, Ultrasound Med. Biol. **26**, 3, 397–403 (2000)

13. Y. Yamakoshi, J. Sato, T. Sato, Ultrasonic imaging of internal vibration of soft tissue under forced vibration, IEEE Trans. Ultrason. Ferroelectr. Freq. Control **37**, 45–53 (1990)
14. E. U. Condon, *Handbook of Physics*, 2nd edn., (McGraw-Hill, New York 1967)
15. M. A. Bernstein, X. J. Zhou, J. A. Polzin, K. F. King, A. Ganin, N. J. Pelc, G. H. Glover, Concomitant gradient terms in phase contrast MR: Analysis and correction, Magn. Res. Med. **39**, 300–308 (1998)
16. H. Knutsson, C.-F. Westin, G. Granlund, Local multiscale frequency and bandwidth estimation, In *Proc. ICIP-94*, Vol. 1, Los Alamitos, CA (IEEE Computer Society 1994) p. 36–40
17. T. E. Oliphant, A. Manduca, R. L. Ehman, J. F. Greenleaf, Complex-valued stiffness reconstruction for magnetic resonance elastography by algebraic inversion of the differential equation, Magn. Res. Med. **45**, 2, 299–310 (2001)

A New Approach for Traveltime Tomography and Migration Without Ray Tracing

Philippe O. Ecoublet[1] and Satish C. Singh[1,2]

[1] Bullard Laboratories, Department of Earth Sciences, University of Cambridge, Madingley Road, Cambridge CB3 0EZ, UK
[2] Laboratoire de Géosciences Marines, Institut de Physique du Globe de Paris, 4 Place Jussieu, 75252 Paris, Cedex 5, France
singh@ipgp.jussieu.fr

Abstract. We present a new method for traveltime tomography. In this method, the traveltime between source and receiver is described by an analytical function, which consists of a series expansion of geometrical coordinates of the source and receiver locations. As the traveltime is derived from the eikonal equation, the analytical function must also satisfy the eikonal equation. This condition imposes a strong constraint on its uniqueness. The coefficients of the series expansion are estimated by minimizing the misfit between the observed and the analytical time function in a least-square sense. Once the coefficients of the series expansion are known, the eikonal equation, which also turns out to be in the form of a series expansion, provides the velocity in the medium. Thus there are two analytical functions, one defining the traveltime and the other defining the slowness, and they can be used for prestack depth migration and velocity model definition. The method can easily be extended to incorporate reflection data and has potential for solving 3-dimensional seismic reflection and global seismology inverse problems.

1 Introduction

The goal of seismic tomography is to determine the large-scale velocity structure of the subsurface of the Earth using traveltime measurements made on the surface. The term "tomography" was first introduced in medical imaging to describe image reconstruction from X-ray line integrals [1,2,3,4]. *Dines* and *Lytle* [5] used the term "computerized geophysical tomography" in analogy to X-ray tomography to determine the velocity distribution between two boreholes from first arrival traveltimes assuming a straight ray path between sources and receivers. The straight ray approximation enables the fast implementation of inversion methods similar to those used in X-ray tomography [6,7], but it does not account for the ray-path dependence on velocity in the medium, which leads to a nonlinear relationship between the data and the model parameters. The nonlinearity can be taken into account to some extent by using an iterative linearized inversion method. A common approach consists of starting from an initial velocity model and iteratively updating this model in such a way that the traveltime data fit the synthetic traveltimes. A review of traveltime tomographic methods is given by *Worthington* [8] and a comparison of various inversion techniques is discussed by *Phillips* and *Fehler* [9].

Ray-based tomography has been applied to various problems [10], but a number of problems remain in the recovery of an accurate and unique image of the Earth from traveltime tomography. Unlike medical imaging, where the measurements of the body that is to be imaged are made from every direction, seismic experiments illuminate the investigated medium from a maximum of three sides, providing nonuniform and incomplete data coverage. Consequently, fewer independent traveltime data than unknown model parameters are available, leading to a nonunique solution, i.e., several different velocity models fit the data [11]. The nonuniqueness of the solution may be constrained by using a priori information for the model parameters [12]. *Berryman* [13,14] used Fermat's principle to constrain the solution. Nevertheless, the computation of traveltime, either using ray-tracing techniques or solving the eikonal equation with a numerical method [15,16], constitutes a time-consuming process in traveltime tomography, particularly in three dimensions.

Alternative approaches to ray-tracing tomography and numerical methods were introduced by *Bates* et al. [17] and were applied to 1-dimensional tomographic inversion by *Enright* et al. [18]. This method deals not with rays but with a functional description of the traveltimes that must satisfy theoretical relationships derived from Fermat's principle. This means that both nonlinearity and nonuniqueness are addressed implicitly by this method. In this paper, we extend this approach to solve a 2-dimensional traveltime tomography problem. In this new method, the traveltime between two points is defined by an analytical function in the form of a series expansion of source and receiver locations. The unknown coefficients of the series expansion are estimated by minimizing the difference between the observed traveltimes and the traveltimes calculated with the analytical function. After estimation of the unknown coefficient, the eikonal equation, derived from the traveltime function, is used to compute the inverse of the square of the velocity. Furthermore, the analytical traveltime function can be used to compute traveltimes between two points in order to implement prestack depth migration.

2 The Traveltime Function

The seismic traveltime between a source, S, and a receiver, R, denoted by $T(R,S)$, is related to the slowness $s(P)$ at the point P along the ray path by the integral

$$T(R,S) = \int_{L(s)} s(P) \mathrm{d}l, \tag{1}$$

where $L(s)$ is the ray path between the source and the receiver, and dl is the arc length along the ray. The traveltime is a solution of the eikonal equation given by

$$\nabla_P T(P,S) \cdot \nabla_P T(P,S) = s^2(P), \tag{2}$$

where ∇_P is the gradient operator at P.

The traveltime $T(R,S)$ can either be computed using a ray-tracing method from (1) or a numerical method from (2) [15]. Here, we assume that $T(R,S)$ can be defined by an analytical function. However, this function must have a number of invariant properties. For example, the first condition requires the traveltime function to vanish for a common source and receiver location S, i.e.,

$$T(S,S) = 0. \tag{3}$$

The reciprocity condition states that the traveltime is invariant when the source and receiver locations are interchanged,

$$T(R,S) = T(S,R). \tag{4}$$

The eikonal equation defined by (2) must remain valid for a common source and receiver point P, i.e.,

$$\nabla_P T(P,P) \cdot \nabla_P T(P,P) = s^2(P), \tag{5}$$

where $\nabla_P T(P,P) = \nabla_P T(P,S)|_{S=P}$.

It should be noted that the left-hand side of (2) is a function of two points, the source location S and the observation point P, whereas the right-hand side of (2), which contains the slowness at the observation point P, is only a function of P. Therefore, taking the gradient with respect to S of the eikonal (2) yields

$$\nabla_S [\nabla_P T(P,S) \cdot \nabla_P T(P,S)] = 0. \tag{6}$$

Equation (6) imposes a uniqueness constraint on the value of slowness at P, which is independent of any source point S and receiver locations. For a common source–receiver point P, (6) gives a boundary condition of the form

$$\nabla_S [\nabla_P T(P,P) \cdot \nabla_P T(P,P)] = 0. \tag{7}$$

Equations (6) and (7) are valid for any points P and S and provide strong constraints on the analytical function.

A distinction can be made between two types of equations: those related to data and those independent of data. We call the first set of equations (3,4,6,7) hard constraints on the analytical function, as they are independent of data, and the second set of equations (1,2,5) soft constraints, as they are dependent on data T.

2.1 Traveltime as a Series Expansion

The formulation of the 2-dimensional problem requires the parameterization which best satisfies the "hard" constraints to be chosen. The following parameters define the source $S(X_S, Z_S)$ and receiver $R(X_R, Z_R)$ locations uniquely in 2-dimensional space (Fig. 1). The coordinates of the source–receiver mid-point, $M(X_M, Z_M)$ are

$$X_M = \frac{X_S + X_R}{2} \quad \text{and} \quad Z_M = \frac{Z_S + Z_R}{2}, \tag{8}$$

the source–receiver distance (d) is

$$d = \left[(X_R - X_S)^2 + (Z_R - Z_S)^2 \right]^{1/2}, \tag{9}$$

and the angle $\theta \left[\frac{-\pi}{2}, \frac{\pi}{2} \right]$ is

$$\theta = \arcsin \left(\frac{Z_R - Z_S}{d} \right). \tag{10}$$

The traveltime function can then be written as a series expansion of these four parameters (X_M, Z_M, d, θ):

$$T(R,S) = \sum_{l=0}^{L} \sum_{m=0}^{M} \sum_{n=-N}^{N} \sum_{p=0}^{P} C_{lmnp} P_l(X_M) P_m(Z_M) \exp(i2n\theta) d^p, \tag{11}$$

where C_{lmnp} are the unknown complex coefficients of the series expansion, P_l is the lth-order Chebyshev polynomial [19], and L, M, N, P are the number of coefficients associated to each term of the series expansion. The term related to symbol L controls lateral variations of velocity and that to symbol M controls the smoothness of velocity variation with depth. Chebyshev polynomials are used in place of a power-series expansion as they are orthogonal in the

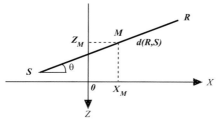

Fig. 1. Geometrical parameters for 2-dimensional tomography. The source–receiver pair (S, R) is uniquely defined by four parameters: X_M and Z_M positions of the mid-point M, the orientation θ with respect to the x-axis, and the source–receiver distance $d(R, S)$

interval $[-1,+1]$. Any other basis function can be used to define the traveltime function. The boundary condition on traveltime for a common source and receiver location, i.e., $d = 0$, (3) yields

$$\sum_{l=0}^{L}\sum_{m=0}^{M}\sum_{n=-N}^{N} C_{lmn0} P_l(X_M) P_m(Z_M) \exp(i2n\theta) = 0, \tag{12}$$

and consequently

$$C_{lmn0} = 0. \tag{13}$$

The condition of traveltime reciprocity (4) is explicitly satisfied by the traveltime series expansion, since the angle θ remains unchanged by swapping over the source and receiver locations.

2.2 The Eikonal Equation

The expression for the eikonal equation at a unique point P (5) takes the form

$$\nabla_P T(P,P) \cdot \nabla_P T(P,P) =$$
$$\left[\sum_{l=0}^{L}\sum_{m=0}^{M}\sum_{n=-N}^{N} C_{lmn1} P_l(X_P) P_m(Z_P) \exp(i2n\theta)\right]^2$$
$$= s^2(P). \tag{14}$$

The condition given by the eikonal equation on the invariance of the slowness at P (7) imposes that (14) be independent of θ. In order to satisfy this condition, the coefficients related to θ must vanish, i.e.,

$$C_{lmn1} = 0, \forall n \neq 0. \tag{15}$$

Incorporating condition (15) into (14), we obtain

$$\nabla_P T(P,P) \cdot \nabla_P T(P,P) = \left(\sum_{l=0}^{L}\sum_{m=0}^{M} C_{lm01} P_l(X_P) P_m(Z_P)\right)^2 = s^2(P). \tag{16}$$

Since the slowness at any point P is always positive, we can say that

$$s(P) = \sum_{l=0}^{L}\sum_{m=0}^{M} C_{lm01} P_l(X_P) P_m(Z_P). \tag{17}$$

The real value of the slowness is ensured by imposing the following conditions on the coefficients C_{lm01}:

$$\Im(C_{lm01}) = 0 \quad \text{and} \quad \Re(C_{lm01}) = R_{lm} \forall l, m. \tag{18}$$

Equation (17) defines the slowness at any point P in the medium as a 2-dimensional series expansion of its coordinates (X_P, Z_P). If the coefficients C_{lmnp} are known, (17) can be used to reconstruct the slowness in the medium. By incorporating the conditions (13), (15) and (18) into the traveltime (11), the traveltime function can be written

$$T(R,S) = \sum_{l=0}^{L} \sum_{m=0}^{M} R_{lm} P_l(X_M) P_m(Z_M) d$$

$$+ \sum_{l=0}^{L} \sum_{m=0}^{M} \sum_{n=-N}^{N} \sum_{p=2}^{P} C_{lmnp} P_l(X_M) P_m(Z_M) \exp(i 2 n \theta) d^p. \quad (19)$$

The substitution of the slowness (17) into (19) gives the traveltime equation:

$$T(R,S) =$$
$$s(M)d + \sum_{l=0}^{L} \sum_{m=0}^{M} \sum_{n=-N}^{N} \sum_{p=2}^{P} C_{lmnp} P_l(X_M) P_m(Z_M) \exp(i 2 n \theta) d^p. \quad (20)$$

Thus the traveltime between a source S and a receiver R is sum of the traveltime through a medium at constant slowness, the slowness of the midpoint $s(M)$, and a power-Fourier series of the geometrical coordinates of the source and receiver. The first term corresponds to a straight ray path approximation, and the second term can be thought of as a correction to this approximation incorporating ray bending in an heterogeneous medium.

2.3 The Equations of Constraint

Condition (6), stating the independence of the eikonal equation with respect to the source location, yields two partial derivative equations:

$$\frac{\partial T(P,S)}{\partial X_P} \frac{\partial}{\partial X_S} \left(\frac{\partial T(P,S)}{\partial X_P} \right) + \frac{\partial T(P,S)}{\partial Z_P} \frac{\partial}{\partial X_S} \left(\frac{\partial T(P,S)}{\partial Z_P} \right) = 0 \quad (21a)$$

and

$$\frac{\partial T(P,S)}{\partial X_P} \frac{\partial}{\partial Z_S} \left(\frac{\partial(P,S)}{\partial X_P} \right) + \frac{\partial T(P,S)}{\partial Z_P} \frac{\partial}{\partial Z_S} \left(\frac{\partial T(P,S)}{\partial Z_P} \right) = 0. \quad (21b)$$

The equation of constraint (21) defines a feasibility domain, independent of the data, in which the traveltime function is a solution of the eikonal equation and therefore complies with the ray theory. The traveltime function defined by (20) does not explicitly satisfy these two partial derivative equations. However, in order that the unknown coefficients of the series expansion satisfy the constraints (21a) and (21b), these equations are generated for a chosen set of S and P points and are solved numerically.

3 Tomography

The unknown coefficients of the series expansion are estimated such that the analytical traveltime function fits the data and satisfies the equations of constraint.

3.1 The Misfit Function

The above is achieved by minimizing a misfit function that accounts for the difference between observed and calculated traveltimes as well as the difference between the equations of constraint and a priori information, for example, information on the velocity from the sonic log. Such a priori information is treated as data. To simplify the notations, we make $\boldsymbol{m}$ a vector containing the unknown coefficients of the series expansion, i.e.,

$$\boldsymbol{m} = (C_{lmnp})^{\mathrm{T}}. \tag{22}$$

We assume that both data and modeling errors follow a Gaussian distribution, and we use a probabilistic definition of the misfit function, where a posteriori probability density function on the model parameters $\mathrm{pdf}(\boldsymbol{m})$ can be written as

$$\mathrm{pdf}(\boldsymbol{m}) = \mathrm{const} \times \exp\left(-\frac{1}{2}\left\{[\boldsymbol{D}_{\mathrm{cal}}(\boldsymbol{m}) - \boldsymbol{D}_{\mathrm{obs}}]^{\mathrm{T}} \mathbf{C}_{\mathrm{D}}^{-1}[\boldsymbol{D}_{\mathrm{cal}}(\boldsymbol{m}) - \boldsymbol{D}_{\mathrm{obs}}] + \mathrm{fc}(\boldsymbol{m})^{\mathrm{T}}\mathbf{C}_{\mathrm{fc}}^{-1}\mathrm{fc}(\boldsymbol{m})\right\}\right). \tag{23}$$

The vector $\boldsymbol{D}_{\mathrm{cal}}(\boldsymbol{m})$ contains N_{t} traveltimes, $\boldsymbol{T}_{\mathrm{cal}}$ defined by (19), and N_{s} slownesses $\boldsymbol{s}_{\mathrm{cal}}$ defined by (17), where

$$\boldsymbol{D}_{\mathrm{cal}}(\boldsymbol{m}) = [\boldsymbol{T}_{\mathrm{cal}}(\boldsymbol{m}), \boldsymbol{s}_{\mathrm{cal}}(\boldsymbol{m})]^{\mathrm{T}}. \tag{24}$$

The data vector $\boldsymbol{D}_{\mathrm{obs}}$ contains the data, N_{t} observed traveltimes $\boldsymbol{T}_{\mathrm{obs}}$, and N_{s} known slowness values $\boldsymbol{s}_{\mathrm{obs}}$, where

$$\boldsymbol{D}_{\mathrm{obs}} = (\boldsymbol{T}_{\mathrm{obs}}, \boldsymbol{s}_{\mathrm{obs}})^{\mathrm{T}}. \tag{25}$$

The vector $\mathrm{fc}(\boldsymbol{m})$ contains N_{fc} equations of constraint (21) defining the invariance of the eikonal equation with respect to the source location.

$$\mathrm{fc}(\boldsymbol{m}) = \{\nabla_{\mathrm{S}}[\nabla_{\mathrm{P}}\boldsymbol{T}_{\mathrm{cal}}(\boldsymbol{m}) \cdot \nabla_{\mathrm{P}}\boldsymbol{T}_{\mathrm{cal}}(\boldsymbol{m})]\}^{\mathrm{T}}. \tag{26}$$

Maximizing the a posteriori probability density function amounts to minimizing the argument of the exponential, that is to say minimizing the misfit function $Q(\boldsymbol{m})$ defined by

$$Q(\boldsymbol{m}) = \left\{[\boldsymbol{D}_{\mathrm{cal}}(\boldsymbol{m}) - \boldsymbol{D}_{\mathrm{obs}}]^{\mathrm{T}} \mathbf{C}_{\mathrm{D}}^{-1}[\boldsymbol{D}_{\mathrm{cal}}(\boldsymbol{m}) - \boldsymbol{D}_{\mathrm{obs}}] \right. \\ \left. + \mathrm{fc}(\boldsymbol{m})^{\mathrm{T}}\mathbf{C}_{\mathrm{fc}}^{-1}\mathrm{fc}(\boldsymbol{m})\right\}. \tag{27}$$

N_D is the total number of data, $N_D = N_t + N_s$, and $\mathbf{C}_D$ is a data error covariance matrix. The diagonal terms of $\mathbf{C}_D$ contain the data variance, denoted by σ_t^2 and σ_s^2 and can be defined as $C_D^{ii} = \sigma_t^{i^2}, \forall i = 1, N_t$, $C_D^{ii} = \sigma_s^{i^2}, \forall i = N_t + 1, N_s$, and $C_D^{ii} = 0, \forall i \neq j$. $\mathbf{C}_{fc}$ is a covariance matrix describing the parameterization errors defined as $C_{fc}^{ii} = \sigma_{fc}^{i^2}, \forall i = 1, N_{fc}$, and $C_{fc}^{ii} = 0, \forall i \neq j$.

3.2 The Initial Model

Although the traveltime series expansion is a linear function of the unknown coefficients, the equations of constraint introduce nonlinear relationships between the coefficients and the data. Consequently, the misfit function becomes nonquadratic; its minimum cannot be reached in a single step, and an iterative optimization process has to be implemented to find the minimum of the misfit $Q(\boldsymbol{m})$. The initial model parameters, denoted $\boldsymbol{m}_0$ for iteration zero, are set to zero except the first coefficient, C_{0001}^0, which is defined as the average slowness $\bar{s}$, i.e.,

$$C_{0001}^0 = \frac{1}{N_t} \sum_{i=1}^{N_t} \frac{T_{obs}^i}{d^i} = \bar{s}, \tag{28a}$$

and

$$C_{lmnp}^0 = 0 \forall (l, m, n, p) \neq (0, 0, 0, 1), \tag{28b}$$

where d^i is the source–receiver distance and C_{lmnp}^0 are the initial coefficients of the traveltime series expansion at iteration zero.

3.3 Optimization

The minimum of the misfit function is obtained by implementing an iterative Gauss–Newton method [20]. The inverse of the Hessian matrix is computed using a singular value decomposition (SVD) [21,22]. The larger eigenvalues are associated with the dominant features of the model parameters, whereas the smaller eigenvalues are dominated by noise. Numerical instabilities during inversion can be avoided by not including the eigenvalues smaller than a cutoff value, λ_{cut}. The minimum of the misfit is reached when the gradient of the misfit function vanishes, i.e., when the increment $\Delta \boldsymbol{m}_N$ tends to zero. The convergence criteria is based on the value of $\Delta \boldsymbol{m}_N$ compared to a threshold. The final solution, denoted $\boldsymbol{m}_*$, is reached when the increment $\Delta \boldsymbol{m}_N$ becomes smaller than a threshold value that can be determined numerically.

3.4 Slowness Image Reconstruction

Once the model parameters, i.e., the unknown coefficients of the series expansion, have been determined, the slowness distribution in the investigated media can be easily obtained using (17). Since the Chebyshev polynomials in (17) cannot represent sharp discontinuities, the final slowness model is always smooth. The degree of smoothness will depend on the number of coefficients defining the series expansion.

4 Error and Resolution Analyses

No inversion is complete without having error and resolution in the formal model. Unlike in conventional tomography, where such analysis is carried out on the final velocity model, the error and resolution analyses need to be first performed on the coefficients of the series expansion. Prior to performing the conventional error analysis, which requires that the a posteriori probability function (PPD) be a Gaussian function, we first need to prove that the PPD we obtain using our method is indeed Gaussian. This requires that, provided the initial model m_0 is a solution of the equations of constraint (i.e., it is within the feasibility region defined by the theoretical constraints), the a posteriori probability density function (23) is Gaussian, and therefore the algorithm converges towards the maximum likelihood solution. An initial model defined by the average constant slowness computed from the traveltime data lies within the feasibility domain and therefore ensures convergence toward the global minimum of the misfit function. Equation (47) (see Appendix B) confirms that the PPD is a Gaussian function, whereas (50) and (51) satisfy the condition required by the initial model.

The result of the stochastic inversion provides statistical information on the estimated model parameters by calculating the a posteriori covariance matrix, denoted $\mathbf{C}_*$, and given by

$$\mathbf{C}_* = \left(\mathbf{G}^T \mathbf{C}_D^{-1} \mathbf{G} + \mathbf{F}_0^T \mathbf{C}_{fc}^{-1} \mathbf{F}_0\right)^{-1} = 2\mathbf{H}_*^{-1}. \tag{29}$$

From the a posteriori covariance matrix, of which a detailed calculation is given in Appendix B, uncertainties of slowness reconstruction and traveltime calculation can be evaluated.

5 Prestack Depth Migration

The traveltime series expansion resulting from the tomographic inversion can be used to compute traveltimes and ray paths between two points. We propose that this result be used to implement a diffraction prestack migration without ray tracing. Kirchoff migration [23] requires intensive traveltime computation of the diffracted wave between each subsurface scattering point and each source and receiver. This operation, usually achieved by shooting rays through the velocity model [24,25], constitutes the most expensive step of the Kirchoff migration. However, two-point traveltime computation can be accurately and rapidly performed using the traveltime series expansion resulting from the tomographic inversion. The smooth and continuous slowness model, defined by the slowness function, is particularly appropriate to perform a prestack migration of seismic data. The computation time required to compute traveltimes with the series expansion depends on the number of coefficients composing the series expansion. Another advantage of our method is that traveltimes can be computed at the time of migration, hence reducing the memory requirement of input/output (I/O) operations.

5.1 Computation of the Incidence Angle of the Ray

Limited-aperture data produce migration smiles which spread the image of the actual scatterers. These artifacts can be reduced by performing a limited aperture migration which requires computation of the ray's incidence angle on arriving at and on leaving each scattering point. Rays whose incidence angles at the scattering point exceed the extreme values controlled by the bounded slope are rejected.

Let us consider a ray emanating from a source S to a point P (Fig. 2). The gradient operator applied to the traveltime $T(P,S)$ gives the ray parameter at P:

$$\nabla_P T(P,S) = \boldsymbol{p} = s(P)\frac{\mathrm{d}\boldsymbol{r}}{\mathrm{d}s}, \tag{30}$$

where $s(P)$ is the slowness at P and $\mathrm{d}\boldsymbol{r}$ is the tangent along the ray with length $\mathrm{d}s$. If $\mathrm{d}x$ and $\mathrm{d}z$ are the projections of $\mathrm{d}\boldsymbol{r}$ along the x- and z-axes, $\mathrm{d}\boldsymbol{r} = (\mathrm{d}x,\mathrm{d}z)^T$, the incidence angle θ_z, defined as the angle between $\mathrm{d}\boldsymbol{r}$ and the z-axis, such that $\mathrm{d}z = \mathrm{d}s\cos(\theta_z)$, is given by

$$\theta_z = \arccos\left(\frac{1}{s(P)}\frac{\partial T(P,S)}{\partial Z_P}\right). \tag{31}$$

The directions of the rays, on arriving at and on leaving each subsurface point, are computed from the traveltime function so that unwanted rays can be eliminated when summing the diffracted energy. Incorporating aperture limitation aims to reduce artifacts and thus to improve the result of the migration.

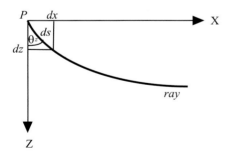

Fig. 2. The ray's incidence angle at $P, \theta z$, defined as the angle between the ray and the vertical axis, is computed from the ray parameter p and the slowness $s(P)$

6 Conclusions

- We have developed a new method for traveltime tomography which does not require ray tracing but takes ray bending into account.
- Our method provides a solution for slowness in a compact analytical form, which can be used to define large-scale models, such as the velocity structure of the Earth.
- It also provides the traveltime function in an analytical form, which we have used for pre-stack depth migration of seismic reflection data. Such an analytical function could also be used to compute traveltimes between any source and receiver efficiently. For example, we could imagine replacing the Jeffrey–Bullen table with such an analytical function, which could be applicable for the 3-dimensional Earth model.

Appendix A: Eikonal Equation from the Traveltime Series Expansion

Here, we develop the expression of the eikonal equation (5) leading to the form given by (14). The gradient of the traveltime function computed at P, $\nabla_P T$ has for components in 2-dimensional space

$$\nabla_P T = \left(\frac{\partial T}{\partial X_P}, \frac{\partial T}{\partial Z_P}\right)^T \tag{32}$$

where the superscript T denotes the transpose. By substituting $\nabla_P T$ with its components in the eikonal equation formulated at a unique point P, (5) can be written as

$$\nabla_P T(P,P) \cdot \nabla_P T(P,P) = \left(\frac{\partial T}{\partial X_P}\right)^2_{d=0} + \left(\frac{\partial T}{\partial Z_P}\right)^2_{d=0} = s^2(P) \tag{33}$$

where d is the source–receiver distance (9). The calculation of the partial derivatives of the traveltime function (11) is straightforward and yields for a null source–receiver distance

$$\left(\frac{\partial T}{\partial X_P}\right)_{d=0} = \sum_{l=0}^{L}\sum_{m=0}^{M}\sum_{n=-N}^{N} C_{lmn1} P_l(X_P) P_m(Z_P) \exp(i2n\theta) \left(\frac{\partial d}{\partial X_P}\right)_{d=0}, \tag{34}$$

and

$$\left(\frac{\partial T}{\partial Z_P}\right)_{d=0} = \sum_{l=0}^{L}\sum_{m=0}^{M}\sum_{n=-N}^{N} C_{lmn1} P_l(X_P) P_m(Z_P) \exp(i2n\theta) \left(\frac{\partial d}{\partial Z_P}\right)_{d=0}. \tag{35}$$

By substituting (34) and (35) into (33) and considering that

$$\left(\frac{\partial d}{\partial X_\mathrm{P}}\right)^2 + \left(\frac{\partial d}{\partial Z_\mathrm{P}}\right)^2 = 1, \tag{36}$$

the eikonal equation (33) becomes

$$\nabla_\mathrm{P} T(P,P) \cdot \nabla_\mathrm{P} T(P,P) =$$
$$\left[\sum_{l=0}^{L}\sum_{m=0}^{M}\sum_{n=-N}^{N} C_{lmn1} P_l(X_\mathrm{P}) P_m(Z_\mathrm{P}) \exp(\mathrm{i}2n\theta)\right]^2$$
$$= s^2(P). \tag{37}$$

Appendix B: Resolution Analysis Posterior After Inversion

A resolution analysis of the inversion result is developed here; we demonstrate that provided the starting model $\boldsymbol{m}_0$ is a solution of the equations of constraint, the a posteriori probability density function (23) is Gaussian, so that the algorithm converges towards the maximum likelihood solution. This analysis provides an expression for the a posteriori covariance matrix that can be used to quantify uncertainties for model parameters and therefore for velocity reconstruction and traveltime computation.

Appendix B.1: The a Posteriori Covariance Matrix

For the case of a linear system, i.e., the data are a linear function of the model parameters, the a posteriori probability density function, $\mathrm{pdf}(\boldsymbol{m})$, is a Gaussian function with a mean model $\boldsymbol{m}$ and a posteriori covariance matrix $\mathbf{C}_*$:

$$\mathrm{pdf}(\boldsymbol{m}) = \mathrm{const.} \times \exp\left(-\frac{1}{2}(\boldsymbol{m}-\boldsymbol{m}_*)^\mathrm{T} \mathbf{C}_*^{-1}(\boldsymbol{m}-\boldsymbol{m}_*)\right). \tag{38}$$

Although the analytical traveltime function is a linear function of the model parameters, the theoretical relationships established by the equations of constraint between the model parameters are nonlinear. Thus, the a posteriori probability density function (38) is not Gaussian and the inversion algorithm may converge towards a local minimum of the misfit function. Nevertheless, we demonstrate that if the initial model satisfies the equations of constraint, i.e., it is contained within the feasibility region defined by the theoretical constraints, the a posteriori probability density function is Gaussian, and therefore the algorithm converges towards the global minimum of the misfit function. Equation (24) can be written as $\boldsymbol{D}_\mathrm{cal}(\boldsymbol{m}) = \mathbf{G}\boldsymbol{m}$, where $\mathbf{G}$ is a matrix containing the geometrical terms of the series expansion. The linearization of the equations of constraint $\boldsymbol{fc}(\boldsymbol{m})$ around an initial model $\boldsymbol{m}_0$ yields, neglecting higher-order terms,

$$\boldsymbol{fc}(\boldsymbol{m}) \approx \boldsymbol{fc}(\boldsymbol{m}_0) + \mathbf{F}_0(\boldsymbol{m} - \boldsymbol{m}_0), \tag{39}$$

where the matrix $\mathbf{F}_0$ contains the partial derivatives of $\boldsymbol{fc}$ with respect to the model parameters computed at $\boldsymbol{m}_0$, i.e.,

$$F_0^{ij} = \frac{\partial fc^i(m_0)}{\partial m_0^j}. \tag{40}$$

By choosing the initial model $\boldsymbol{m}_0$ such that it is a solution of the equations of constraint, which requires

$$\boldsymbol{fc}(\boldsymbol{m}_0) = 0, \tag{41}$$

the misfit function takes the following form:

$$Q(\boldsymbol{m}) \approx (\mathbf{G}\boldsymbol{m} - \boldsymbol{D}_{\text{obs}})^{\mathrm{T}} \mathbf{C}_{\mathbf{D}}^{-1} (\mathbf{G}\boldsymbol{m} - \boldsymbol{D}_{\text{obs}}) + (\boldsymbol{m} - \boldsymbol{m}_0)^{\mathrm{T}} \mathbf{F}_0^{\mathrm{T}} \mathbf{C}_{\boldsymbol{fc}}^{-1} \mathbf{F}_0 (\boldsymbol{m} - \boldsymbol{m}_0). \tag{42}$$

At the maximum likelihood solution $\boldsymbol{m}_*$, the gradient of the misfit $\boldsymbol{g}$ vanishes, $\boldsymbol{g} = \nabla_{\mathrm{m}} Q(\boldsymbol{m}_*) = 0$, so that

$$\mathbf{G}^{\mathrm{T}} \mathbf{C}_D^{-1} (\underline{\mathbf{G}} \boldsymbol{m}_* - \boldsymbol{D}_{\text{obs}}) + \mathbf{F}_0^{\mathrm{T}} \mathbf{C}_{\boldsymbol{fc}}^{-1} \mathbf{F}_0 (\boldsymbol{m}_* - \boldsymbol{m}_0) = 0. \tag{43}$$

Note that, since the calculated data are linear functions of the model $\boldsymbol{m}$, the matrix $\mathbf{G}$ is independent of $\boldsymbol{m}$. After some manipulations [26], the misfit function $Q(\boldsymbol{m})$ can be written in the form

$$Q(\boldsymbol{m}) \approx \frac{1}{2}(\boldsymbol{m} - \boldsymbol{m}_*)^{\mathrm{T}} \mathbf{H}_*(\boldsymbol{m} - \boldsymbol{m}_*) - \frac{1}{2} \boldsymbol{m}_*^{\mathrm{T}} \mathbf{H}_* \boldsymbol{m}_* + \boldsymbol{D}_{\text{obs}}^{\mathrm{T}} \mathbf{C}_D^{-1} \boldsymbol{D}_{\text{obs}} + \boldsymbol{m}_0^{\mathrm{T}} \mathbf{F}_0^{\mathrm{T}} \mathbf{C}_{\boldsymbol{fc}}^{-1} \mathbf{F}_0 \boldsymbol{m}_0, \tag{44}$$

where $\mathbf{H}$ is the Hessian of the misfit function calculated at the maximum likelihood model $\boldsymbol{m}_*$, with

$$\mathbf{H}_* = \nabla_m \nabla_m Q(\boldsymbol{m}_*) = 2 \left(\mathbf{G}^{\mathrm{T}} \mathbf{C}_D^{-1} \mathbf{G} + \mathbf{F}_0^{\mathrm{T}} \mathbf{C}_{\boldsymbol{fc}}^{-1} \mathbf{F}_0 \right). \tag{45}$$

The last three terms of the right-hand side of (44) are constant, i.e., independent of $\boldsymbol{m}$; the first term, however, is not. So

$$Q(\boldsymbol{m}) \approx \frac{1}{2}(\boldsymbol{m} - \boldsymbol{m}_*)^{\mathrm{T}} \mathbf{H}_*(\boldsymbol{m} - \boldsymbol{m}_*) + \text{const}. \tag{46}$$

By replacing the misfit function (46) into the probability density function pdf($\boldsymbol{m}$) and moving the constant terms in the factor of the exponential, we obtain

$$\text{pdf}(\boldsymbol{m}) = \text{const} \times \exp\left(-\frac{1}{4}(\boldsymbol{m} - \boldsymbol{m}_*)^{\mathrm{T}} \mathbf{H}_*(\boldsymbol{m} - \boldsymbol{m}_*) \right). \tag{47}$$

Equation (47) shows that the a posteriori probability density function is also Gaussian with its center at the maximum likelihood model $\boldsymbol{m}_*$ and a posteriori covariance matrix, denoted $\mathbf{C}_*$, given by

$$\mathbf{C}_* = \left(\mathbf{G}^{\mathrm{T}}\mathbf{C}_{\mathrm{D}}^{-1}\mathbf{G} + \mathbf{F}_0^{\mathrm{T}}\mathbf{C}_{fc}^{-1}\mathbf{F}_0\right)^{-1} = 2\mathbf{H}_*^{-1}. \tag{48}$$

This result demonstrates that, provided the initial model satisfies the equations of constraint, the a posteriori probability density function is Gaussian. This implies that the minimum of the misfit function is at the maximum likelihood solution $\boldsymbol{m}_*$.

We now demonstrate that the initial model, which is defined as the average slowness $\bar{s}$ in (28), is a solution of the equations of constraint and therefore lies in the feasibility domain. The initial calculated traveltime function $T_{\mathrm{cal}}^i(\boldsymbol{m}_0)$ reduces to a single term:

$$T_{\mathrm{cal}}^i(\boldsymbol{m}_0) = C_{0001}^0 d^i = \bar{s} d^i, \tag{49}$$

where d^i is the source–receiver distance for the ith calculated traveltime. The eikonal equation, computed for the initial model $\boldsymbol{m}_0$, takes the form

$$\nabla_{\mathrm{P}} T_{\mathrm{cal}}^i(\boldsymbol{m}_0) \cdot \nabla_{\mathrm{P}} T_{\mathrm{cal}}^i(\boldsymbol{m}_0) = \bar{s}^2, \tag{50}$$

and, consequently,

$$\nabla_{\mathrm{S}} \left[\nabla_{\mathrm{P}} T_{\mathrm{cal}}^i(\boldsymbol{m}_0) \cdot \nabla_{\mathrm{P}} T_{\mathrm{cal}}^i(\boldsymbol{m}_0)\right] = 0, \tag{51}$$

which suggests that the initial model satisfies the condition for the equations of constraint, and hence the probability density function is a Gaussian.

Appendix B.2: Estimation of Uncertainty

The a posteriori covariance matrix, which is twice the inverse of the Hessian matrix (48), provides information on the uncertainty of the model parameters. The physical meaning of the Hessian matrix is given by the curvature of the misfit function. A high curvature of the misfit function in the neighborhood of its minimum implies large values for the Hessian matrix's components, and therefore a small covariance. The diagonal elements of the covariance matrix contain the variance of the model parameters. As the variance of a sum is the sum of the variance of each coefficient, the uncertainty in the final slowness $s_{\mathrm{cal}}(P)$ will be the square root of the sum of the variance of each term comprising the series expansion. If $\sigma_{s_{\mathrm{cal}}(P)}$ is the uncertainty of the calculated slowness at P, then

$$\sigma_{s_{\mathrm{cal}}(P)} = \left(\sum_{l=0}^{L} \sum_{m=0}^{M} \sigma_{C_{lm01}}^2 P_l(X_{\mathrm{P}}) P_m(Z_{\mathrm{P}})\right)^{1/2}, \tag{52}$$

where $\sigma_{C_{lm01}}$ is the standard deviation of the C_{lm01} coefficients of the series expansion. The uncertainty in the estimated velocity is defined as the ratio of the slowness uncertainty and the slowness squared. If $v_{\text{cal}}(P)$ is the calculated velocity at P, then the uncertainty in the calculated velocity $\sigma_{v_{\text{cal}}(P)}$ is

$$\sigma_{v_{\text{cal}}(P)} = \frac{\sigma_{s_{\text{cal}}(P)}}{s_{\text{cal}}^2(P)}. \tag{53}$$

Similarly the uncertainty in the calculated traveltime between a source S and a receiver R, $\sigma_{T_{\text{cal}}(R,S)}$, is given by

$$\sigma_{T_{\text{cal}}(R,S)} = \left(\sum_{l=0}^{L} \sum_{m=0}^{M} \sigma_{C_{lm01}}^2 P_l(X_M) P_m(Z_M) d(R,S) \right.$$
$$\left. \sum_{l=0}^{L} \sum_{m=0}^{M} \sum_{n=-N}^{N} \sum_{p=2}^{P} \sigma_{C_{lmnp}}^2 P_l(X_M) P_m(Z_M) \exp(\text{i}2n\theta) d^p(R,S) \right)^{1/2}, \tag{54}$$

where $\sigma_{C_{lmnp}}$ is the standard deviation of the C_{lmnp} coefficient of the series expansion.

Acknowledgements

This project was supported by a PhD studentship by the Geosciences Research Centre - Elf UK to P.E.E. The British Institutions Reflection Profiling Syndicate (BIRPS) is funded by the Natural Environment Research Council and BIRPS Industrial Associates [Amerada-Hess Ltd., BP Exploration Co. Ltd., Chevron UK Ltd., Conoco (UK) Ltd., Lasmo North Sea Plc., Mobil North Sea Ltd., Shell UK Exploration and Production].

References

1. K. Tanabe, Projection method for solving a singular system of linear equations and its applications, Num. Math. **17**, 203–214 (1971)
2. R. M. Mersereau, Direct Fourier transform techniques in 3-D image reconstruction, Comput. Biol. Med. **6**, 247–258 (1976)
3. A. C. Kak, Computerized tomography with X-ray, emission, and ultrasound sources, Proc. IEEE **67**, 1245–1272 (1979)
4. A. K. Louis, F. Natterer, Mathematical problems of computerized tomography, Proc. IEEE **71**, 379–389 (1983)
5. K. A. Dines, R. J. Lytle, Computerized geophysical tomography, Proc. IEEE **67**, 1065–1073 (1979)
6. R. Gordon, A tutorial on ART, IEEE Trans. Nucl. Sci. **21**, 78–93 (1974)
7. Y. Censor, Finite series-expansion reconstruction methods, Proc. IEEE **71**, 409–419 (1983)
8. M. H. Worthington, An introduction to geophysical tomography, First Break, 20–26, Nov. (1984)
9. W. S. Phillips, M. C. Fehler, Traveltime tomography: A comparison of popular methods, Geophys. **56**, 1639–1649 (1991)
10. R. P. Bording, A. Gersztenkorn, L. R. Lines, J. A. Scales, S. Treitel, Applications of seismic travel-time tomography, Geophys. J. R. Astr. Soc. **90**, 285–303 (1987)

11. P. Carrion, Dual tomography for imaging complex structures, Geophys. **56**, 1395–1404 (1991)
12. D. D. Jackson, The use of a priori data to resolve non-uniqueness in linear inversion, Geophys. J. R. Astron. Soc. **57**, 137–157 (1979)
13. J. G. Berryman, Fermat's principle and non-linear traveltime tomography, Phys. Rev. Lett. **62**, 2953–2956 (1989)
14. J. G. Berryman, Stable iterative reconstruction algorithm for nonlinear traveltime tomography, Inv. Prob. **6**, 21–42 (1990)
15. J. E. Vidale, Finite difference calculation of traveltimes in three dimensions, Geophys. **55**, 521–526 (1990)
16. T. J. Moser, Shortest path calculation of seismic rays, Geophys. **56**, 59–67 (1991)
17. R. H. T. Bates, V. A. Smith, R. D. Murch, Manageable multidimensional inverse scattering theory, Phys. Rep. **201**, 185–277 (1991)
18. S. A. Enright, S. M. Dale, V. A. Smith, R. D. Murch, R. H. T. Bates, Towards solving the bent-ray tomographic problem, Inv. Prob. **8**, 83–94 (1992)
19. G. Arfken, *Mathematical Methods for Physicists*, 3rd ed. (Academic, New York 1985)
20. J. A. Scales, *Introduction to Nonlinear Optimization* (Macmillan, New York 1985)
21. G. Golub, C. Reinsch, Singular value decomposition and least squares solution, Num. Math. **14**, 403–420 (1970)
22. W. H. Press, S. A. Teukolsky, W. T. Vetterling, B. P. Flannery, *Numerical Recipes in Fortran — the Art of Scientific Computing*, 2nd ed. (Cambridge Univ. Press, Cambridge 1992)
23. Ö. Yilmaz, *Seismic Data Processing* (Society Exploration Geophysicists, Tulsa, Okla. 1987)
24. W. Schneider, Integral formulation for migration in two and three dimensions, Geophys. **43**, 49–76 (1978)
25. J. R. Berryhill, Wave-equation datuming, Geophys. **44**, 1329–1333 (1979)
26. A. Tarantola, *Inverse Problem Theory; Methods for Data Fitting and Model Parameter Estimation* (Elsevier, Amsterdam 1987)
27. G. M. Jackson, F. Pawlak, Interactive tomography for VSP migration velocity models, 56th Annu. Int. Meeting EAEG, Extend. Abstr. (1994)
28. G. M. Jackson, Experiences with anisotropic well seismic migration – field data example, 57th Annu. Int. Meeting EAEG, Extend. Abstr. (1995)
29. R. J. Michelena, J. M. Harris, Tomographic traveltime inversion using natural pixels, Geophys. **56**, 635–644 (1991)
30. P. Podvin, I. Lecomte, Finite difference computation of traveltimes in very contrasted velocity models — a massively parallel approach and its associated tools, Geophys. J. Int. **105**, 271–284 (1991)
31. A. Tarantola, B. Valette, Generalized nonlinear inverse problems solved using the least square criterion, Rev. Geophys. Space Phys. **20**, 219–232 (1982)
32. J. G. Berryman, Weighted least-squares criteria for seismic traveltime tomography, IEEE Trans. Geosci. Remote Sensing **27**, 302–309 (1989)
33. P. O. Ecoublet, Bent-ray traveltime tomography and migration without ray tracing, Ph.D. thesis, Department of Earth Sciences, Cambridge University (1995)

Simple Models in the Mechanics of Earthquake Rupture

Shamita Das

Department of Earth Sciences, University of Oxford, Parks Road, Oxford,
OX1 3PR, UK
das@earth.ox.ac.uk

Abstract. Starting from Navier's equations of motion in elastodynamics, it is shown that the slip on an earthquake fault can be written as a convolution of the stress changes on the fault with the half-space Green's function. An alternate representation relation in which the fault slip is a convolution of the fault slip on other parts of the fault at previous times with a different kernel is also derived. Using either of these (equivalent) expressions, the behavior of the fault slip as a function of space and time as a fault grows at a constant speed from a point to a finite circular shape (the "interior" crack problem) is obtained. Spontaneous faults, in which the fault speed is not assumed a priori but determined from some fracture criterion, are considered, and the limiting rupture speeds in different cases obtained. The growth of a pre-existing circular crack is determined. Finally, the rupture of a zone surrounded by already ruptured zones (the "exterior" crack problem) is studied. In the last two cases, it is shown that the rupture process is so complex that the terms "rupture front" and "rupture velocity" are no longer meaningful.

1 Introduction

The primary goal of the study of the mechanics of earthquake faulting is to understand its underlying physical processes, with a long-term view of developing the rational capability to reduce the damage caused by earthquakes.

Earthquakes represent the release of elastic strain energy, accumulated in the Earth's crust, due to the process of plate tectonics, by fracturing of crustal material. As earthquakes occur, the elastic waves generated by them propagate throughout the Earth, causing damage locally, and are recorded globally on seismometers. These instruments are now so sensitive that they can record ground motions the size of an atom. In fact, such instruments are too sensitive to be used on the Earth, where many other sources of disturbance exist (such as Earth tides, atmospheric disturbances and cultural noise), but have been used on the Moon. The time series recorded by the seismometers are analyzed to obtain information on the earthquake, starting from the most basic parameters, such as the location and occurrence time of the earthquake and its size, to far more complex information on the details of the rupture process. Examples of the latter are the speed and path of the rupture; the amount of slip that occurred on different parts of the fault; the rate at which the two sides of the fault slip past one another; and the amount of stress released. This is the classical inverse problem. That is, given a set of seismograms, determine the details listed above. But before we can study the inverse problem, we need to study the forward problem in order to obtain insight into the physics of the problem. Furthermore, the inverse problem needs additional constraints to obtain stable solutions [1], and these can be obtained only by studying the forward problem.

From an academic point of view, an earthquake is a source of information, the acquisition and analysis of which comprises the subject of seismology. The study of the earthquake source is of considerable importance, as it provides information about regions in the Earth where earthquakes occur, regions which are quite inaccessible to us due to the fact that earthquakes occur from about 5–670 km depth in the Earth, with most earthquakes occurring under water. Though we ourselves cannot sample such regions, the earthquake fault samples it for us by itself passing through these regions. The information is coded into the waves generated by the earthquake, so that the entire subject of modern seismology consists of deciphering this coded information from the seismogram. Even today, the mechanics of earthquakes deep in the Earth remains poorly understood. The situation is better for earthquakes at shallow depths such as 5–20 km, where the earthquakes are understood to be due to brittle fracture of the Earth's crust. In this chapter, we confine ourselves to these earthquakes.

2 Brief Derivation of the Underlying Equations

The three modes of fracture are illustrated in Fig. 1. The solid arrows indicate the direction of fault displacement, with the open arrows showing the direction of the fracture propagation. The difference in the three modes is the relative direction of the fault displacement with respect to the direction of rupture propagation. The first mode is called the "tensile" mode or "opening" mode (Mode I). The other two are the shear modes. Mode II is the in-plane shearing mode, and Mode III the antiplane shearing mode or "tearing" mode. In reality, all three modes can be present in an earthquake, but we shall only consider the simple case when the opening mode is absent and

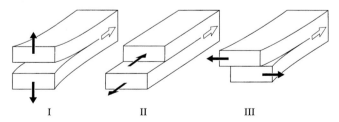

Fig. 1. The different modes of fracture

the earthquake is pure shear. We shall use the terms "fault", "fracture" and "crack" interchangeably in this paper.

We shall show that the fault displacement history and distribution is a convolution of the impulse response of the medium (Green's function) with the fault slip rate in space and time. The development of the basic theory of elastodynamics is based on standard texts such as *Love* [2] and *Morse* and *Feshbach* [3], with the applications to earthquake studies based on *Aki* and *Richards* [4] and *Das* [5]. The notation of *Aki* and *Richards* [4] is generally used. We start with the well-known equations of motion in an elastic medium, developed by Navier in 1821:

$$\tau_{ij,j} + f_i = \rho \ddot{u}_i, \tag{1}$$

where τ_{ij} is the stress tensor $= c_{ijkl} e_{kl}$, $i, j, k, l = 1, 2, 3$ and $\tau_{ij} = \tau_{ji}$; c's are the elastic constants, $e_{kl} = \frac{1}{2}(u_{i,j} + u_{j,i})$ is the strain tensor, u_i is the component of displacement in the i-direction, f_i is the component of the body force in the i-direction, ρ is the density, comma denotes spatial derivatives, dots denote time derivatives and Einstein's summation convention for tensors is used. The set of three equations given by (1) give the equations of motion in any general elastic medium. For an isotropic, elastic medium, $c_{ijkl} = \lambda \delta_{ij} \delta_{kl} + \mu(\delta_{ik}\delta_{jl} + \delta_{il}\delta_{jk})$, where δ's are the Kronecker deltas and λ and μ are the Lamé parameters.

A Green's function is the impulse response of an elastic solid, that is, the displacement generated in the body by the application of a unit force $f_i = \delta(\boldsymbol{X})\delta(t)$, where, $\boldsymbol{X} = (X_1, X_2, X_3)$ and t is time. If the unit impulse is applied at $\boldsymbol{X} = \boldsymbol{X}'$ and $t = t'$ in the n-direction in a volume V, then we denote the ith component of displacement at general $(\boldsymbol{X}, t)$ by $G_{in}(\boldsymbol{X}, t; \boldsymbol{X}', t')$, where $i, n = 1, 2, 3$. Then $G_{in}(\boldsymbol{X}, t; \boldsymbol{X}', t')$ satisfies the equations of motion (6), that is,

$$\rho \frac{\partial^2}{\partial t^2} G_{in} = \delta_{in} \delta(\boldsymbol{X} - \boldsymbol{X}') \delta(t - t') + \frac{\partial}{\partial x_j} \left(c_{ijkl} \frac{\partial}{\partial X_l} G_{kn} \right) \tag{2}$$

everywhere in the volume V. We shall use the initial condition that $G_{in}(\boldsymbol{X}, t; \boldsymbol{X}', t')$ and its time derivative vanish for $t \leq t'$ and $\boldsymbol{X} \neq \boldsymbol{X}'$. Reciprocity relations in space and time, such as

$$G_{in}(\boldsymbol{X}, t; \boldsymbol{X}', t') = G_{ni}(\boldsymbol{X}', -t'; \boldsymbol{X}, -t), \tag{3}$$

can be established. If the boundary conditions are homogeneous and independent of time and the medium is homogeneous, then it can be shown that

$$G_{in}(\boldsymbol{X}, t; \boldsymbol{X}', t') = G_{in}(\boldsymbol{X} - \boldsymbol{X}', t - t') = G_{ni}(\boldsymbol{X} - \boldsymbol{X}', t - t'), \tag{4}$$

so that in this case the Green's functions depend only on *differences* in space and time coordinates, rather than on the coordinates themselves. This property is very useful when solving problems numerically, where we need to determine and store the Green's functions on a computer.

The representation theorem of classical wave mechanics for an elastic medium is,

$$u_k(\boldsymbol{X},t) = \int_{-\infty}^{\infty} dt' \iiint_V f_i(\boldsymbol{X}',t')G_{ki}(\boldsymbol{X}-\boldsymbol{X}',t-t')dV(\boldsymbol{X}')$$

$$+ \int_{-\infty}^{\infty} dt' \iint_\Sigma \left[G_{ki}(\boldsymbol{X}-\boldsymbol{X}',t-t')\tau_{ij}n_j - u_i(\boldsymbol{X}',t')c_{ijkl}(\boldsymbol{X}')n_j \right.$$

$$\left. \times G_{kn,l}(\boldsymbol{X}-\boldsymbol{X}',t-t') \right] d\Sigma(\boldsymbol{X}') \tag{5}$$

for a volume V surrounded by a surface Σ, where n_j is the jth component of the unit (outward) normal to Σ. This representation is exactly the same, term by term, as that of electrodynamics or water waves, with the difference that here we have two wave velocities (for compressional and shear waves). Equation (5) states the way in which displacement $\boldsymbol{u}$ at some point in the medium $(\boldsymbol{X},t)$ is made up of contributions due to the force $\boldsymbol{f}$ throughout V, together with contributions due to the traction $\tau_{ij}n_j$ and the displacement $\boldsymbol{u}$ itself on Σ. The point $(\boldsymbol{X}',t')$ is called the "source point," and the point $(\boldsymbol{X},t)$ is called the "receiver point" (where the observer is located). The relation (5) is a very powerful relation since it is true for the Green's function obtained using *any* boundary condition on Σ. The boundary conditions for $\boldsymbol{G}$ can be chosen in *any* way that turns out to be useful. We give two examples below which are widely used in seismological applications.

Case When G Is Obtained with Σ as a Rigid Boundary

Let us call this Green's function $\boldsymbol{G}^{\text{rigid}}$. $G_{ik}^{\text{rigid}}(\boldsymbol{X}-\boldsymbol{X}',t-t') = 0$ for $\boldsymbol{X}'$ in Σ (Dirichlet condition). Then, the second term of (5) vanishes. If body forces are absent, we have

$$u_k(\boldsymbol{X},t) = -\int_{-\infty}^{\infty} dt' \iint_\Sigma u_i(\boldsymbol{X}',t')c_{ijkl}n_j \frac{\partial}{\partial \xi_l} G_{kn}^{\text{rigid}}(\boldsymbol{X}-\boldsymbol{X}',t-t')d\Sigma. \tag{6}$$

Case When G Is Obtained with Σ as a Free Surface

Let us call this Green's function $\boldsymbol{G}^{\text{free}}$. The traction $c_{ijkl}n_j(\partial/\partial \xi_l)G_{kn}^{\text{free}}(\boldsymbol{X}-\boldsymbol{X}',t-t') = 0$ for $\boldsymbol{X}'$ in Σ (Neumann condition), and the third term of (5) vanishes. If body forces are absent, we obtain

$$u_k(\boldsymbol{X},t) = \int_{-\infty}^{\infty} dt' \iint_\Sigma G_{ki}^{\text{free}}(\boldsymbol{X}-\boldsymbol{X}',t-t')\tau_{ij}n_j d\Sigma. \tag{7}$$

Equations (5,6,7) are different forms of the representation theorem. Taken together, they appear to imply a contradiction as to whether $u(\boldsymbol{X},t)$ depends upon the displacement on Σ (6), the traction (7) or both (5). But traction and displacement cannot be assigned independently, so there is no contradiction.

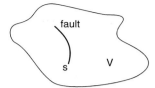

Fig. 2. A volume V bounded by a surface Σ and containing a fault S

Next, let us consider that the volume V has a fault in it (Fig. 2). Then, the surface Σ consists of two parts, the outer surface of V and the fault surface S. So we can now replace the integrals over Σ by integrals over $(\Sigma+S)$. It can be shown that that the integrals over Σ vanish when $\Sigma \to \infty$, and we are simply left with the integrals over the fault surface S in the representation relations above; the "displacement" u_i under the integral signs in the representation relations above can now be replaced by the displacement discontinuity across the fault surface, and the representation relation (6) becomes

$$u_k(\boldsymbol{X},t) = - \int_{-\infty}^{\infty} \mathrm{d}t' \iint_S [u_i(\boldsymbol{X}',t')] \, c_{ijpq}\nu_j \frac{\partial}{\partial \xi_q} G_{kp}(\boldsymbol{X}-\boldsymbol{X}', t-t') \mathrm{d}S \,, \quad (8)$$

where $\boldsymbol{\nu}$ is the normal on the fault surface and $[\ldots]$ denotes displacement discontinuity. This is a very important relation in seismology, and we will return to it in a later section.

When the form of the representation relation given by (6) is used, we obtain the kinematic problem of the seismic source – we call it the "dislocation" model. That is, given the displacements $\boldsymbol{u}$ on the fault surface S, we find the displacement at some other point in the medium. On the other hand, when the form of the representation relation given by (7) is used, we obtain the dynamic model of the seismic source. That is, given the forces and physical laws governing the motions, we have to determine the motion itself. In seismology, this is now called the "crack" model. This approach has to take into account the inertia terms in the equations of motion, namely, the waves that are radiated as the fault propagates. In the "static" approach, the acceleration terms are omitted and the waves ignored. If a fault propagates very slowly, then the static approach is valid. In the analysis of geodetic data (GPS studies, nowadays), the static solutions are used to analyze the data. We shall see later how much error is introduced by this.

The kernels in (6) and (7) are nonintegrable, for the general case of a fault that initiates, propagates and comes to a stop, as earthquakes do. This situation motivated the development of sophisticated numerical techniques for the solution of dynamic crack problems. As a vehicle for illustration of a numerical method of solution of such dynamic faulting problem, we shall consider the numerical boundary-integral equation (BIE) method. We next briefly derive the representation relations for earthquake dynamics. In order to reduce the earthquake source problem to the solution of a BIE, we must use relations between stress and displacement fields throughout the medium and the displacement discontinuity or the traction perturbations on the fault plane, given by the representation relations. Together with boundary conditions, these relations comprise the integral equations, the solution of which give the the displacement and stress fields due to a dynamically propagating fault. We model the earthquake source as a propagating plane shear crack in an infinite medium which is homogeneous and linearly elastic everywhere *off the crack plane*. (Note that earthquakes cannot occur in a medium that is homogeneous everywhere!) As the fault propagates on the planar surface, waves are radiated out in three spatial dimensions. The geometry of the problem is shown in Fig. 3. The plane $X_3 = 0$ is taken as the fault plane, that is the plane across which slip occurs during the earthquake, and denoted by $S(t)$, since the broken part of the fault grows with time t. Initially, the infinite body is under a uniform state of stress, σ_{ij}^0. The stress on the fault plane $X_3 = 0$ can be separated into the normal stress, σ_{33}^0, and a shear stress, $\sigma_{13}^0 = \sigma^0$, say. The component σ_{23}^0 can be taken as zero by taking the coordinate axis X_1 in the direction of the maximum initial shear (without loss of generality). Let us assume that the initial shear stress is increased sufficiently to initiate a fault which then propagates on the $X_3 = 0$ plane. The normal stress σ_{33}^0 over the fault plane remains constant throughout the rupture process, for a planar fault. Let us take the origin of time $t = 0$ as the time when the fault initiates and starts extending. We study the case when the fault propagation speed is rapid enough to generate elastic waves. The fault edges may move at some pre-assigned speed or the position of the fault edge may be found as a function of time, using some fracture criterion. The latter is called "spontaneous" propagation in seismology. Let us consider the former case only for the moment, that is, the fault propagation speed in all directions on the fault

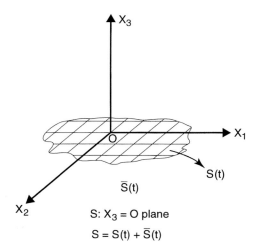

Fig. 3. Geometry of the fault plane

plane is known. As the fault propagates, there is relative motion between the two faces of the fault – between the regions $X_3 < 0$ and $X_3 > 0$ – and a displacement discontinuity appears across the broken region of the fault plane. This discontinuity is a function only of the coordinates X_1 and X_2 and time t. The shear stress on the fault surface is zero if there is complete stress release. Alternatively, it can be equal to the the frictional stress σ on the fault faces, given by $\sigma = \mu \sigma_{33}^0$, where μ is the coefficient of friction; μ may be taken constant, but in reality it may be a function of space and time. Let the incremental stresses due to the displacement $\boldsymbol{u}$ from its initial configuration be τ_{ij}, so that $\sigma_{ij} = \sigma_{ij}^0 + \tau_{ij}$, that is, τ_{ij} is the stress change due to the motion.

There exist symmetries in the problem which we can exploit. For planar shear cracks, the solution can be shown to be antisymmetric in X_3, that is, the displacement components u_1, u_2 and traction perturbation τ_{33} are odd in X_3, while u_3, τ_{13} and τ_{23} are even in X_3, and it is sufficient to solve the problem for the upper half-space $X_3 \geq 0$. Further, from the continuity of tractions across $X_3 = 0$, it follows that $\tau_{33} = 0$ everywhere on $X_3 = 0$. Then, the representation relation (7) gives

$$u_k(\boldsymbol{X}, t) = \int_{-\infty}^{\infty} \mathrm{d}t' \iint_S G_{ki}(\boldsymbol{X} - \boldsymbol{X}', t - t') \, \tau_{k3}(\boldsymbol{X}', t') \, \mathrm{d}S, \tag{9}$$

where $k = 1, 2, 3$ and S is the infinite fault plane $X_3 = 0$. Letting $X_3 \to 0$ and accounting for the symmetry of the displacement components, the required representation relation is obtained to be

$$u_k(\boldsymbol{X}, t) = \int_{-\infty}^{\infty} \mathrm{d}t' \iint_S G_{ki}(\boldsymbol{X} - \boldsymbol{X}', t - t') \, \tau_{k3}(\boldsymbol{X}', t') \, \mathrm{d}S, \tag{10}$$

where $\boldsymbol{X}$ and $\boldsymbol{X}'$ are 2-dimensional vectors on S now. The total slip across the fault is twice that given by (10). The required components of the Green's functions G are the solution to Lamb's problem and can be expressed in terms of elementary functions. The analytical expressions for G_{ki} for the 2- and 3-dimensional problems are given in Appendix I of [1]. The kernel G possesses only weak singularities and can be directly discretized for numerical computation, as we shall discuss later.

Writing the system of integral (9) fully, we have

$$\begin{aligned} u_1(X_1, X_2, 0, t) = \int_{-\infty}^{\infty} \mathrm{d}t' \iint_S [&G_{11}(X_1 - X_1', X_2 - X_2', 0, t - t')\tau_{13}(X_1', X_2', 0, t') \\ &+ G_{12}(X_1 - X_1', X_2 - X_2', 0, t - t')\tau_{23}(X_1', X_2', 0, t') \\ &+ G_{13}(X_1 - X_1', X_2 - X_2', 0, t - t')\tau_{33}(X_1', X_2', 0, t')] \\ &\mathrm{d}X_1' \mathrm{d}X_2', \end{aligned} \tag{11}$$

$$u_2(X_1, X_2, 0, t) = \int_{-\infty}^{\infty} dt' \iint_S [G_{21}(X_1 - X_1', X_2 - X_2', 0, t - t')\tau_{13}(X_1', X_2', 0, t')$$
$$+ G_{22}(X_1 - X_1', X_2 - X_2', 0, t - t')\tau_{23}(X_1', X_2', 0, t')$$
$$+ G_{23}(X_1 - X_1', X_2 - X_2', 0, t - t')\tau_{33}(X_1', X_2', 0, t')]$$
$$dX_1' dX_2', \tag{12}$$

$$u_3(X_1, X_2, 0, t) = \int_{-\infty}^{\infty} dt' \iint_S [G_{31}(X_1 - X_1', X_2 - X_2', 0, t - t')\tau_{13}(X_1', X_2', 0, t')$$
$$+ G_{32}(X_1 - X_1', X_2 - X_2', 0, t - t')\tau_{23}(X_1', X_2', 0, t')$$
$$+ G_{33}(X_1 - X_1', X_2 - X_2', 0, t - t')\tau_{33}(X_1', X_2', 0, t')]$$
$$dX_1' dX_2'. \tag{13}$$

The terms with τ_{33}, which appear in the complete expressions above, disappear for pure planar shear cracks, and the displacement component $u_3 = 0$. For shear cracks with a tensional component of motion (for example, due to shearing across nonplanar faults) the terms in τ_{33} must be kept, and $u_3 \neq 0$. For purely tension cracks, the shear-stress components and u_1 and u_2 vanish, so that the above expressions are applicable for that problem as well. In the rest of this discussion, we shall consider only pure planar shear faulting.

The integration domain for the integral equations (11,12,13) covers all points influenced by disturbances which propagate with the fastest wave velocity of the problem (the compressional wave speed of the medium for general 3-dimensional problems). That is, it is defined by the causality condition, which gives the region over which the Green's functions are nonzero. On the fault plane $X_3 = 0$, this is

$$v_P(t - t') - \sqrt{(X_1 - X_1')^2 + (X_2 - X_2')^2} \geq 0, \quad t \geq t' \geq 0, \tag{14}$$

that is, it is the conical region defined by

$$v_P^2(t - t')^2 - (X_1 - X_1')^2 - (X_2 - X_2')^2 \geq 0, \quad t \geq t' \geq 0. \tag{15}$$

This cone is called the "cone of dependence," and the surface of the cone is called the "characteristic surface." The integrations in (11,12,13) extend only over this region given by (15). For the sake of clarity, the cone of dependence and the "characteristic lines" for the 2-dimensional problem is shown in Fig. 4a. If we further assume that the fault initiates at a point and extends at a speed not exceeding the fastest wave speed of the medium (the compressional wave speed), then the region of integration is further reduced to the "cone of influence." Again, we illustrate this for the 2-dimensional problem for clarity, in Fig. 4b. For the 3-dimensional problems under consideration here, the Green's functions are nonzero only in the time interval between the arrivals of the P wave and the Rayleigh wave, further reducing the region of integration. The volume in two spatial coordinates and in time over which the integrations are to be carried out is shown in Fig. 4b.

To find the displacement and stress field on the entire shear fault plane, we have to solve the system of two integral equations (11,12) under the initial and boundary conditions discussed above. The region of integration (15) includes the unbroken region, say, $\bar{S}(t)$, where the stress components τ_{13} and τ_{23} are unknown. So, before we can carry out the integrations in (11,12), we must first determine these two components of stress. As long as the fault-edge position is known, this can be done, by using the property that the displacement outside the fault is zero, and setting $u_1, u_2 = 0$, and solving for τ_{13} and τ_{23} [5]. The computations are carried out by time-stepping, so that at each time to calculate displacements, say, all necessary stress components have already been determined (Fig. 4c).

An alternate representation relation, which has some practical uses such as computational efficiency, can be obtained by using the representation relation (6) [1]. Discretization of that kernel is far more complex. It cannot be discretized directly as above, as it contains nonintegrable singularities. *Das* and *Kostrov* [6] found one way of overcoming this problem indirectly [1]. Very recently, *Cochard* and *Madariaga* [7] have developed a way of performing this discretization directly for 3-dimensional problems. The advantage of this representation is that the integrations are confined only to the broken part of the fault plane, whereas in (7) all parts of the fault plane where a P wave from the point of fault initiation has arrived contribute to the integral. Since most faults rupture at speeds less than the P-wave speed, considerable computational efficiency is obtained from using (6).

The numerical method developed using (11,12) was tested by *Das* [5] for the problem of a fault that initiated at a point and started propagating at a constant velocity in all directions but never stopped; it was tested by comparing it with the corresponding analytical solution for this problem, which was given by *Kostrov* [8]. This is the only dynamic faulting problem for which an analytical solution exists. This comparison is shown in Fig. 5 for the case when the circular fault speed is half the compressional wave speed of the medium, using two different grid sizes.

It is important to point out here a very fundamental property of the integral equations (11,12,13), namely, that the slip components u_k on the fault depend only on the Green's functions and on the stress *changes* on the fault; they are *independent of the total stress level*. Thus, doubling of the stress drop will double the fault slip, and so on. However, since the final stress on the fault (say, the frictional stress) does depend on the total stress on the body, the stress drop does depend on the stress *but only very indirectly*, so that it is very difficult, if not impossible in reality, to find the ambient stress level in the Earth from analyzing seismograms. Occasional papers claiming to be able to do this must be read keeping this in mind!

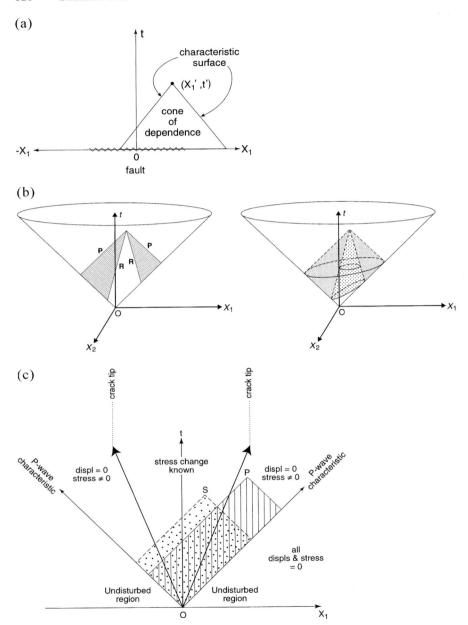

Fig. 4. (a) The cone of dependence and the characteristic lines, illustrated for the 2-dimensional problem. (b) The domain of integration in the $(X_1 - t)$ plane, illustrated for the 2-dimensional problem. (c) The volume of integration, illustrated for the 3-dimensional problem. At P, the displacement is zero. At S, the stress drop is known

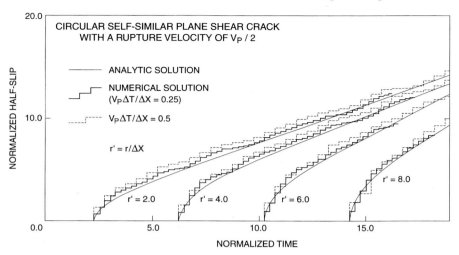

Fig. 5. Comparison of the analytical with the numerical solutions

Next, we consider specific dynamic faulting problems of interest in seismology.

3 The Finite Circular Shear Fault

This is a good model for moderate-sized earthquakes, which are more or less equidimensional. Moreover, it is an excellent illustrative example. We consider the problem of a circular shear fault which initiates at a point, propagates at a preassigned constant velocity v, say, and stops when it reaches some finite radius, r say (Fig. 6). Let the stress drop on the fault be assigned constant, $\Delta\sigma$ say, and directed in the X_1-direction. The fault region $S(t)$ is defined by

$$S(t): X_1^2 + X_2^2 \leq v^2 t^2 \text{ for } vt \leq r \ ; \ X_1^2 + X_2^2 = r \text{ for } vt > r \,. \tag{16}$$

Then we have the mixed boundary value problem

$$\tau_1 = \Delta\sigma, \ \tau_2 = 0 \text{ on } S(t) \ ; \ a_\alpha = 0 \text{ on } \bar{S}(t) \,. \tag{17}$$

Even this relatively simple problem of a finite fault propagating at a constant velocity and stopping cannot be solved analytically.

We consider an instantaneously appearing crack of diameter $3\Delta X$, which grows to a final diameter of $41\Delta X$ at a speed $v = v_P/2$. We shall allow backslip to occur on the crack in this particular example. The half-slip $a_1/2$ is plotted against time in Fig. 7 and is normalized by $(r\Delta\sigma)/3\mu$ for this problem. The slip on the crack is found to be essentially in the direction of the stress drop even after the crack has stopped, and it coincides with the solution of the corresponding self-similar problem until the first diffracted

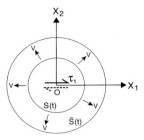

Fig. 6. Geometry of the finite circular shear fault

waves from the crack edge arrive and decrease the slip rate. The rise time and slip at the centre are larger than at the crack edge. Figure 7 shows that the maximum slip is reached at or soon after the time when the Rayleigh wave arrives from the crack edge. The final offset at any point on the circular fault at a radius r' is given by

$$u_1(r') = (1.23)\frac{12}{7\pi}\left(\frac{\tau_{13}}{\mu}\right)\sqrt{r^2 - r'^2}. \qquad (18)$$

The static solution for this problem is the same as above, with the factor 1.23 replaced by 1. Thus, the dynamic slip overshoots the static value by 23% for this fault rupture velocity. Estimates of this overshoot, obtained by using different numerical methods and/or different grid sizes are found to lie between 20 and 27%. The dynamic overshoot of slip in the interior of the crack may of course be interpreted as the overshoot of the static stress drop there. In other words, the dynamic stress drop is larger than the static stress drop. *Kostrov* and *Das* [1] gave a general formula for estimating the expected overshoot for any fault rupture speed and showed that the dynamic overshoot increases with increasing fault rupture speed v.

For more complex situations, such as cracks with friction etc., we refer the reader to *Kostrov* and *Das* [1].

4 Spontaneous Faults

In the Earth, faults initiate and start to propagate but not at a constant speed. *Kostrov* [9] showed that infinite-fault rupture speeds go through a phase in which they accelerate, until they reach their terminal rupture velocity, in material where the fault has constant strength along the fault plane, and then continue to propagate forever at this terminal speed. Finite faults of course eventually stop, but the initial phase of rupturing is similar to that of the infinite fault. The real case is even more complex, where the fault may propagate through different rocks with different elastic properties and strengths. So the problem of determining the velocity of rupture propagation is an important one in seismology. It was shown by *Madariaga* [10] that the changes of speed of the fault edges generates high-frequency wave radiation

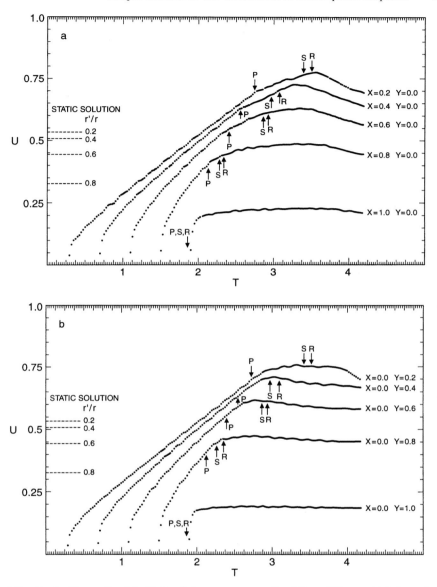

Fig. 7a,b. Normalized half slip (u) versus normalized time ($T = v_P t/r$) for a finite circular crack for (**a**) points along the X_1-axis and (**b**) points along the X_2-axis. $X = X_1/r$, $Y = X_2/r$, r being the final crack radius and $v_P \Delta t / \Delta X = 0.5$. The theoretical arrival times of the diffracted P, S, and Rayleigh waves from the crack edge are indicated on each curve by *arrows*. The static solutions for the points $r'/r = 0.2, 0.4, 0.6$, and 0.8, where r' is the distance of the point from the center of the crack, are given by $u = 0.55\sqrt{1 - r'^2/r^2}$ and marked along the abscissae. [Note that 0.55 is the factor $12/(7\pi)$]

and hence are the source of some of the damaging waves radiated from a fault. In order to study this changing of the fault front rupture speed, we must use a "fracture criterion" to determine the rupture speed, rather than assume the speed a priori.

4.1 Fracture Criterion

It can be shown that at the edge of the crack, large stresses exist. Full explanations of this are given in text books on fracture mechanics (though most such books do not deal with the shear fault problem that we are interested in). A simple way to understand it is to use the definition of the strain at a point. Strain is the change in length (area or volume) divided by the original length (area or volume) before any forces were applied, and it is often expressed as a percentage. At the edge of the crack, one goes from the region $S(t)$, where the fault faces are discontinuous, to the region $\bar{S}(t)$, where the fault faces are continuous. Then, the linear strain at the fault edge (and hence the stress) is infinite! In reality, since this transition from broken to unbroken is not sharp, the strains and stresses are not infinite but are large. As soon as one moves further away from the crack edge into the unbroken region, the stresses very quickly become finite. This fall-off can be shown to be an inverse square-root function of the distance to the fault edge [11]. Now the state of a point (broken or unbroken) depends on the state of nearby points. Thus, the criteria for determining whether a point will be broken or not is essentially nonlocal, that is, it is a global criterion. Numerically, this is impossible to use.

So we now consider a truly local fracture criteria and its use in numerical formulations. *Das* and *Aki* [12] introduced such a criterion, in which a point ahead of the fault edge is allowed to break if the stress concentration there exceeds some critical level. For some simple, 2-dimensional problems, they showed that this simulates the well-known Irwin fracture criterion well and also gives results that qualitatively are close to that obtained by using the global Griffith fracture criterion. (We do not discuss these criteria here but refer the reader to *Das* and *Aki* [12]). This numerical criterion is truly local in the sense that the state of a grid at a given time does not depend on the state of adjacent grids at the *same* time! The criterion is called the "critical stress level" fracture criterion nowadays. We next consider several cases of spontaneous cracks propagating under the "critical stress level" failure criterion, which are of interest in seismology.

Das [13] studied infinite and finite spontaneous rupture. The most significant result was that if the strength of the material through which the fault propagates is low relative to the stress drop, the terminal rupture speed in the purely in-plane direction of fracture can reach v_P while that in the pure anti-plane fracture direction is $0.57v_P \approx v_S$ (for a Poisson solid and within the numerical resolution). In all other directions, the rupture is of mixed mode (II and III), and the terminal velocity lies between the two above limits [13]. Thus, an initially equidimensional crack will become elongated in the direction of the applied stress on such fault planes. What is important is that such studies showed that there is no physical restriction for in-plane shear cracks to propagate at speeds exceeding the Rayleigh wave speed of the medium, as had been believed earlier. For higher values of the relative fault strength, a remarkable change in these terminal velocities is encountered. The terminal fracture speeds in the purely in-plane and purely anti-plane directions become practically the same, and equal to $\approx 0.5 v_P$, for some strengths. For even higher values of strength, the terminal rupture velocity remains unchanged, though the time required to reach this velocity increases with increasing fault strength. Thus a sudden transition occurs in crack behavior going from lower to higher values strength. Faults of other shapes have been considered in many studies, and *Kostrov* and *Das* [1] give complete references. Models with more complex frictional properties on the fault faces have recently been developed. *Andrews* [14] has used a slip-weakening friction law; *Cochard* and *Madariaga* [15] used rate-dependent frictional properties. The reader is referred to these studies.

Next we consider the fracture of a pre-existing circular fault on an infinite plane of constant strength, the more realistic situation. In this case, the initial conditions are different from the cases discussed above, where a small fault appeared suddenly and starts propagating. Due to this difference in the initial stress, the initial part of the fracture process is very different. The initial stress on the fault plane is not uniform now, and its deviation from the uniform applied stress is given by the stress distribution around the initial crack. The expressions for the initial stress distribution for a circular fault have been written by *Eshelby* [16] and were evaluated numerically and illustrated by *Kostrov* and *Das* [17]. Let us consider the problem of a shear crack which is initially of diameter $9\Delta X$. Let this initial crack be denoted by $S(0)$. The direction of the initial applied stress σ_α^0 is assumed to be constant. For an initially circular crack, we may assume this stress to be applied in the X_1-direction without loss of generality. Let the constant prescribed stress drop $\Delta \sigma$ on $S(0)$ be in the X_1-direction. Then the discrete problem we must solve is

$$\tau_{1ijk} = \Delta \sigma, \tau_{2ijk} = 0 \text{ for } S_{(0)} : (i\Delta X)^2 + (j\Delta X)^2 \leq (4.5\Delta X)^2,$$
$$a_{\alpha ijk} = 0 \text{ for } \bar{S}_{(0)} : (i\Delta X)^2 + (j\Delta X)^2 > (4.5\Delta X)^2.$$

The stress varies smoothly along the crack edge from the end of one axis to the other. The numerically calculated stresses show the inverse square-root behavior near the crack edge.

We may now use this σ_α^0 and solve the following dynamic problem:

$$\tau_{\alpha ijk} = 0 \text{ on } S_{(0)},$$
$$\tau_{1ijk} = \sigma_1^0 - \Delta \sigma \; ; \; \tau_{2ijk} = 0 \text{ on } (S_{(k)} - S_{(0)}),$$
$$a_{\alpha ijk} = 0 \text{ on } \bar{S}_{(k)}.$$

Dynamic fracture is initiated by artificially releasing one of the two grid points along the crack edge with the highest stress concentration, which lie along the X_1-axis for this problem. Figure 8 shows the fractured regions of the crack as a function of normalized time for the case when $S = 0.75$ for the region outside $S_{(0)}$. The fracture process is symmetric about the X_1-axis. The fracture commences propagation from the initially broken area at $X_1 = 5.5\Delta X$, $X_2 = 0$ along the perimeter of the initial crack, that is, it follows the stress concentration in the vicinity of this area. At the same time the disturbance generated by the fracturing starts spreading out over the $X_3 = 0$ plane, and at the normalized time $v_P t/r_0 = 1.7$, where r_0 is the initial crack radius, it initiates fracturing close to the point which is diametrically opposite the fracture initiation point. This is the time of arrival of the S wave at this point from the initially fractured point. A second area of fracture then starts spreading out from this point, the propagation still occurring along the edge of the initial crack. Until time $v_P t/r_0 = 5.5$, fracture propagation occurs mainly in the in-plane and mixed modes. At this time the fracture finally commences in the purely anti-plane mode as well, that is, the stress concentration in the purely anti-plane direction finally reaches the critical stress level required for fracture. The calculations are carried out until $v_P t/r_0 = 15$. By time

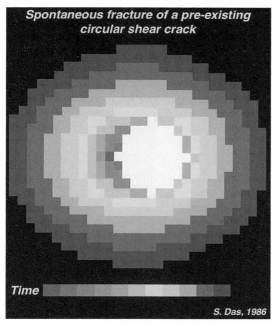

Fig. 8. Fracture process of a pre-existing circular crack shown by contours at the fixed normalized time intervals of $v_P t/r_0 = 4$. The initial crack is white and the grid that is released to initiate dynamic fracture is the darkest shade

$v_P t/r_0 = 8.8$, the fracture front has accelerated to its terminal velocity in all directions, this velocity being $v_P/2$, which is at about the Rayleigh wave speed of the medium; the crack front is plotted only up to this time in the figure. From time $v_P t/r_0 \sim > 10$, the difference in its dimension in the X_1- and X_2-directions become negligible compared to its actual dimensions.

Finally, we consider a model for an aftershock, where most of the fault has ruptured, leaving behind some unbroken regions. If we consider one such region on a very large fault, we can model it as an "exterior" crack problem, all the problems discussed up to now having been "interior" crack problems. We consider fracture of a circular "asperity" on an infinite fault; the results are valid, even in the general case of a finite fault, up to the time when the effects of other asperities on the fault or the main crack edge are felt at the particular asperity under consideration. In a dynamic problem, this is the time when waves diffracted from the edges of other asperities or the main fault edge return to the asperity under study and modify the solution there.

Let us consider the 3-dimensional problem of a pre-existing circular asperity of radius r_0 on an infinite plane $X_3 = 0$. At infinity, the two half-spaces are shifted by an amount u_0, say, as shown in Fig. 9. Let the initial asperity area (unbroken) be denoted by $S(0)$ and $\bar{S}(0)$ be its complement. As before, we shall use σ^u to denote either the fracture strength σ^Y or the frictional strength σ^{stat} on $S(0)$. On $\bar{S}(0)$, the frictional traction is assumed constant. Let us also assume that the displacement of the two half-spaces at infinity occurred quasistatically and that the broken region of the fault plane slipped quasistatically in response and resulted in concentrating stresses along the boundary of the asperity. The necessary static shear traction components

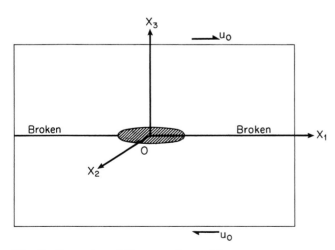

Fig. 9. Geometry of the asperity problem

on $X_3 = 0$ can be obtained for a Poisson solid from the well-known solution of *Mindlin* [18] as

$$\left. \begin{array}{l} \sigma_{13}^{\text{static}}(\boldsymbol{r}) = 16\mu u_0/(7\pi\sqrt{r_0^2 - r^2}), \\ \sigma_{23}^{\text{static}}(\boldsymbol{r}) = 0 \end{array} \right\} \text{ on } S(0): \quad r < r_0,$$

$$\sigma_{i3}^{\text{static}}(\boldsymbol{r}) = 0, \qquad \text{on } \bar{S}(0): \quad r > r_0,$$

where $\boldsymbol{r} = (X_1, X_2)$. A property of $\sigma_{13}^{\text{static}}$ from the above expression is that its value at the centre of the asperity is half its average value over the entire asperity.

Next, we must discretize the analytic form of $\sigma_{13}^{\text{static}}$ for use in the numerical algorithm. We may average $\sigma_{13}^{\text{static}}$ by averaging over grid areas, but the method of this discretization is crucial as it must conserve the important properties of the solution. In this example, the traction distribution on the asperity is radially symmetric but a straightforward discretization of the analytical initial stress distribution over cartesian grids centered at $(i\Delta X, j\Delta X)$ destroys this property along the bounding edge of the asperity. The radial symmetry can be preserved by averaging the static traction distribution over polar grids having areas $(\Delta X)^2$ and centered at $(i\Delta X, j\Delta X)$. The normalized averaged values of $\sigma_{13}^{\text{static}}$ across any diameter of the asperity is shown in Fig. 10; the normalization factor being $(16\mu u_0)/7\pi$ for the case when $r_0 = 5.5\Delta X$ (there are 97 grids on the asperity in this case). The expected inverse square-root behavior of the tractions near the edge is seen to be well preserved. The property that the value of $\sigma_{13}^{\text{static}}$ at the center is half its average value is found to be conserved automatically. It must be re-iterated here that some other method of discretization may lead to a different fracturing process. If, however, the most important property, that is, the radial symmetry of the initial stress, is preserved, then the most important features of the resulting fracture process will be the same, though obviously the details will differ.

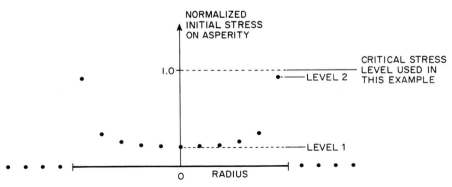

Fig. 10. Discretized initial stress (= negative of the stress drop) distribution shown along the diameter of the circular asperity (after [19])

This problem is specially suited for use in the algorithm of (6). In particular, if there is no backslip on the fault (and this is the *only* assumption we make regarding the nature of the solution), there is no traction change on $\bar{S}(0)$ during fracture, so that the fracture process calculations require that the summations in the discute form of (6) extend only over the initially unbroken asperity $S(0)$. Further, the calculations need be performed only over the as-yet unbroken part of the asperity $S(t)$, which *decreases* as time increases. Also, once the radius of the Rayleigh cone exceeds the initial asperity diameter $2r_0$, further calculations need be carried out over only a fixed number of time-steps. We clarify these statements using Fig. 11, which shows a cross-section through the region of integration in (X_1, X_2, t) space, the unbroken region of the asperity being indicated by stippling. To determine the stresses at representative points A and B, it is necessary to carry out the convolutions in the region of intersection of the P-wave cone (marked P) having vertex at A or B, with the cylindrical region of radius r_0. Since in 3-dimensional problems the Green's function is zero once the Rayleigh wave has passed (the Rayleigh cone is marked R in the figure), the regions over which the convolutions actually have to be carried out are even smaller and are indicated by hatching. Once the radius of the Rayleigh cone exceeds the initial asperity diameter, the figure shows that there is no contribution to the convolution from lower time levels (illustrated for point B). This also implies that to solve this problem the Green's function need only be determined for a fixed time interval related to the initial asperity size and not up to the time

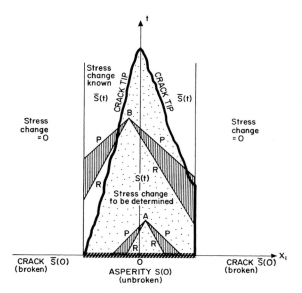

Fig. 11. A section through the volume of integration in (X_1, X_2, t) space for the asperity problem (after [19])

level for which the solution is desired, thereby resulting in further economy in computation.

The spontaneous fracture process is to be determined for some given critical stress level on the asperity. If this level was chosen to be below level 1 of Fig. 10, the whole asperity would break at once. If this critical stress level were between levels 1 and 2, all grids along the circumference would break immediately and dynamic fracturing of the remainder of the asperity would occur. The shape of the region which would break immediately would be an annulus and its width would depend on the particular value of σ^u that was chosen. If the critical stress level were above level 2, the asperity would remain unbroken. If, however, a few points on the asperity were relaxed, the resulting additional stress concentrations may (or may not) cause dynamic fracture to commence. We shall consider the last case, where dynamic fracture does occur. The critical stress level is assumed constant over the asperity in this example. Different distributions of σ^u over the asperity can be easily considered, if so desired. To initiate dynamic fracture, one or more of the most highly stressed grids along the edge of the asperity may be released. Let us release the grid closest to the asperity edge on the X_1-axis, say. The slip on the asperity is found to be mainly in the X_1-direction, the a_2 component of slip being found to be virtually zero, and the traction perturbation component $\tau_{23} \approx 0$.

The fracture process for this case is shown in Fig. 12. The sequence of fracturing is keyed to the normalized times $v_\mathrm{P} t/\Delta X$, shown in the figure by different levels of shading. The fracture propagation is symmetric about the X_2-axis, as is expected for this case. The fracture propagates along the edge of the asperity in two directions and completely encircles the asperity.

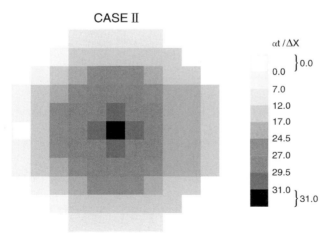

Fig. 12. Fractured areas as a function of normalized times. The key shows the different areas fractured during the normalized time intervals marked (after [19])

The fracture then propagates inwards from all directions and finally fractures the central grid. Such a fracture process was termed the "double encircling pincer" process by *Das* and *Kostrov* [19], in analogy with the term for the classic military maneuver. To characterize the rate of fracturing, one may divide the total fracture time by the asperity diameter, which in this case gives $0.35v_P$. This value is quite unrelated to the fracture speed locally on the asperity, which is seen in Fig. 12 to be higher along the asperity edge. A more useful measure of the time history of fracture in such a case may be the number of grid elements fractured at a given time. The normalized half-slip $a_1/2u_0$ as a function of time is shown for several points on the asperity in Fig. 11, the location of these points being indicated in the upper part of this figure. The rise time is found to be shortest for points near the center (which fracture later) and longest for points along the edge (which fracture earlier).

By releasing grids at other points along the circumference of the asperity, an essentially similar fracturing process was obtained [19]. *Das* and *Kostrov* [20] have also studied the fracturing process for an elliptical asperity.

References

1. B. V. Kostrov, S. Das, *Principles of Earthquake Source Mechanics*, Appl. Math. Mech. Ser. (Cambridge Univ. Press, New York 1988)
2. A. E. H. Love, *A Treatise on the Mathematical Theory of Elasticity* (Dover, New York 1926)
3. P. M. Morse, H. Feshbach, *Methods of Theoretical Physics* (McGraw-Hill, New York 1953)
4. K. Aki, P. G. Richards, *Quantitative Seismology: Theory and Methods* (Freeman, San Francisco 1980)
5. S. Das, A Numerical Method for Determination of Source Time Functions for General Three-Dimensional Rupture Propagation, Geophys. J. R. Astron. Soc. **62**, 591–604 (1980)
6. S. Das, B. V. Kostrov, On the numerical boundary integral equation method for three-dimensional dynamic shear crack problems, J. Appl. Mech. **54**, 99–104 (1987)
7. A. Cochard, R. Madariaga, Dynamic faulting under rate-dependent friction, Pure Appl. Geophys. **142**, 419–445 (1994)
8. B. V. Kostrov, Selfsimilar Problems of Propagation of Shear Cracks, J. Appl. Math. Mech. **28**, 1077–1087 (1964)
9. B. V. Kostrov, Unsteady Propagation of Longitudinal Shear Cracks, J. Appl. Math. Mech. **30**, 1241–1248 (1966)
10. R. Madariaga, Dynamics of an Expanding Circular Fault, Bull. Seismol. Soc. Am. **66**, 639–666 (1976)
11. B. R. Lawn, T. R. Wilshaw, *Fracture of Brittle solids* (Cambridge Univ. Press, New York 1975)
12. S. Das, K. Aki, A Numerical Study of Two-Dimensional Rupture Propagation, Geophys. J. R. Astron. Soc. **50**, 643–668 (1977)

13. S. Das, Three-dimensional spontaneous rupturre propagation and implications for earthquake source mechanism, Geophys. J. R. Astr. Soc. **67**, 375–393 (1981)
14. D. J. Andrews, Dynamic plane-strain shear rupture with a slip-weakening friction law calculated by a boundary integral method, Bull. Seismol. Soc. Am. **75**, 1–22 (1985)
15. A. Cochard, R. Madariaga, Complexity of seismicity due to highly rate-dependent friction, J. Geophys. Res. **101**, 25321–25336 (1996)
16. J. D. Eshelby, The Determination of the Elastic Field of an Ellipsoidal Inclusion, and Related Problems, Proc. R. Soc. Lond. A **241**, 376–396 (1957)
17. B. V. Kostrov, S. Das, Evaluation of Stress and Displacement Fields due to an Elliptical Plane Shear Crack, Geophys. J. R. Astron. Soc. **78**, 19–33 (1984)
18. R. D. Mindlin, Compliance of Elastic Bodies in Contact, J. Appl. Mech. **16**, 259–268 (1949)
19. S. Das, B. V. Kostrov, Breaking of a single asperity: Rupture process and seismic radiation, J. Geophys. Res. **88**, 4277–4288 (1983)
20. S. Das, B. V. Kostrov, An elliptical asperity in shear: Fracture process and seismic radiation, Geophys. J. R. Astron. Soc. **80**, 725–742 (1985)

Index

A-scan, 169
– -driven approach, 172
aberrating medium, 108
absorption, 79
adiabatic-mode theory, 52
ambiguity function, 58
anisotropy, 138, 183, 192, 202, 242, 253
– azimuthal, 204
– polarization (or radial anisotropy), 204
– seismic, 203
antenna, 173
apodization, 29, 161
array
– convex and linear, 140
– factor, 173
– linear, 114, 139
– piezoelectric transducer, 18
– transducer, 21, 56, 135
attenuation, 48, 143, 234
– coefficient, 263
average
– frequency, 26
– medium, 84
– spatial, 26
– time, 26

B-mode images, 135
B-scan, 168
back propagation, 66
backscattering, 234
– coefficient, 239
ballistic wave, 8
beam pattern, 161
beamforming, 135, 265
– electronic, 142
– plane-wave, 56
beamtracking, 163

Born approximation, 97, 236
– inverse, 253
boundary-integral equation, 315
broadband
– seismology, 193
– seismometer, 193
bulk modulus, 262

C-scan, 168
causality, 97
chaos
– quantum, 5
chaotic, 7
– cavities, 32
Chebyshev polynomials, 298
coda, 8, 201
coherent, 85
– Green's function, 86
compression
– temporal, 28
– time, 22
cone of dependence, 318
convection
– mantle, 219
correlation
– function, 84
– – spatial, 26
– length, 26, 217
covariance
– function, 213
– matrix, 213
crack, 313
– model, 315
crystallite, 234

D.O.R.T. method, 107
data space, 212
defects, 40

deformation tensor, 201
detection, 107
– process, 116
diffraction limits, 20
diffusive, 6
directivity pattern, 26, 29
Dirichlet condition, 314
dislocation model, 315
Doppler, 259

earthquake, 201, 311
eigenfrequency, 198
eigenfunction, 198
eikonal equation, 296
elastic parameters, 258
elasto-dynamics equation, 198
elastodynamic finite integration technique, 169
elastography, 258, 277
– magnetic resonance, 258, 277
emission
– acoustic, 268
ergodic, 32

fault, 313
field
– far, 144, 269
– near, 144, 149
figure-of-merit, 239
flaw, 168, 234
– detection, 234
flow imaging, 142
focal spot, 20
focus, 114
focusing, 97, 107, 142
– selective, 112
– spatial, 22, 60
– variable-range, 71, 74
force
– acoustic radiation, 258
– radiation methods, 258
– static radiation, 268
forward
– problem, 212, 312
– scattering problem, 98
fracture, 313
fracture criterion
– Griffith, 324
– Irwin, 324

Fresnel zone, 8

geometrical regime, 241
grain, 234
Green's function, 85, 313
group velocity, 197

Hanning window, 161
hardness, 262
Helmholtz equation, 53
Huygens' principle, 150

image
– theorem, 29
– vibro-acoustic, 272
imaging
– elasticity, 258
– medical, 107, 277
– phase-sensitive magnetic resonance, 263
– process, 116
– ultrasound, 135
impedance
– mechanical, 269
impulse response, 109, 113, 149, 313
– function, 98
– spatial, 152
incoherent field, 85
internal wave, 83
inverse
– problem, 212, 216, 312
– scattering, 171, 278
– scattering problem, 100
– scattering theory, 97
inversion
– direct, 286

kernel distribution, 19

Lamé parameters, 280, 313
lobe
– grating, 21, 144
longitudinal, 183
lossy media, 48
Lyapunov
– coefficient, 7
– exponent, 4

Marchenko equation, 97
matched filter, 25, 29, 117

matched-field
— processing, 43, 57
— tomography, 59
mean free path, 6
microstructure
— duplex, 246
migration
— prestack, 303
mode
— normal, 49
— — theory, 53
— opening, 312
— tearing, 312
— tensile, 312

Neumann condition, 314
Newton–Marchenko equation, 97, 98

ocean acoustics, 43

parabolic
— equation, 49
— equation model, 91
parameter space, 212, 215
perturbation theory, 205
— first-order, 211
phase conjugation, 17, 30, 36, 43, 60
pixel-driven approach, 172
plasma wave equation, 98, 99
polycrystal
— metal, 234
potential, 99
pulse–echo, 168, 178
— measurement, 107
— mode, 234

random, 83
random media, 17
ray
— -tracing algorithm, 183
ray theory, 49
Rayleigh, 260
— integral, 151
— regime, 240
— wave, 197, 318
reciprocity, 26, 60, 103, 113, 143, 207, 313
— theorem, 153
resolution, 22, 170, 272

SAFT algorithm, 171
scanner
— ultrasound, 135
scattered wave, 201
scattering, 21, 79, 234, 236
— direct
— — problem, 99
— sound, 83
scattering cross-section, 6
seismic
— moment tensor, 201
— record, 195
seismic tomography, 218, 295
seismogram, 312
— synthetic, 201
seismology, 107
seismometer, 192, 311
self-averaging, 36
self-focusing, 108
— process, 20
shallow water, 30, 88
shear mode, 312
— anti-plane, 312
— in-plane, 312
shear modulus, 262, 286
shear-wave methods, 259
sidelobes, 144
signal-to-noise ratio, 172, 175, 272
singular value decomposition, 111, 113, 302
slip, 311
source location, 116
speckle pattern, 164
spectral method, 49
split-step
— algorithm, 91
— Padé algorithm, 93
splitting, 203
steel
— austenitic, 168
— austenitic stainless, 187
— ferritic, 168
stiffness
— complex-valued, 278
— tensor, 280
— tissue, 277
stochastic regime, 240
strain

– tensor, 198, 313
– tissue, 278
stratification wave, 83
stress
– critical level, 324
– field, 268
– tensor, 279, 313
stripping
– mode, 72
subzone
– overlapping
– – technique, 288
surface, 58
– characteristic, 318
synthetic aperture
– focusing technique, 168
– – formalism, 170
– reconstruction, 253

tectonics, 311
texture, 241
thermocline, 44
time
– domain, 97
– reversal, 1, 97
time-of-flight diffraction technique, 177
time-reversal
– cavities, 104
– focus, 98
– invariance, 1, 17
– mirror, 43, 60, 104, 108, 117
– operator, 108
time-reversed wavefield, 2
titanium alloys, 40
tomographic models, 191

tomography, 43, 212
transducer
– array, 56
– multi-element, 135
– planar, 234
transfer function, 28
– medium, 269
turbulence wave, 83

underwater acoustics, 30, 107

variational method, 288
vibro-acoustography, 259
viscoelastic parameters, 262
viscosity, 270
– tensor, 280

wafer, 32
wave
– body, 195
– circumferential, 124
– creeping, 169
– elastic, 233
– internal, 32
– Lamb, 32, 123
– Love, 196
– multiple-scattered, 7
– P and S, 191, 318
– shear, 183
– shock, 262
– surface, 32, 85, 118, 167, 195
waveguide, 43, 116
– ultrasonic, 27
weld, 175, 177

Young's modulus, 270

Topics in Applied Physics

67 **Hydrogen in Intermetallic Compounds II**
Surface and Dynamic Properties, Applications
By L. Schlapbach (Ed.) 1992. 126 figs. XIV, 328 pages

68 **Light Scattering in Solids VI**
Recent Results, Including High-T_c Superconductivity
By M. Cardona and G. Güntherodt (Eds.) 1991. 267 figs., 31 tabs., XIV, 526 pages

69 **Unoccupied Electronic States**
Fundamentals for XANES, EELS, IPS and BIS
By J. C. Fuggle and J. E. Inglesfield (Eds.) 1992. 175 figs. XIV, 359 pages

70 **Dye Lasers: 25 Years**
By M. Stuke (Ed.) 1992. 151 figs. XVI, 247 pages

71 **The Monte Carlo Method in Condensed Matter Physics**
By K. Binder (Ed.) 2nd edn. 1995. 83 figs. XX, 418 pages

72 **Glassy Metals III**
Amorphization Techniques, Catalysis, Electronic and Ionic Structure
By H. Beck and H.-J. Güntherodt (Eds.) 1994. 145 figs. XI, 259 pages

73 **Hydrogen in Metals III**
Properties and Applications
By H. Wipf (Ed.) 1997. 117 figs. XV, 348 pages

74 **Millimeter and Submillimeter Wave Spectroscopy of Solids**
By G. Grüner (Ed.) 1998. 173 figs. XI, 286 pages

75 **Light Scattering in Solids VII**
Christal-Field and Magnetic Excitations
By M. Cardona and G. Güntherodt (Eds.) 1999. 96 figs. X, 310 pages

76 **Light Scattering in Solids VIII**
C60, Semiconductor Surfaces, Coherent Phonons
By M. Cardona and G. Güntherodt (Eds.) 1999. 86 figs. XII, 228 pages

77 **Photomechanics**
By P. K. Rastogi (Ed.) 2000, 314 Figs. XVI, 472 pages

78 **High-Power Diode Lasers**
By R. Diehl (Ed.) 2000, 260 Figs. XIV, 416 pages

79 **Frequency Measurement and Control**
Advanced Techniques and Future Trends
By A. N. Luiten (Ed.) 2001, 169 Figs. XIV, 394 pages

80 **Carbon Nanotubes**
Synthesis, Structure, Properties, and Applications
By M. S. Dresselhaus, G. Dresselhaus, Ph. Avouris (Eds.) 2001, 235 Figs. XVI, 448 pages

81 **Near-Field Optics and Surface Plasmon Polaritons**
By S. Kawata (Ed.) 2001, 136 Figs. X, 210 pages

82 **Optical Properties of Nanostructured Random Media**
By Vladimir M. Shalaev (Ed.) 2002, 185 Figs. XIV, 450 pages

83 **Spin Dynamics in Confined Magnetic Structures**
By B. Hillebrands and K. Ounadjela (Eds.) 2002, 166 Figs. XVI, 336 pages

84 **Imaing of Complex Media with Acoustic and Seismic Waves**
By M. Fink, W. A. Kuperman, J.-P. Montagner, A. Tourin (Eds.) 2002, 162 Figs. XII, 336 pages

You are one click away from a world of physics information!

Come and visit Springer's
Physics Online Library

Books
- Search the Springer website catalogue
- Subscribe to our free alerting service for new books
- Look through the book series profiles

You want to order? Email to: orders@springer.de

Journals
- Get abstracts, ToC´s free of charge to everyone
- Use our powerful search engine LINK Search
- Subscribe to our free alerting service LINK *Alert*
- Read full-text articles (available only to subscribers of the paper version of a journal)

You want to subscribe? Email to: subscriptions@springer.de

Electronic Media
- Get more information on our software and CD-ROMs

You have a question on an electronic product? Email to: helpdesk-em@springer.de

••••••••• Bookmark now:

http://www.springer.de/phys/

 Springer

Springer · Customer Service
Haberstr. 7 · 69126 Heidelberg, Germany
Tel: +49 (0) 6221 - 345 - 217/8
Fax: +49 (0) 6221 - 345 - 229 · e-mail: orders@springer.de
d&p · 6437.MNT/SFb

Druck: Strauss Offsetdruck, Mörlenbach
Verarbeitung: Schäffer, Grünstadt